OF

Tourism and Tourism Spaces

Tourism and Tourism Spaces

Gareth Shaw and
Allan M. Williams

SAGE Publications
London • Thousand Oaks • New Delhi

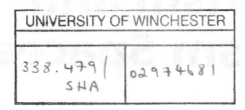
 SAGE Publications Ltd
1 Oliver's Yard
55 City Road
London EC1Y 1SP

SAGE Publications Inc.
2455 Teller Road
Thousand Oaks, California 91320

SAGE Publications India Pvt Ltd
B-42, Panchsheel Enclave
Post Box 4109
New Delhi 100 017

British Library Cataloguing in Publication data

A catalogue record for this book is available from the British Library

ISBN 0 7619 6991 8
ISBN 0 7619 6992 6 (pbk)

Library of Congress Control Number 2003115350

Typeset by TW Typesetting, Plymouth, Devon
Printed in Great Britain by Athenaeum Press Ltd, Gateshead

Contents

List of boxes

List of figures

List of tables

Acknowledgements

The authors would like to give their thanks to the following people at the University of Exeter: Terry Bacon and Sue Rouillard for their cartographic skills in drawing all the diagrams, and Cathy Aggett for all the long hours she spent at the word processor.

1   *Introduction*

Tourism is increasingly the centre of popular and policy discourses. It is both demonized and idealized, as a destroyer and a creator, whether of valued environments, social and cultural practices, or wealth. One of the roles of tourism researchers should be to provide a greater understanding of the underlying processes that shape the emerging tourism landscape. We have argued elsewhere (Shaw and Williams 2002) that such research has at best been uneven, and at worst has failed to respond to these challenges. Although there are a growing number of exceptions, tourism research is still often descriptive, atheoretical, and chaotically conceptualized in being abstracted from broader social relationships. The task of remedying these deficiencies lies beyond the scope of this, or probably any other, book. Rather, it requires a collective endeavour by tourism researchers. This book aims to make a small contribution to such a reorientation, but the enormity of even this task will become clear as gaps in our knowledge and understanding appear, as the volume unfolds.

In this Introduction we begin by seeking to conceptualize tourism in terms of broader debates about mobility. Tourism is intertwined with many other forms of mobility, such as labour and retirement migration, or knowledge and capital transfers. In some senses, however, these share a common structure, characterized by what Urry (2000) terms 'scapes and flows'. These are mutually informing, and constantly shifting over time. One of our key interests is in how globalization – in its different meanings – shapes tourism flows, and indeed spaces and places. It is only within this context that we can start to understand the complex opportunities, risks and constraints that accompany tourism practices. Having briefly outlined the above theme, the remainder of the Introduction sets out the rationale and structure of this volume.

TOURISM AND MOBILITY

Barbara Cassani, Chief Executive of Go – the UK budget scheduled airline – has been quoted as saying that 'Everyone in the population wants to travel. It's something that people feel positive about and find liberating' (*Financial Times*, 11 August 2001). This is necessarily an exaggeration, which ignores both the complexity of travel motivations and the unequal access to travel. But it does emphasize the importance of travel and tourism in the contemporary world. It is not only essential to the system of trade and production, and a dynamic element of consumption, but also a cultural icon and an increasingly important constituent of cultural capital. There has been an immense increase in mobility – whether in economic, cultural, political or environmental terms – and tourism has been at the forefront of such changes.

Mobility encompasses many forms: goods, information, services, and financial transactions are all mobile over space, as are people (corporeal mobility). These forms are necessarily interrelated. For example, the mobility of tourists across space is inevitably accompanied by mobility of goods (for consumption by the tourists on holiday or after returning home), information about destinations, services provided by travel intermediaries, and significant financial transactions between places of origin and destination. This is a reminder that tourism cannot be understood as a unique and independent social phenomenon, but is intricately woven into the fabrics of daily lives, the constitution of communities, and the functioning of social and natural systems.

The character of tourism as corporeal mobility has been changing, especially in the way it interacts with the provision of services and the social changes resulting from the development of the internet, including the potential for virtual tourism. But, despite the growth of virtual travel via the internet, there is no convincing evidence that this is replacing corporeal travel (Urry 2001: 1). Urry (2002) explains this in terms of 'the obligations of co-presence' (Boden and Molotch 1994). Co-presence (of individuals) is essential to some forms of social intercourse, and from this stems the need for mobility. It follows that where mobility involves a significant corporeal displacement from the usual place of residence, then various forms of tourism will result (Table 1.1).

Table 1.1 **Tourism and co-presence**

Basis of co-presence	Activity requiring co-presence	Tourism implication
Legal, economic and family obligations	To work, to attend family events, to visit public institutions	Business and VFR tourism
Social obligations	To meet face to face, to develop trust, to note body language	Business and VFR tourism
Time obligations	To spend quality time with family, partners or friends	Leisure tourism
Place obligations	To sense a place directly, through embodied experiences	Leisure, heritage and cultural tourism
Live obligations	To experience a particular live sporting, political or cultural event	Sports and cultural tourism
Object obligations	To work on objects that have a particular physical location	Business tourism

Note: VFR = visiting friends and relatives tourism.
Source: main bases of co-presence are based on Urry (2002), with the present authors' elaboration of tourism implications.

Tourism therefore remains one of the most significant forms of mobility. One of the greatest challenges for researchers is to understand how it both shapes and is shaped by wider societal processes. This requires a perspective on change, and for this we can draw on the concept of scapes and flows (Urry 2000), understood as 'networks of machines, technologies, organizations, texts and actors that constitute various interconnected nodes along which the flows can be relayed. Such scapes reconfigure the dimensions of time and space' (p. 35). Flows of people, information, images, money etc move along these scapes. 'Such flows generate for late twentieth-century people, new opportunities and desires, as well as new risks' (p. 36).

It is useful to think of the 'landscapes' of tourism as being composed of scapes and flows. Tourism space is deeply structured by scapes – motorways, flight routes, airports, etc., which facilitate and channel movement. Scapes have blurred rather than sharp boundaries, and much travel occurs outside them, but they are fundamental to understanding the massing of tourism flows along particular routes – for example, the charter-flight routes from northern to southern Europe, from Japan to Hawaii, or from the northern states of the USA to Florida. They are more than transport routes, for these scapes also consist of the material investments in hotels, restaurants and other services that facilitate travel. They are also invested with the tourist imagination: the tourist gaze (Urry 1990) is signposted along the scapes, informed by diverse media imaging of not only the destinations, but also the routes themselves. The sea route into the harbours of New York, the Trans-Canada railway route, and the flight pathways over the poles are all romanticized scapes, where images and imaginations reinforce the attraction of travel and tourism. Mass tourism – drawing on the economies of scale inherent in such scapes – is particularly influenced by their structures. These flows are channelled along scapes or routes which tend 'to wall them in' (Urry 2000: 38), for there are fragmented and uneven flows of people, ideas, and information across space (Urry 2000: 38–9). Many other forms of tourism (e.g. ecotourism, minority sports tourism) are also shaped by these scapes, even if only through actively seeking out the interstices between the major channels.

Scapes are significant because they contribute to the ascendancy of relative location over absolute location. The direct distance between potential points of origin and destination no longer matters. Instead, scapes create inequalities in tourist and related flows as they bypass some areas, while connecting others with channels enriched with transport and tourism facilities. Scapes are characterized by inertia, resulting from the technology, fixed capital investments and knowledge embedded within them (Williams 2002), but they are not immutable. Instead, they are constantly being revised and reconstructed in context of tourism globalization. They are contested – for example, by cities such as Barcelona, Glasgow and Baltimore – which have sought to reposition themselves in tourism markets (Law 2002). This represents an attempt to generate new flows which, in time, will lead to the repositioning of scapes. The resulting shifts distribute and redistribute not only scapes and flows, but also economic, cultural and environmental costs and benefits.

Scapes and flows are useful concepts because they highlight the relationships and tensions between structures and processes, and they provide insights into what can be termed the path-contingent nature of change. This provides a starting point for addressing some of the central concerns of this volume:

- How are tourism structures and flows created? This leads to a central concern with questions of *production and consumption*, and the interrelationships between these.

- How are tourism structures and flows reproduced? This raises questions about *regulation*.

- Who and what areas benefit from, or incur costs resulting from, these structures and flows, and when and how are these manifested? This poses questions about *welfare*, but also about *cultural and environmental impacts*.

- How do tourists, host communities and other participants experience these flows and structures? This problematizes issues relating to *authenticity and cultural adaptation*.

- To what extent are individuals, communities and states able to contest their relative locations in these structures and flows, and the distribution of cost and benefits that stem from them? This highlights questions about the *state, government, and governance*.

We develop a broad political-economy perspective, which helps to address these questions, but which is also informed by behavioural and cultural research.

The above questions are assuming increasing significance and urgency in the face of growing, and increasingly globalized, mobility, which is the theme of the next section.

TOURISM AND GLOBALIZATION

Globalization is a widely used, perhaps misused, term. There is widespread popular and political acceptance of globalization: financial services, currencies, corporations, markets, governments and governance, cultures and corporeal mobility are no longer constrained by national spaces. Instead, Urry's (2000) concept of scapes and flows transcends national boundaries, so that 'geographies seem to be shrinking even disappearing' (Amin and Thrift 1997: 147). This is a process with long roots, and there have been significant trans-border flows of capital for at least two centuries, and of corporeal mobility for much longer. But globalization is distinguished from earlier types of internationalization, because it involves 'qualitative' and not only 'quantitative' changes (Dicken et al. 1997).

In what way is globalization qualitatively different? Held provides an insight:

> ... (the) explosion of travel, migration, fighting, and economic interchange provided an enormous impetus to the transformation of the form and shape of human communities; for the latter increasingly became enmeshed in networks and systems of interchange – a new era of regional and global movement of people, goods, information and microbes was established. (2000: 1)

The key concept here is 'networks and systems of interchange' or the interconnectedness of places and spaces. Globalization is distinguished (from internationalization) by increases in both the geographical reach and intensity of interconnectedness. There has been what Harvey (1989a) famously described as 'time–space compression'. Urry also writes of time–space shrinkage resulting from the impacts of new technologies, which 'carry people, information, money, images and risks' so that there are 'new fluidities of astonishing speed and scale' (2000: 33).

Time–space compression, and changes in the scope and intensity of interconnections, have been neither homogeneous nor worked out on an isotropic, featureless surface. Instead, intensified flows across space were, and are, highly structured. The introduction of new technologies, such as the internet and satellite television, have not eliminated spatial differences, nor destroyed the tyranny of (relative) location, but have instead created new inequalities, in terms of how individuals and places use these differently. There are differences not only in the degree to which places are connected to new and old scapes (in, for example, internet cabling and air transport routes), but also in the extent to which they are able to harness these and influence the resulting distribution of benefits and costs.

The question of how places can influence globalization takes us to the debate about the location of power. Box 1.1 summarizes the salient points. In essence: 'globalists' focus on erosion of the power of national states; 'traditionalists' emphasize the intensification of long-established internationalization processes but with national states remaining key sites of regulation; and 'transformationalists' emphasize the changes brought about by increased interconnections, without predicting an inevitable outcome in terms of the power of national states. Assessing the competing claims of these theories is problematic. For example, Dicken et al. (1997) emphasize that there are no entirely globalized corporations, because even those with the greatest global reach still tend to have a high degree of embeddedness in their home country, so that national states continue to have policy leverage over them. There is also a parallel debate on global–local relationships, and the two are necessarily interlinked, as the national state is often a key intermediary for the local. Perhaps the most significant outcome of this debate is the question of whether the local is largely passive in the face of global changes, or whether localities can contest and use globalization (see Chapters 8 and 10).

The general debate about globalization has a number of implications for understanding tourism. These are evident in terms of production,

---

Box 1.1   Competing theories of globalization

There are three main competing theories of globalization, and they are distin-guished mainly by their interpretation of the changing relationships between national states and global phenomena.

- *Globalists* argue that large scale economic, social and political changes have significantly diminished the power of national states, ushering in a 'borderless world' (Ohmae 1990). This development is viewed positively by commentators such as Ohmae, while others – for example, Robinson (1996) – consider that it has unleashed a 'savage capitalism', which is destructive of cultures, local economies and environmental systems.

- *Traditionalists* recognize that there have been important changes, but regard them as constituting no more than an intensification of long-established internationalization processes. In their view, national states remain the key sites for the regulation of economies, not least because capital and economic activities remain mainly oriented to national markets. This view is typified by Hirst and Thompson (1996), who back up their arguments with an empirical analysis of trade and investment flows within and beyond national boundaries.

- *Transformationalists* argue that globalization leads to a new political, economic and social framework, which necessarily has transformed the nation state. However, unlike the other theorists, they predict neither its demise nor its continuing supremacy. Instead, they argue that 'the social-spatial context of states is being altered and, along with it, the nature, form and operations of states' (Held 2000: 3).

Source: Held (2000); Cochrane and Pain (2000)

---

consumption and the location of power to control or regulate tourism (see also Chapters 2 and 7). Some of the key aspects of the globalization of tourism are set out below.

First, we note that *tourism is highly implicated in globalization*. It is, of course, influenced by globalization, as evident in the way globalized information flows impinge on tourist decision-making or in the way that globalized investment flows are creating international hotel chains and tourist attractions (see the discussion of 'McDisneyization' in Chapter 5) which foster global tourist flows. But tourism is also helping to create, recreate and distribute images and objects around the world (souvenirs, photographs and informa-tion leaflets transported home by tourists, or imported goods for consumption in holiday destinations).

Second, tourism has been subject to *the intensification of interconnections*, and, as noted earlier, this has two dimensions. There has been geographical stretching of tourism flows, evident above all in the growth of long-haul holidays, but also in the opening of up new destinations such as Prague, on the doorstep of the major western European market. And there has also been greater intensity in these interconnections. Places are linked not just by larger flows of tourists but by qualitatively different flows. For example, increasing numbers of tourists engage in circulation (cycles of movement, or of comings and goings) rather than simple one-off holidays. This is exemplified by the circulation between first and second homes, or by multiple and repeated visits to particular destinations, perhaps visiting friends and relatives; this trend is particularly evident among the retired and early retired populations of the more developed countries (Williams et al. 2000). The flows have also been intensified by associated changes such as the increased flows of credit and electronic financial transactions accompanying tourists, as they take advantage of the globalization of financial services. The Thomas Cook travellers cheque, an icon of an earlier era of tourism, has long since been replaced by the credit card as an icon of contemporary tourism, at least in most developed societies. Together, these changes have contributed to the globalization of tourism.

Third, while globalization has contributed to the creation of new structures, along which tourists flow to new or distant locations, these have been *grafted onto* what may be termed *existing tourism landscapes*. Globalization has reinforced the pivotal tourism role of world cities such as London, New York, Rome and Paris. Moreover, the map of tourism flows remains highly regionalized, with intensification evident within three main regional foci: Europe, North America/Caribbean, and East Asia (Shaw and Williams 2002: Chapter 2). The impacts of tourism, whether in these macro regions, or in countries such as the Czech Republic, Belize or Thailand, have remained highly uneven, contingent upon the highly focused nature of the structures along which tourists flow.

Fourth, globalization does not concern only production and consumption. It also has *resonance for identities and meanings*, for there is a process of interpenetration, as distant cultures are increasingly brought into direct contact with one another (Cochrane and Pain 2000). Giddens (1996) reminds us that globalization is not simply an 'out there' phenomenon, bound up in the emergence of world systems, but it also refers to transformations in 'the texture' of everyday life. It is an 'in here' phenomenon, with profound implications for the emergence of new forms of local cultural identity and self-expression. Tourist experiences – which are increasingly globalized – do contribute to identity formation, if only in reinforcing awareness of difference. But tourism is also a vehicle for the transfer of ideas and artefacts, which contribute to the reshaping of identities. Values and identities are unlikely to remain untouched in homes filled with souvenirs from trips around the Caribbean or the hills of Chianti. To varying degrees then – and this partly depends on tourist motivations and experiences (see Chapter 6) – tourism contributes to the emergence of hybrid identities. And this applies to hosts as well as guests. These issues are explored in this volume in relation to the concept of authenticity.

Fifth, *globalization has modified the location of power and the nature of tourism dependency* – understood as unequal relationships and external control (Britton

1991) – without challenging fundamental inequalities. The intensification of interconnections has created new opportunities for exploitative relationships for capital, in terms of where and how profits are extracted from tourists. But local communities and countries are not simply passive in the face of globalization. They do contest their position, although interpretation of their capacity for effective action depends on which of the major theories of globalization is adhered to (Box 1.1). In this volume we take the view that national states continue to be significant sites for regulating national economic spaces, including those in which tourism occurs (see Chapter 2). Figure 1.1 provides a highly idealized model of the relationships between globalization and the location of power. In practice, of course, the strength of national regulation is uneven among developed countries (contrast the United States and Switzerland, for example), let alone between these and the less developed economies.

Even if the roles of national states are in question, there is general consensus that globalization has reinforced the significance of place, and that localities can contest their place in the world (Chapter 10). Dicken et al. (1997) comment that globalization processes are not abstract but 'are realized in institutionally, historically and geographically specific sites'. The global–local nexus is not necessarily a threat. Rather it offers opportunities to combine global and local potentialities. But these have to be seized through appropriate local actions. This theme is considered further in the following section, and in more detail in later chapters.

TOURISM OPPORTUNITIES, RISKS AND CONSTRAINTS

The role of tourism in the global economy is frequently mythologized. On the one hand, proponents of tourism, such as the World Tourism Organization, quote statistics which seek to demonstrate that not only is tourism the largest industry in the world, but it is also one of the more rapidly growing (see also Williams and Shaw 1998a, on the approach of the EU to tourism). Currently some 600–700 million international trips are made each year, but the World Tourism Organization predicts that by 2020 this will have increased to 1.5 billion. In contrast, there is a phalanx of critiques of tourism, which variously present it as exploitative, short-termist, and destructive of social and natural systems (see Chapter 7). For example, Mathieson and Wall's classic study of tourism impacts (1982) set out a framework which highlighted the problems inherent in analysing and evaluating these. There is also a more emotive literature, much of it in the popular domain, which presents tourism and tourists as a blight or a plague on unsuspecting communities. The truth inevitably lies between these two poles.

Tourism certainly is a large and expanding set of activities. There are, of course, problems in its quantification, and there is a well-rehearsed debate as to the conceptual and empirical difficulties involved in defining tourism, or the tourism industry (see Leiper 1990; Smith 1998; Wilson 1998). Rather than rehearsing what is often an arid debate about definitions, it is more useful to

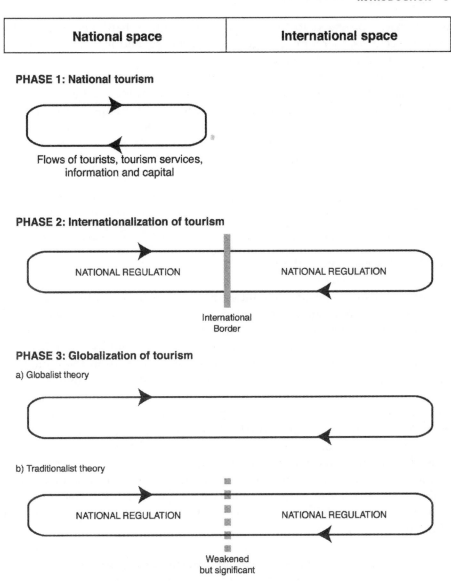

National space	International space

PHASE 1: National tourism

Flows of tourists, tourism services,
information and capital

PHASE 2: Internationalization of tourism

NATIONAL REGULATION   NATIONAL REGULATION

International
Border

PHASE 3: Globalization of tourism

a) Globalist theory

b) Traditionalist theory

NATIONAL REGULATION   NATIONAL REGULATION

Weakened
but significant

Figure 1.1   **Tourism, globalization and national regulation**

think of tourism as a complex set, or bundle, of economic, political, socio-cultural and environmental processes related to tourist activities. Tourist activities can be defined in terms of either location (involving a stay away from home of at least one night) or experience (leisure and recreational activities undertaken away from the home and the immediate neighbourhood, but not necessarily involving an overnight stay). There is a core of activities, such as providing accommodation and long-distance travel, which is devoted almost entirely to tourism, but this is surrounded by activities such as catering or running tourist attractions, wherein tourism activities are intermingled with services provided to local markets. Drawing a hard line around the tourism sector is, therefore, a largely futile exercise, and it is more useful to recognize that it has blurred boundaries.

The blurring of boundaries is one of the keys to interpreting the opportunities, risks and constraints associated with tourism. These can be understood only if they are analysed in the context of tourist and non-tourist relationships. Tourism is only one of the many flows through which communities and individuals are related, but a highly significant one. As Schiller et al. observe, the constant and various flows of:

> ... goods and activities have embedded within them relationships between people. These social relations take on meaning within the flow and fabric of daily life, as linkages between different societies are maintained, renewed, and reconstituted in the context of families, of institutions, of economic investments, business, and finance and of political organisations and structures including nation states. (1992: 11)

Business tourism is embedded in inter-firm relationships, visiting friends and relatives (VFR) tourism is embedded in geographically stretched family relationships, and leisure tourism is embedded in the meanings people make of, and impose on, their use of time, or their working and free-time lives. In short, the question therefore, is 'How does tourism relate to the economic, political and cultural processes emanating from non-tourism activities?'

The difficulties of disentangling tourism from its wider context mean that we should be less concerned with trying to quantify or document total impacts on an area, or an individual or household, and instead more focused on analysing their contingent nature – that is, the opportunities, risks and constraints in particular places. This is exemplified by a brief review of some of the sociocultural, economic and environmental relationships of tourism (Table 1.2). With some simplification, it is possible to systematize the opportunities, risks and constraints generated by tourism in respect of each of these.

Table 1.2 **Perspectives on tourism opportunities, risks and constraints**

	Economic	**Sociocultural**	**Environmental**
Opportunities			
Growth	Dynamism	Cultural-exchange	Funding
Diversification	Risk-reduction	Alternatives	Biodiversity
Risks			
Uncertainty	Non-predictability	Uncontrollability	Non-regulation
Dependency	Vulnerability	Imitation	Uniformity
Homogeneity	Competition	McDisneyization	Parkification
Constraints			
Assets	Free goods	Community	Nature
Intrarelationships	Non-trust	Conflicts	Non-conformism
Interrelationships	Competition	Irritation	Non-conformism

Opportunities. Tourism provides two forms of opportunities, which can be generalized in terms of growth and diversification. These are most clearly and directly evident in economic terms. Tourism provides an opportunity to stimulate modern capitalist growth in less developed economies, or to boost ailing economies, providing a dynamic basis for restructuring. It can also contribute, via diversification, to risk reduction, whether in agricultural or mature industrial communities. In sociocultural terms, tourism can intensify the interconnections between places; if communities have previously been relatively isolated, this provides greater cultural alternatives in lifestyles, and potentially greater multiculturalism (although this is a contested goal). The opportunities for the environment lie in harnessing the revenue generated by tourism for environmental improvements, including taking positive actions to promote biodiversity.

Risks. Tourism presents three major risks to places: uncertainty (resulting from fluctuations in demand curves that are highly sensitive to perceptions of risk), over-dependency on a single activity, and greater homogeneity, in part related to the globalization of consumption. The economic risks associated with lack of predictability in demand, and vulnerability to short- and long-term fluctuations in demand, are clearly expressed through changes in visitor numbers and expenditure. Increased homogeneity – as typified by the Mediterranean mass tourism product – results in intense competition between places, for a largely undifferentiated demand, mostly on the basis of price (Williams 2001). In such circumstances growth can be associated with highly exploitative relationships (e.g. local hotels squeezing wages in response to pressures from tour operators to reduce prices). The sociocultural risks lie in the lack of local control over relatively open systems characterized by high mobility levels (characteristic of tourism). Dependency on particular market segments may also lead to strong demonstration effects – especially among younger people – evident in imitative social or cultural practices which may not be viable, or may be a source of conflict in the host community. There is also a tendency for tourism to contribute to the homogenization of culture. In part, this results from people increasingly wanting their tourist experiences to be as 'McDonaldized' as everyday life (Ritzer and Liska 1997). This 'McDisneyization' and 'McDonaldization' of the tourist industry on a global scale produces 'homogeneous, calculable and safe experiences wherever they are to be consumed' (Urry 2000: 38). But at the same time, the demands made on local communities – in terms of services provided, languages spoken, behaviours accepted or imitated – lead to greater cultural uniformity (see the discussion of authenticity in Chapter 7). The risks for the environment from unregulated tourism are obvious, but there are also risks in the production of topographical uniformity (importing sand, building sea breaks, draining marshes) to provide the environments demanded by some forms of tourism. In extremis, this produces what we term 'parkification', as particular landscapes, such as golf courses, are created for tourism.

Constraints. We identify three generic types of constraints on tourism development, which have been termed assets, intrarelationships (within tourism), and interrelationships (between tourism and other sectors). One of the economic assets of tourism is that many tourist attractions – sea views, townscapes, fresh air, clean water – are free goods, which do not have to be paid for. But because it is difficult, if not impossible, to establish private property rights over such goods, it is also difficult to establish responsibility for maintaining these. Another constraint is that the fragmentation of tourism tends to contribute to a lack of trust and confidence among firms, and therefore there is weak capacity for building the types of coalitions that are necessary for effective local partnerships to advance tourism (Bramwell and Shurma 1999). It may also be difficult to resolve competition between tourism and other sectors over land or labour, which may drive up prices and reduce the competitiveness of both activities. One of the assets of tourism may be the local sociocultural system, either because of its receptivity to tourists, or because folkloric practices are attractions in their own rights. Yet tourism growth potentially may undermine both aspects, leading to social cleavages between those sections of the tourism industry serving different market segments, and to growing host–guest irritation (Chapter 7) among many sections of the wider community. In terms of the environment, nature is often a major tourist attraction, but many of the most attractive 'natures' are also highly fragile environments. In these settings, the practices of different groups of tourists, and of tourists versus locals, may be non-conforming; they may be mutually exclusive in terms of enjoyment of the same place, while also threatening the viability of the local ecosystem.

The above synopsis provides no more than an introductory framework for the study of tourism. It is necessarily a simplification, as the economic, sociocultural and environmental categories are not discrete. For example, nature is in part socially constructed, while nature is also incorporated into social, cultural and economic practices. Moreover, whether particular tourism developments constitute opportunities, risks or constraints is highly contingent. For example, rural tourism in one place may provide welcome diversification given agricultural decline, and the potential for generating income to support environmental programmes in what had been a bleak landscape of modern farming. But, elsewhere, similar developments may compete for scarce labour, and be weakly regulated, threatening to destroy the 'rural' characteristics that initially attracted tourists.

TOURISM AND TOURISM SPACES: AN APPROACH

The central concern of this book is to explore the relationship between tourism and tourism places and spaces, while also deepening our theoretical perspectives on this relationship. Our starting point is political economy, while recognizing the importance of the cultural dimension. This does not mean that we believe that everything can be reduced to material relationships, but we see these as fundamental to the shaping of tourism. As such, we aim to build on Britton's (1991) seminal contribution to theorizing the political economy of tourism as a system of production, consumption, and circulation, which needs to be understood within an essentially political context.

In common with Britton, our approach is influenced by what has become known as the 'cultural turn' (see Lee and Wills 1997). In essence, this recognizes that economic relationships are infused with culturally symbolic processes, which are expressed differently in different cultural systems, and which are therefore necessarily territorially embedded (Thrift and Olds 1996). One of the central tenets of the book, therefore, is the need to take into account how tourism processes are place- and time-contingent, as well as place-shaping. This applies as much to those chapters that focus on economic relationships (Chapters 2–4), as those which consider, motivations, and experiences and authenticity (Chapters 5–7).

Another theme associated with the 'cultural turn' is the need to look at the interplay of production and consumption (Gregson 1995). Consumers – in this case, tourists – are not just passive objects; rather, they explore and experience sites of consumption (Jackson 1995), and their practices contribute to the ways in which places are constituted. More explicitly, tourists contribute to tourism experiences; they actively create these for themselves and for other tourists: the atmosphere of a tourism site, and the experiences of tourists, are often dependent on the co-presence of other tourists. At the same time, their interrelationships with host communities actively shape places, as is explored later in this volume in the case of commodification and authenticity (Chapter 7).

Ateljevic argues that there is a need to conceptualize tourism 'as a nexus of circuits operating within production–consumption dialectics enabled by the processes of negotiated (re)production' (2000: 371). These dialectics are perhaps most clearly evident in respect of tourist attractions. Ritzer sees shopping malls, theme parks, and casinos, among other tourist venues, as 'cathedrals of consumption', where there is a 'dizzying proliferation of settings that allow, encourage, and even compel us to consume . . . goods and services' (1999: 2). In order to attract increasing numbers of consumers (tourists), these cathedrals of consumption 'need to offer, or at least appear to offer, increasingly magical, fantastic, and enchanted settings in which to consume' (Ritzer 1999: 8). Consumption informs production, as much as production shapes consumption in these settings (Chapters 9 and 10).

Consumers actively contribute to the consumption experiences at such 'cathedrals of consumption'. Indeed, for Pine and Gilmore (1999) there is the growing hegemony of 'the experience economy', which they consider to be the main source of value extraction in the modern economy (Chapter 5). The experience economy is different from the service economy because consumers have to be engaged to produce and experience the event (see the discussion of labour as performance in Chapter 3). Hence, one of our interests in this book is the need to understand the tourist experience (Chapter 6). The emphasis on the cultural dimension of these experiences does not mean that they can be dissociated from the underlying material relationships that shape them. Rather, we agree with Ateljevic that:

> Recognising the importance of the economy while not analytically separating out culture or consumption, has given rise to theories which attempt to account for both the individual's material condition and specific experience but at the same time situate the individuals into the political and economic structures of power, conflict and resistance. (2000: 373)

Thus, East and South-East Asian tourism, and the scapes through which such tourist flows move, provide an illustration of some of the cultural aspects of tourism production and consumption (Box 1.2). Huong and King (2002) emphasize how Confucian ideas pervade the Vietnamese approach to travel. Lew and Wong's (2002) work on tourism and the overseas Chinese illustrates the cultural basis of the economic relationships of tourism. And Timothy (2002) discusses the growth of diasporic Chinese communities in west-coast America as tourist attractions for the dominant white population of the United States.

The structure of the book

Chapters 2–4 of the book focus largely on production, albeit in the context of a nexus of production–consumption relationships. In the following three chapters, the scale of analysis shifts more to the individual, the tourist group and the community, and the themes focus more on culture and consumption. The final section concentrates on tourism spaces and places, and their relationships to tourism processes.

Chapter 2 addresses the fundamental question of how tourism is commodified, and this leads to consideration of the nature of tourism in a system of capitalist relationships. Particular emphasis is placed on the value of regulation theory as a way of interpreting tourism production and consumption.

Not all tourism production and consumption occurs in capitalist economies, but – with a few exceptions such as North Korea – it is the dominant system. This has become even more marked following the post-1989 transition in central and eastern Europe, and the former Soviet Union, and the continuing embedding of less developed economies into the system of world trade. The remarkable economic transformation in China provides strong evidence of this, although the transition to capitalism in that country is highly uneven. It is also evident that not all production is commodified in the same way, and perhaps the strongest example of this is the persistence of Visiting Friends and Relatives (VFR) tourism. There are a number of other examples of tourism production/consumption where commodification is less well developed, such as the use of second homes, self-catering (use of own labour in some aspects of producing the tourism experience), or walking and camping in remote areas. But even in these instances, of course, tourism occurs in context of the broader economic system, and the tourists do enter into market relationships in buying transport to the destination, visiting tourist attractions, or simply buying camping equipment or food.

Box 1.2 Tourism and the cultural turn in economic analysis: Asian tourism and migration

Vietnamese values and travel

Behaviour and consumption theories have largely been developed in Western capitalist societies, and there has been only limited research in other cultural settings. Vietnamese tourists, for example, are influenced by Confucianism, Taoist teaching on harmony, and Buddhist direction on personal behaviour (Nguyen and King 2002). Some of the key values of Confucianism are social stability, respect for authority, meritorious social mobility, and clearly recognized mutual obligations. In respect of tourism, this means that, whereas Western travel tends to be formally regulated by rules and laws, mutuality and obligation are more significant in Vietnamese society. This is evident in the importance of VFR tourism, and in the solidarity among members of tourist groups.

Source: based on Huong and King (2002)

The overseas Chinese and tourism investment

'Global tribes' (with shared world views) are increasingly important in economic activities (Kotkin 1993), as is exemplified by the Chinese diaspora, for whom language and shared notions of racial identities are particularly important. This is manifested in both their travel to, and investment in, tourism activities in China. Overseas investment in tourism has been significant, ever since 1978, when the doors to tourism were first opened. It is estimated that more than a third of all direct foreign joint-venture investments in China have been in property development, including hotels, luxury resorts, and golf courses, as well as in housing and offices. The economic experience, knowledge, and material capital of the overseas Chinese have played a critical role in modernizing tourism, but at the same time this has been facilitated by what we can term the social capital of the Chinese 'global tribe'. This global phenomenon has particular place implications. For example, Hong Kong Chinese mostly speak the same language as neighbouring Guangdong Province in China, which has been the focus of 80% of their investments in China.

Source: based on Lew and Wong (2002)

'Chinatowns' as tourist attractions

Chinese migration to the United States was driven primarily by the needs of capitalist production, but, in an unusual example of production–consumption circuits, the resultant migrant enclaves – Chinatowns – have become tourist attractions. For example, San Francisco's original Chinatown was generally considered a 'mysterious' slum, but this image also created a basis for tourism commodification. 'Advertisements promised that white tourists visiting Chinatown would experience the 'sounds, the sights, and the smells of Canton', and they could imagine themselves in 'some hoary Mongolian city in the distant land of Cathay' (Takaki 1994: 53–4). In due course, tourism income and investment transformed the alleyways of the slums into picturesque lanes.

Source: based on Timothy (2002)

But even when we focus on tourism in its commodified form, this is not to say that there is some monolithic world capitalist system and that, therefore, a one-fit theory will do. There may be broadly similar underlying capitalist relationships in different societies but these are culturally and socially contingent, at different scales, whether the national or the local. Institutions also differ among places and countries. One of the strengths of regulation theory is that it directs us to such differences not only in the forms of capitalist accumulation, but also in the way such economies are (culturally and politically contingently) regulated. This is evident, for example, in different approaches to privatization, or to the operating environment of small firms, which are considered later in this book. Moreover, the particularities of tourism production have significance for how tourism is commodified.

Having established a broad framework for the analysis of production–consumption in Chapter 2, the next two chapters investigate production issues in greater detail. Chapter 3 focuses on tourism firms and outlines some of the primary issues in their operations. The discussion revolves around consideration of the two polar forms of organization, the micro firm and the transnational company, but we emphasize that in reality there is a plurality of organizational forms. Thereafter, we focus on what is considered to be the key relationship in the firm, the labour process. This is understood as the organization of labour within and between firms, and how it is combined with, and shaped by, particular technologies and capital investment. Labour accounts for a particularly large proportion of production costs in many, and probably most, sectors of tourism, but is significant in all sub-sectors, even those with relatively high fixed capital costs, such as air transport. Attention is therefore given to how, in the light of the particularities of tourism production, firms have different strategies to reduce labour costs. The chapter ends with a brief review of labour as performance, emphasizing the experiential nature of tourism, and the active engagement of both tourists and tourism workers in the co-production of tourism experiences.

In Chapter 4, the focus shifts from the individual firm to the relationships between firms. Drawing on Schumpeter's (1919, 1939) distinctions between weak and strong competition, we divide the discussion into three parts, First, we consider repetitive competition, that is within existing paradigms (mainly but not entirely on the basis of cost). Second, we turn to disruptive competition, that is within changing parameters, and consider the influence of changes in technology, products, and markets. Such competition is 'disruptive' of existing markets and firm operations, and partly in response to this, many firms have sought to develop inter-firm collaboration. In the third part of the analysis, therefore, we consider various forms of inter-firm relationships, ranging from weak partnerships, through strategic alliances, to merger and acquisition activity. The central argument here is the need to conceptualize firms as having blurred rather than discrete boundaries, and therefore to focus on the multiple and changing forms of their relationships with other firms, and economic agents.

Chapter 5 examines the nature of tourism consumption. It begins by considering the overall significance of consumption in capitalist societies, drawing on the works of Saunders (1981) on access to consumption, and Bourdieu (1984) on cultural capital. This more socially and culturally oriented discussion needs to be read alongside the regulation-theory perspective on consumption, outlined in Chapter 2. More direct links to the discussion of regimes of accumulation are provided in the next section of the chapter, which considers the thesis that there has been a shift from Fordist mass consumption to more individualized and segmented post-Fordist consumption. The arguments revolve, in particular, around the role of the new middle class as signifiers of valued forms of consumption, and around flexibility. The chapter concludes with a critical examination of some of the forms of 'new tourism', which are supposed to constitute post-Fordism. Heritage tourism and ecotourism provide useful vehicles to explore some of these themes. The discussion inevitably points to the complexities of new forms of consumption, and to the ambiguities that characterize much of the voluminous literature on this subject. Ritzer's (1998) concept of 'McDonaldization', relating to the application of technologies, and control over consumption processes, provides one way to reconcile some of the apparent contradictions.

In Chapter 6 the focus shifts to the tourist experience, which is one of the keys to understanding not only consumption, but also production; this is captured in Pine and Gilmore's (1999) treatise on 'the experience economy'. Tourist experiences are examined through the lens of authenticity, which is to be understood as socially constructed. The concept provides insights into both the tourist and the host communities, as well as the interaction between these. This leads to a consideration of tourist behaviour, which is necessarily time- and place-contingent, despite the prevalence of globalization tendencies. Authenticity also raises critical questions about the nature of Fordist, as opposed to post-Fordist, consumption. The question is posed as to whether, for example, backpacker tourism along some of the more favoured tourism scapes has become constituted as a form of post-Fordist tourism, in both the production and consumption of the associated tourism experiences.

Chapter 7 begins with a review of the nature of the impact of tourism on communities. This leads to an exploration of the processes of commodification. Building on some of the ideas outlined in Chapter 2 (which focuses more on the commodification of exchange relationships), Chapter 7 concentrates more on local social systems and cultures, as a way of assessing the impacts of tourism on host communities. In line with our view that individuals and communities are not passive in the face of global tourism, but can contest how they are situated in relation to it, we examine the nature of reactions from host societies towards tourism and tourists. The practices of the host communities also contribute to shaping wider tourism processes. Such discourses are significant, in that the battlegrounds of globalization are becoming less political and economic, and far more cultural. As Ateljevic argues:

> . . . producers and consumers communicate and negotiate between each other in the economic, social, political and cultural (con)texts they create, constitute and (re)produce, which result in the construction of common sense understanding, so-called hegemony. (2000: 376)

The third part of this volume focuses on tourism places. In Chapter 8, we set out a broad framework for examining the relationships between tourism and place. A simple model is outlined, based on the degree of tourism dependency and the extent to which tourism is leading to economic diversification. In reality, of course, tourism places are not (re)made on a blank page. Instead, we argue, it is necessary to consider how regimes of accumulation and modes of regulation are articulated in, and with, particular places. Some of the key themes considered are the concentration of production and consumption, changes over time in the product cycle and the notion of the resort life cycle, how local labour markets are constituted, and the embeddedness of capital. The local character of regulation systems is also discussed, especially the role of interest groups, the concept of governance, and the role of the local state. Against this background, the subsequent two chapters explore a number of examples of how places have been made and remade through their relationships with tourism. The final chapter returns to the themes set out in the introduction in relation to structures and flows, and also explores some of the major challenges being faced by tourism, tourism places and indeed tourism researchers.

This book does not aim to provide a comprehensive survey of its subject matter. That is beyond the scope of this, and probably of any, book. Rather, this book provides what is inevitably a snapshot, capturing – and hopefully advancing – some aspects of our understanding of the rapidly changing phenomenon of tourism. In places, it reviews pertinent debates about tourism, and elsewhere it seeks to provide new insights into how we read these. It is, however, guided by several overarching goals. The first is to help clarify some of the issues inherent in the questions concerning structures and flows set out earlier in this chapter. The second is to contribute to the already increasing interchange between tourism studies and critical social theory (see also Shaw and Williams 2002, preface and Chapter 1). The third is to explore some of the relationships between tourism and tourism places and spaces. Whilst the fourth is to consider globalization, not as some all-pervading process, but as a series of – often incoherent – processes, which create opportunities and risks for tourism, while exposing some of the constraints within which tourism relationships are developed. This firmly directs our attention to the contingencies of time and, especially in this volume, place.

SUMMARY

Much of tourism research has been atheoretical and has abstracted tourism from the broader social and economic relationships within which it is set. This volume aims to make a contribution to the reorientation of such research, by considering a number of theoretical perspectives, some of which, such as political economy, have been relatively underdeveloped in tourism studies. The introduction sets out some of the key reference points for the remainder of the book, as summarized below.

- Tourism constitutes one form of mobility, and it is interwoven with other types such as migration, capital flows, and knowledge flows along IT networks.

- Urry's concept of 'scapes and flows' provides a useful perspective on the tensions between tourism structures and flows.

- Changes in mobilities have to be analysed in context of globalization which, in turn, is understood as increases in the geographical reach and intensity of interconnections.

- Globalization also involves the erosion of the power of national states to shape tourism within their boundaries. However, the extent of such shifts is contested, and there are major differences between three theoretical positions – 'globalist', 'traditionalist', and 'transformationalist'.

- The role of tourism tends to be mythologized but it is characterized by opportunities, risks and constraints on different scales, from the individual to the global.

- The central concern of this book is to explore the relationship between tourism and tourism places and spaces, drawing on a number of theoretical perspectives.

- The book is divided into three parts, which focus, in turn, on: (a) the political economy of tourism, especially of production; (b) tourism consumption and tourist experiences, and the impacts of tourism on communities; and (c) the relationships among tourism places, which are explored through case studies of how places are made and remade through engagement with tourism.

PART 1 PRODUCTION, REGULATION AND COMPETITION

---

2   *Production and Regulation*

THE NATURE OF TOURISM PRODUCTION

There are a number of distinctive features of tourism that play a major role in shaping its production and consumption. This chapter outlines some of them and then proceeds to examine the following themes: the nature of tourism commodification; the extent to which tourism production and consumption are determined by capitalist relationships, with illustrations of how these vary, between mature liberal market economies and less developed and emerging market economies; the insights from applying regulation theory to tourism; and the extent to which globalization undermines the utility of regulation theory, given that the latter tends to assume the primacy of the national state. Below we consider the distinctiveness of tourism production.

Tourism is, above all, place-specific, and it is consumed *in situ*, so that it is strongly entangled with the making and remaking of local communities and nature. Some salient features are summarized below.

- Tourism is *conditional on the production and consumption of a bundle of services, goods and ultimately experiences*. Some forms of tourism experiences, therefore, cannot exist unless particular combinations of services and goods are provided. Inclusive tours are one of the most obvious manifestations of this (the selling of a package of holiday services), but all tourism experiences are dependent on the availability of particular combinations of travel, hospitality services, and tourist attractions. This is why it is essential to study inter-firm relationships in order to understand the production of tourism (Chapter 4). It also accounts for the inherent difficulties faced by policy-makers who seek to change the tourism trajectory of specific places. For example, attempts to enhance tourist experiences of a particular place through improvements to tourist attractions will be constrained if tourists continue to experience poor accommodation in, or inadequate travel to, this place.

- *Property rights – in the sense of establishing the right to use and extract income from a particular asset – are problematic in tourism.* Hann summarizes some of the key issues:

  > [First] property relations exist not between persons and things but between people in respect of things. Second, these relationships are 'multi-stranded' and involve membership of various overlapping groups, based on kinship, the local community, religion etc. Third, property rights can be thought of as forming a bundle, which it is instructive to disaggregate. Rights to regulate and control are usually distinct from rights to use and exploit economically. (2000: 1)

  There are implications here for tourism. First, there are no inherent property rights over tourism assets; rather, these are socially constructed through the relationships among people – either through practices or through formal contracts. Second, ability to extract income/benefits from tourism property depends on membership of various groups. For example, control of the income from a family tourism business is subject to gender and other relationships within that household. Third, while the right to commodify a tourism asset (such as a particular beach) may rest with those with the title deeds to that land, the right to regulate and control it may be vested in the public authorities on behalf of the community. In addition, we should note that many tourism experiences – particular views of landscapes or townscapes, or relaxation in a warm climate – are very difficult to establish property rights over. This raises questions about free-riding and 'public goods', which we discuss later.

- *Tourism is characterized by temporality and spatiality.* Tourists essentially consume tourism experiences at particular sites – although we also acknowledge that anticipation before, and recollection afterwards, are part of the total tourism experience. There is therefore spatial fixity, which gives rise to a number of implications (Shaw and Williams 2002, Chapter 1): strong potential for spatial polarization, necessity for host–guest relationships, direct environmental impacts of tourists, and the need to travel to the tourism site. Moreover, tourism is characterized by perishability. Tourism experiences have to be consumed at particular times and cannot be deferred, e.g. enjoyment of winter or sun tourism in particular places, or events such as the Olympic Games or the World Cup (and, in economic terms, bed spaces in those places have to be filled on particular nights). Consequently, temporal polarization tends to reinforce spatial polarization. Temporality and spatiality are of course relational rather than absolute, and can be modified, e.g. through investment in indoor facilities, snow-making equipment, or new services at previously little-used beaches, in order to extend the tourist season or create new sites of tourism production/consumption.

- Tourism is part of the experience economy (see also Chapter 5), so that *production is incorporated into the tourism experience.* Pine and Gilmore emphasize that 'companies stage an experience whenever they engage customers, connecting with them in a personal, memorable way' (1999: 3). Many traditional services have tried to cash in on this, for example, with restaurants that have become themed, where 'the food functions as a prop' (p. 3) for the entertainment experience. 'Front of house' service employment – for example, as receptionist, waiter, tour guide – involves a strong element of 'tourism performance' (Crang 1994, 1997; Coleman and Crang 2002; see also Chapter 3). The tourism experience is, in fact, likely to be 'multiply-conditional': it depends not only on the performance of a number of producers, but also on that of the individual tourist, and other tourists present at the site of the experience. This set of interactions produces the particular tourism experience, in terms of the atmosphere created, the entertaining repartee while ordering dinner in a restaurant, or watching and even joining in the performance of a local cultural festival.

- Many forms of tourism are deeply *entangled with a socially-constructed Nature.* Urry asks, 'What indeed is nature and should it not include the social as well as the apparently physical environment' (1995: 28). The idea of Nature, and often the physicality of Nature, are socially determined. There are relatively few 'natural landscapes' or 'natures' left in any of the developed countries, especially in Europe. Moreover, there are many different 'readings' of Nature, and these are highly contingent on the individual's cultural lense (Macnaghten and Urry 1998). Tourism has partly been shaped by wider discourses over the significance and meaning of Nature, especially by the way that 'expertise' in Nature has been appropriated by the 'new middle classes' in advanced capitalist societies (see Chapter 5). The natures that tourists wish to experience (whether visually or in any other sensory form) are socially signposted. In turn, tourists – through their presence at particular sites – contribute to the signposting of valued natures. There are also major sustainability issues inherent in the often contradictory practices of tourists seeking to consume valued sites of nature.

- With the exception of a few highly isolated enclaves, or landscapes which are effectively devoid of human settlement (Antarctica, higher mountain ranges, etc.), the mobility inherent in tourism means that *tourists interact with local host communities.* These interactions both inform and constrain tourism experiences, and they punctuate the scapes and flows of tourism. The 'performance' of the community, consciously or otherwise, has major implications for the tourism experience. Tourists experience local communities and cultures – either actively seeking them out, or more passively as backcloths to their own practices (Chapter 7). But the community may variously be irritated by, or adapt its practices to, the existence of tourism; the latter raises issues of authenticity and commodification for local communities (see Chapter 7). Communities may resist or embrace, or simply be overwhelmed by, the influences of the tourists. These host–guest relationships are central to tourism experiences and tourism impacts.

The features outlined above are not unique to tourism. For example, the symbolic combination of production and consumption in performance characterizes many other practices, such as attending the theatre, or going to a shopping mall. Similarly, many other forms of service activity display strong temporal and spatial polarization. *But tourism is distinctive because of its particular combination of production, consumption and experiential characteristics.* Of course, the combination is highly variable among different forms of tourism, even within capitalist societies. We address this issue in the following section, in respect of commodification.

TOURISM AND COMMODIFICATION

Serageldin (1999) provides a useful economic framework for understanding tourism as a source of value. Although designed specifically to address urban heritage, this has a more general application. He contends that heritage provides three sources of value. Extractive use value derives from goods that can be extracted from the heritage site: for example, payment of entry fees, or trading from that site. Non-extractive use value is derived from the services that support the site. These are complex, and not all generate income directly. If the tourist passes through a locale without spending any money on services, there may be no economic non-extractive income. However, a number of service outlets – shops, restaurants, hotels, etc. – will probably extract income from tourists, or what Serageldin terms 'recreational use value'. Finally, historic buildings have non-use value, or what may be termed 'existence value': the simple existence of the historic building yields value to those who would feel impoverished if it was destroyed. This is the value attributed to the Taj Mahal by most people – who have not, and probably will not, visit the site. In the commodification of tourism, we are concerned with the first two types of value.

All modern societies are characterized by commodification, as part of the process of (re)producing their material conditions of existence (Watson and Kopachevsky 1994). In capitalist societies, commodification is based essentially on the values allocated to goods, services and experiences by market-exchange mechanisms. These reflect the level and type of demand for tourism, as well as the conditions of tourism production. Unlike some goods and services, the commodification of tourism is based not only on the labour, capital and natural resources used in production, but also on 'the sign value' or symbolic value of the tourism experience (see Chapter 7). For MacCannell (1976) tourism, as a commodity, is an expression of 'the semiotics of capitalist production'. Tourism commodities can become a means to achieve particular cultural or social goals: the purchase of tourism experiences also represents the purchase of a lifestyle, a statement of taste, or a signifier of status. As a result, some tourism commodities may become 'fetishized' (Watson and Kopachevsky 1994) – which means that they seem to assume a life of their own, and become transformed into 'the sacred'. The exchange values (or prices) of such commodities – for example, visits to famous hotels or exotic resorts – may become detached from the actual costs of production.

Watson and Kopachevsky (1994) argue that the attribution of symbolic value to products is especially pronounced compared with most other types of consumer behaviour. Power over attribution is highly uneven and changes over time. Different social classes, at different periods, have been hegemonic in the attribution of values. Power to signify has shifted over time from the aristocracy (notably associated with the Grand Tour) to the new middle class in recent decades (Chapter 5). The latter group has played a pivotal role in signifying that knowledge of, and engagement with, rural and heritage tourism is an important source of cultural capital. Advertising is especially powerful in signposting, that is in ascribing and expanding the values attached to tourism commodities.

While tourism has been extensively commodified in capitalist economies, there are limits to this. Two of the characteristics of tourism, noted in the first part of this chapter, are especially important here. First, the opaqueness of property rights circumscribes the capacity to extract income from tourism practices. It is no more possible to extract income *directly* from tourists who are wandering around London, Paris, or New York, gazing on their townscapes, than it is from the tourist visually absorbing seascapes or mountainscapes (unless a way has been found of 'gating' these – e.g. via entry charges or the sale of permits providing access to national parks). Second, the fact that tourism has to be consumed *in situ* does, however, provide a means for local service establishments (hotels, restaurants, shops, etc.) to extract income from what are relatively predictable flows of tourists.

Based on these two factors, commodification processes in capitalist societies can be classified as follows:

- *Direct* commodification of the tourism experience, e.g. charges for using a 'gated' tourist site. Examples can include gated beaches and rural sites, as well as theme parks, museums and other tourist attractions charging entrance fees.

- *Indirect* commodification of the tourism experience, that is selling services which are essential to support, or add to, the tourism experience. Examples include travel, accommodation, meals, souvenir sales, and 'accidental or purposive' general shopping. These services tend to cluster around particular tourist attractions in order to capture tourist expenditures at these sites. They are usually shaped by the scapes of tourism, and become embedded into these. In some cases, the owners of the tourist attractions may also be the owners of some, or even all, service outlets. Disney, for example, provides a range of accommodation and eating experiences at its theme parks.

- *Part commodification* of tourism experiences. Examples include self-catering accommodation and car hire – where the tourist provides labour that is not costed at market price, although it may involve considerable personal cost. In second homes, the tourist also provides the fixed capital asset (house), as well as the labour input required to deliver the tourism experience.

• *Non-commodification* of tourism experiences. This includes accommodation, catering, guiding and other services provided by friends and family. Other manifestations include walking around cities, rambling or hiking in rural areas and visiting other valued, but non-gated, tourism sites.

This classification is necessarily generalized. For example, for some tourists, staying in particular notable hotels (e.g. Raffles Hotel in Singapore, or the Savoy in London) can become the objective of the tourism experience, rather than the means of supporting it; in other words, these hotels become tourist attractions in their own right. In this case, those who hold property rights in the hotel are engaged in direct rather than indirect commodification.

Any tourism destination is likely to be the site of all four types of commodification processes, but it is the exact combination of these which, in large part, determines place characteristics. These, of course, are likely to change over time, although not in any pre-ordained sequence. Figure 2.1 outlines some of the possible forms of commodification associated with different types of tourism. These are necessarily simplified, but they illustrate

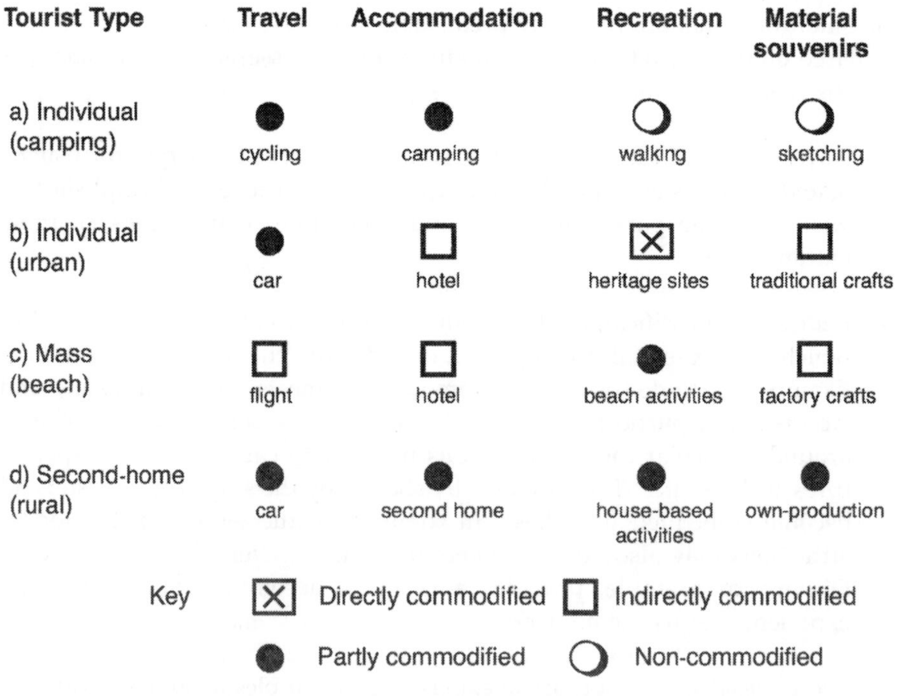

Tourist Type	Travel	Accommodation	Recreation	Material souvenirs
a) Individual (camping)	● cycling	● camping	◗ walking	◗ sketching
b) Individual (urban)	● car	☐ hotel	☒ heritage sites	☐ traditional crafts
c) Mass (beach)	☐ flight	☐ hotel	● beach activities	☐ factory crafts
d) Second-home (rural)	● car	● second home	● house-based activities	● own-production

Key   ☒ Directly commodified   ☐ Indirectly commodified

● Partly commodified   ◯ Non-commodified

Figure 2.1 **Idealized forms of tourism commodification**

the different economic processes involved, and hint at the different economic impacts for local businesses, other economic agents, and the tourists themselves.

While it is not possible to predict the sequence of commodification processes at an individual site, three generalized tendencies can be noted, and these are, to some extent, contradictory. First, there has been a general tendency towards the increased commodification of tourism. Britton, for example, noted that tourism has been 'increasingly commodified as a culture of consumption has evolved' (1991: 453). Second, while there are limitations to the number of gated sites that individual tourists can be attracted to while on holiday, there has been greater scope for the growth of ancillary services that indirectly commodify the tourist experience, such as additional services available in hotels, more themed bars and enhanced retailing. Third, the growing demand for more flexible holidays, combined with an emphasis on privatized consumption, has encouraged the expansion of part-commodification of tourism experiences, via second home growth, and the use of rented cars rather than guided coach excursions.

Material souvenirs provide an interesting example of the various forms of tourism commodification. Souvenir production can be non-commodified – for example, collecting attractive or unusual rocks from a volcano or a beach. It is rarer for souvenirs to be partly commodified, but there are some examples: a tourist may buy raw materials from a local market (perhaps some aromatic dried flowers or herbs) and then, through his/her own labour, add value to these (perhaps packaging them attractively to give to friends as presents). However, most souvenir production falls into the category of indirect commodification, through persuading visitors at a tourism site to purchase objects that they believe will add to the total tourism experience – miniature Big Bens, Leaning Towers of Pisa, and Eiffel Towers epitomize this, as much as the ceramics, leather goods, cheeses or cured hams bought in Italy or France. In some cases, purchase of the souvenirs can become the central object of the tourist experience, where these have been appropriately signposted; for example, glass objects from Murano in the Venetian lagoon. In this case, the production and sale of the souvenirs becomes a form of direct commodification of the tourism experience.

Souvenirs illustrate how commodification mediates the economic relationship between tourism and place. Traditional handicraft products in tourist destinations are often sought out by tourists as souvenirs. Where tourist demand is sufficient, this may lead to the commodification of production, which previously had been for domestic use only. For example, the making and selling of simple cooking utensils to tourists constitutes a source of income for those engaging in this form of indirect commodification of the tourism experience. However, critics contend that commodification destroys authenticity, because – in response to expanding demand – simpler items, or those more appealing to tourists are produced (Greenwood 1977; MacCannell 1973). But handicraft production for sale as souvenirs does not necessarily lead to degeneration of traditional crafts. Cohen (1988), for example, argues that such sales can support existing crafts (creating a market, at the time when factory-produced goods threaten to replace domestic or local production), or even add new ones to meet tourist demands for what they perceive to be authentic items. This 'perceived' authenticity tends to acquire greater symbolic meaning from viewing production in context, rather than simply buying crafts in a shop. However, authenticity is a 'negotiable concept' (see Chapter 6): some tourists may be content with what they know to be mass-produced 'false' craft goods, or with newly introduced but visible crafts, while others demand goods that are historically authenticated.

In practice, there are often complex relationships between commodification and souvenir production, as is illustrated in Markwick's (2001) case study of handicrafts in Malta (Box 2.1). This example demonstrates not only the economic consequences of commodification, but also that this is a culturally imbued process. The purchase of souvenirs can be one of the principal meeting grounds between tourists and the host community. Evans sums this up eloquently:

> Markets are often the prime source of souvenir and artefact, the closest that many tourists get to local interaction beyond the hospitality industry. They do offer a compromise, a meeting ground between both [*sic*] tourist, broker and host community, that is both authentic and staged, since they can be located and housed in strategic sites, away from sensitive and private areas. (2000: 129)

This is a timely reminder that we need to consider economic relationships in their cultural context, and not simply as a set of market mechanisms.

TOURISM, CAPITALIST RELATIONSHIPS AND REGULATION THEORY

The dominant mode of tourism production is capitalist in at least two senses: first, in that most tourism services are produced for markets, and second, even where they are non-commodified, they are produced in societies that are capitalist. Thus, an individual in a developed country may go on a walking holiday departing by foot from his/her house, and could camp on common land – that is a virtually non-commodified tourism experience. But the free time for this experience is generated by working in a capitalist economy, while the Nature that he/she is experiencing is conditional on capitalist relationships (as to which areas are cultivated, and how they are cultivated).

As indicated earlier, the extent of commodification is constantly changing between, and within, societies, and tourism experiences have generally become more embedded in capitalist relationships. For example, in the first half of the twentieth century, rural tourism in southern Europe was characterized mostly by urban migrants making return journeys to visit their families in the countryside, and such visits were often associated with providing free labour to the family farm at harvest time (Cavaco 1995). In the second half of the twentieth century, this was increasingly replaced by the commodified rural tourism of the urban middle classes. Of course, even VFR

Box 2.1 Handicraft production and tourism: two faces of commodification in Malta

There are two organizational forms of craft production in Malta: a 'cottage' industry (production within the home) and small and medium-sized enterprises (SMEs). They differ in their scale, marketing, capacity for investment and spatial distribution (SMEs are often located in 'craft villages', which have been specifically developed as tourist attractions). Both types of craft production have been influenced strongly by tourism development. Not only have tourist numbers increased (from 12,000 in 1959 to 1 million in 1998), but the tourism market has become increasingly diversified, with the growth of cultural tourism.

Two contrasting case-studies

There have been two main forms of commodification of craft work for the production and sale of souvenirs to tourists:

- *Spontaneous commercialization* of traditional household production, for example lace. Tourism demand has helped revive the craft production of lace goods, which had been under threat from mass production and international imports. The result has been product diversification. There is still some production 'as art rather than a craft' of highly skilled pieces, for use in churches, or at weddings and other special occasions. But most production is of simpler pieces, for sale to tourists. Lace-working remains a cottage industry, but with a subtle change: most of the work is now undertaken outdoors in order to bring what had been the 'backstage' of production before the tourists. While tourists are willing to pay relatively high prices for these goods, the producers' lack of marketing knowledge, and the micro-scale of their operations, means they are dependent on commercial intermediaries.

- *Sponsored commercialization* has occurred through the introduction of crafts by foreign entrepreneurs, responding to potential tourist demand. These crafts, as in the case of glass-blowing introduced in the 1960s by British entrepreneurs, are largely unrelated to local culture. Tourist meaning rests more in the process of production than the products, and organized and individual visits to view glass blowing are key economic and cultural relationships. These small firms have sufficient knowledge and capital to engage in both product innovation and direct marketing and sales.

These contrasting forms of indirect commodification involve not only different types of economic and cultural relationships, but also contrasting distributions of tourism-generated income.

Source: based on Markwick (2001)

tourism involves some commodification, such as transport, and local purchases of food and other goods by the host family. Similarly, the shift from serviced accommodation to second homes or to self-catering in many mature tourism markets may reduce and change the form of commodification (see

Figure 2.1), but does not eliminate it. Expenditure on the maintenance of the second home replaces paying for hotel rooms, and food for cooking at home is bought rather than meals in restaurants. But although this involves significant displacement of expenditure socially, and in time and space, tourism is experienced within a framework of capitalist relationships.

Other than producing for markets, the key feature of capitalist production is the relationship between capital and labour, and this is a wage relationship. In this sense labour is a commodity bought and sold in the market place. Class relations are constructed around this relationship and remain central to the functioning of capitalist societies, albeit we have to see these in broad terms. As Wright states, at the concrete – rather than the abstract – level, there is 'no longer necessarily a simple coincidence of material interests, lived experience and collective capacity' (1989: 295–6). Labour is also different to all other commodities, because people are not simply passive elements in production, but are reflective and actively engage in the process. We return to this theme in Chapter 3, in relation to tourism employment as a form of performance. But here we wish to explore further the notion of the centrality of wage and capitalist relationships, by considering the experiences of capitalism in the less developed countries and the emerging market economies in central and eastern europe (CEE).

The reintroduction of market economies in CEE led to a fundamental change in capital–labour relationships, which were increasingly mediated through markets rather than state socialist property ownership and central planning. However, the experiences of CEE have demonstrated that markets are not simply 'market places' where goods are bought and sold in a social and legal vacuum. Therefore, the introduction of price liberalization and privatization of property rights was not sufficient for the creation of effective markets (Williams and Balaz 2000a). Instead, it created a type of raw capitalism in CEE in the early 1990s that has probably never existed in western capitalism, and certainly not in living memory. In the West, markets are a set of institutions and practices (Daviddi 1995: 2) that have evolved over a long time. The institutions include shared values, accepted business practices, and the institutions of stock exchanges and investment banks, among others, in routinizing capitalist relationships. The lack of formal regulation – either via voluntary codes or state legislation – exacerbated the difficulties in CEE, and further hampered the development of tourism in the transition phase, at least in the short term.

The less developed countries (LDCs) also illustrate the complexities of capitalist relationships. Harrison (1992) considered the contributions of modernization theories and underdevelopment theories to understanding tourism in LDCs. According to modernization theories, the LDCs are becoming more like the developed countries (DCs) in their internal structures. Tourism can be seen as an instrument of modernization, for it facilitates the transfer of knowledge, capital and values from the DCs. In contrast, underdevelopment theory stresses that development and underdevelopment are linked in a world economic system, which is characterized by unequal exchange – evident in the relative prices of exports from the LDCs compared with exports from the DCs. In this perspective, the prices paid to economic agents in the LDCs (for tourism, for example) are 'unequal' compared with the prices they pay for imports from the DCs. 'Unequal' can be interpreted in different ways: for example, in relation to the labour inputs required to produce given outputs, or in terms of movement of relative prices over time. In either case, the unequal prices received for tourism services influence the wage relationship between the owners of tourism capital and their workers. The critical point, however, is that unequal exchange mediates but does not alter the fundamental importance of wage relationships.

One further and critical qualification needs to be stated. Although we have discussed DCs, emerging market economies in CEE, and the LDCs in general terms, more concrete analysis requires examination of the conditions of production and consumption in particular countries. Simple predictive theories have limited value. Instead, there is a need for 'situational analysis', whereby researchers situate tourism in relation to key elements in individual countries (Dieke 2000), including the development stage of the country, the roles of the public versus the private sector, and institutional and regulatory frameworks. Although Dieke was writing specifically about LDCs, these remarks apply equally to advanced capitalist economies. Regulation theory provides an useful conceptual framework for situational analysis.

Regulation theory

There are many variants of regulation theory, but here we rely on the interpretation of French writings as summarized in Dunford (1990). Economies are characterized as having a regime of accumulation and a mode of regulation, which are defined as follows:

> A regime of accumulation is a systematic organisation of production, income distribution, exchange of the social product, and consumption. (p. 305)

And a mode of regulation is:

> ... a specific local and historical collection of structural forms or institutional arrangements within which individual and collective behaviour unfolds and a particular configuration of market adjustments through which privately made decisions are co-ordinated and which give rise to elements of regularity in economic life. (p. 306)

In effect, the regime of accumulation is how production, distribution and consumption are organized. As noted earlier, capitalist regimes are distinguished by the centrality of wage and property relations. One of the problems in any regime of production is that while individual decision-makers may be rational in pursuing their own (short-term) goals, they are incapable of ensuring the reproduction of the economic system as a whole, for example by providing the health and education services necessary for an effective labour force. In other words, there is a need for a regulatory system to mediate and normalize the crisis tendencies that are inherent in capitalist accumulation (Tickell and Peck 1992). There are many different foci of regulation, including the monetary system, wage relations and working conditions, competition, provision of collective services (health, education, housing, security, etc.), and international relations.

At any one time, and in any one country, there tends to be a hegemonic (dominant) structure, which is the dominant political, economic and institutional strategy. It is contended that, in advanced capitalist societies, there have been two main hegemonic structures in the last century: Fordism/Keynesianism and neo-Fordism/neo-liberalism. As the Fordism/Keynesianism model became exhausted, and was no longer able to contain recurring crises (evident in declining profits, falling investment, labour unrest, etc.), it was supplanted by neo-Fordism/neo-liberalism. This argument is, however, theoretically and empirically contested.

The main features of the two hegemonic structures are set out in Table 2.1. In short, it is contended that there has been a shift in the regime of accumulation from standardized mass production and consumption, to more flexible markets, segmented along lifestyle lines (Abercrombie 1991). The Keynesian model of state management of economic crises through macro-economic instruments, welfare policies and territorial (urban, regional, etc.) policies have also proven untenable, in the face of globalization and economic liberalization pressures to enhance international competitiveness. In part, this is because of the 'success' of the Fordist/Keynesian model, a version of which was exported to the newly industrialized countries from the 1960s, although with a much stronger role for the state. This contributed to the massive increase in global competition in succeeding decades, which challenged the hegemony of the Fordist/Keynesian model.

These two models are idealized, and concrete analyses reveal varying degrees of liberalism versus interventionism. Therefore, it may be more useful

Table 2.1 **Hegemonic structures: Fordism/Keynesianism versus Neo-Fordism/neo-liberalism**

Fordism/Keynesianism	Neo-Fordism/neo-liberalism
Regime of accumulation	
Mass production/consumption	Flexible production (assisted by IT)
Standardization	Smaller-scale production
Large-volume sales	Increased market segmentation
Economies of scale	More individualized consumption
Mode of regulation	
Strong state intervention	Globalization weakens national
Welfare state (sustains consumption)	regulation
Keynesian economic management (government spending instrumental in countering cyclical and other crises in the economy)	'Rolling back frontiers of the state' Privatization

to think that – at any one time – there are not two dichotomous economic types, but varying composites of regimes of production. There are also significant differences among countries in terms of the hegemonic mode of regulation. For example, Esping Andersen (1990) has identified three distinctive strands of welfare capitalism in western Europe in the 'golden age' of post-1945 Keynesian regulation, which he terms the Scandinavian, the liberal Anglo-Saxon, and the corporate Bismarckian or Rhineland model. Each involves different levels and types of state regulation, and relationships between the state, capital and labour.

Since the 1970s, all these models of welfare capitalism have been under intense pressure from globalization, competition, and increasingly mobile capital. The UK and the Netherlands have been most responsive to neo-liberal agendas, but all the European welfare models have faced similar challenges, and responded to them in similar fashion, to some degree. Neo-liberalism has been strongest in the United States, however. Elsewhere, the 1998-99 East Asia/South-East Asia economic crisis revealed a number of weaknesses in the economic models of the newly industrialized countries, including lack of transparency ('crony' capitalism) and lack of international competition. Subsequently, economic liberalization has gradually led to greater openness to imports and foreign investment, and to privatization. Many less developed countries have also adopted neo-liberal modes of regulation, often as a result of their imposition by the International Monetary Fund. For example, Desforges (2000) reports that in Peru, until the 1990s, the state was the 'main engine of development' (via state-sponsored enterprises and subsidies to the private sector). By the early 1990s, state expenditure had led to massive unsustainable debts, and the response of the Fujimori government was to adopt neo-liberal policies, reduce state expenditure and privatize state-owned companies.

In summary, regulation theory provides a useful level of abstraction about production and consumption, and how these are regulated. It is not a deterministic theoretical framework. Instead, it directs our attention to the analysis of national differences. In the next section, we consider its value for the study of tourism.

Tourism and regulation theory

Most tourism research from a political-economy perspective has been focused on LDCs and their relationships to the DCs through the world economic system (Bianchi 2002). In contrast, there has been less work on the political economy of tourism in the advanced capitalist economies, and surprisingly little research on regulation theory, although a number of authors signal its relevance (Ateljevic 2000; Williams and Shaw 1999). In general, the regime of accumulation has received more attention than the mode of regulation, and Ioannides and Debbage (1998) have produced a thoughtful review of the prevalence of Fordist and post-Fordist regimes in different branches of tourism (Table 2.2). They consider that the central question is whether:

Table 2.2 **Fordist versus Post-Fordist production in tourism**

'Fordist' production in tourism (1950s to 1990s)	'Post-Fordism' and 'flexibility' in tourism (1990s to future)
The production process	
Economies of scale	Economies of scale *and* scope
Mass, standardized and rigidly packaged holidays	Emergence of specialized operators, tailor-made holidays
Packaged tours, charter flights	Market niching
Narrow range of standardized travel products	System of information technologies (SIT) (CRS technology, teleconferencing, videotext, videobrochures, satellite printers, etc.) front- and back-office automation, internet, World Wide Web
Holding of holidays 'just in case'	
Tour-industry determined quality and type of product	Custom-designed flexible holidays
Industrial concentration (horizontal, and to lesser extent vertical, integration)	Tourist-determined product type
	Horizontal integration, subcontracting (e.g. the hotel industry externalizing laundry operations or specialized kitchen activities)
	Adoption of regionally based, integrated, computer information systems and strategic network alliances in the airline industry
Labour practices	
Low labour (functional) flexibility	Functionally flexible (skilled) year-round employees flanked by peripheral, numerically flexible, unskilled workers
High labour turnover, labour is seasonal, low wages	
Mostly unskilled labour force	
The consumption process	
Mass tourists	Independent tourists
Tourists as psychocentrics (inexperienced, predictable), sun-lust seekers, motivated by price	Experienced, independent, flexible (sun-plus) travellers
	Fewer repeat visits
	Demand for 'green tourism' or other alternative forms (e.g. ecotourism)

Source: Based on Ioannides and Debbage (1998)

> . . . trends indicate that parts of the travel and tourism supply system experience various degrees of flexibility in terms of production and labour practices. A key question is to which sectors of the travel and tourism economy can the theorised shift from Fordism to flexible production be most easily applied. (Ioannides and Debbage 1998: 106)

In reality, of course, concrete analysis reveals that in a sector as amorphous as the travel industry, 'with so many permeable boundaries and so many diverse linkage arrangements to exploit, a polyglot of coexisting multiple incarnations has evolved, displaying varying traits of flexibility' (Ioannides and Debbage 1998: 108). Nevertheless, they discern identifiable tendencies:

- *Pre-Fordist* artisanal and craft production is typical of many souvenir shops, small restaurants and lodging houses, which typically are 'mom and pop' businesses. These tend to be small-scale, in independent ownership, weakly managed and reliant on (or exploitative of) family labour. Long hours are combined with flexible working practices. There are mostly low levels of technology, with exceptions such as microwaves, dishwashers and, increasingly, information technology (IT) for websites and email.

- *Fordist* mass consumption and production is typical of large hotels, airlines, tour companies, and cruise ships, for example. These benefit from economies of scale, and the industry is characterized by concentration, and horizontal and vertical integration. The use of IT and various other forms of technology is widespread. Recently, they have engaged in limited product differentiation, such as the 'brand super segmentation' of hotel chains such as Accor (with its targeted Sofitel, Novotel and Ibis hotels).

- *Neo-Fordist* production is increasingly evident, displaying increased flexibility of production and consumption. This takes many forms, including outsourcing to reduce overhead and inventory costs: for example, aircraft maintenance and cleaning, and restaurant and hotel contract-catering (Bull and Church 1994). There is also greater reliance on inter-firm networks and alliances (see Chapter 4).

There is also evidence of post-Fordist trends in consumption, and the possible demise of mass tourism consumption (Urry 1995). There are more flexible holidays, including the growth of shorter breaks and more specialized holidays (Chapter 5). Greater flexibility is also being built into package holidays, with shifts from full board to room and breakfast only, and from serviced to self-catering accommodation. It is, however, debatable whether the latter constitutes a shift to post-Fordism. Instead, Ritzer (1998) argues that tourists still want 'McDonaldized' holidays: these are predictable, highly efficient (value for money), calculable in terms of cost, and controlled (in terms of risk, host encounters, etc.). Fully inclusive tours originally epitomized such products. Recent shifts to more flexible holidays have only been possible because of the growing 'McDonaldization' of host societies. The standardization of many aspects of these local societies – in terms of the types of restaurant, shops, etc. on offer – means that tourists have become less reliant on highly packaged holidays with inclusive services and more willing to eat outside their hotels. But their consumption retains its mass character.

There are other reasons to regard the demise of mass tourism as greatly exaggerated (Shaw and Williams 2002: 239–43). First, mass tourism is still being extended to new markets, both socially (to lower-income market segments) and geographically (to the emerging market economies). Second, many commentaries use evidence of decline in individual resorts (which undoubtedly exists) as evidence of generic decline in this type of tourism, which is questionable. Third, we concur with Hudson's (1997) view that rather than there being a decline in mass production, there has been a shift from mass, standardized to high-volume tourism production with greater flexibility. Although Hudson's comments were directed at manufacturing, they apply equally well to mass package tourism.

While there has been little detailed research on the changing regime of accumulation in tourism, there has been a deafening silence about the mode of regulation. Yet the general mode of regulation, as well as elements specific to tourism, have considerable significance. The role of the national state is critical, although the degree and extent of engagement with tourism is variable among countries and through time. For example, tourism has had a relatively weak voice within the national state in the UK, as was made starkly evident by government reactions to the 2001 foot and mouth crisis in agriculture. The government response, which included effectively closing down mobility into, and within, large areas of the countryside, was predicated on the economic interests of farming, ignoring the greater weight of tourism in many rural economies. In contrast, tourism is more strongly represented in the national state in other countries, especially where it is a key element of the economy, as in Spain (Valenzuela 1998) or many smaller Caribbean islands (Wilkinson 1997).

Table 2.3 lists some key roles of the state in this respect. Although not comprehensive, it illustrates the range of ways in which the state – directly or indirectly – regulates tourism. Moreover, in several areas of intervention, there is evidence that neo-liberalism is impacting on tourism. International tourism mobility has generally been liberalized (although there are exceptions) as has international capital mobility, while reduced welfare expenditure and reduced taxation levels have implications for the distribution of the disposable income available for holidays. Social investment has generally been reduced by national states but there is contradictory evidence that local and regional states have become more involved in economic intervention. In reality, there is no simple dichotomy of the interventionist versus the liberal state, but rather a continuum between these poles. Below we consider each of the roles of the state in turn.

First, the national state *mediates relations with the global economy*, through exercizing control over the mobility of people, goods and capital. This role should not be underestimated, for the OECD, as reported in Ascher (1984), has identified a number of obstacles in international tourism relating:

- to individuals intending to travel (currency restrictions, visas, etc.)

- to companies providing services to facilitate travel (e.g. the rights of tour operators and travel agents), including restrictive requirements in respect of qualifications, rights of establishment and trading rights

- to companies providing transportation (landing, berthing, crossing rights, etc.)

- to companies providing 'reception' or hospitality facilities (controls over imports, limits to foreign ownership, etc.)

National states also influence exchange rates to varying degrees (Vellas and Bécherel 1995: 82–83). Exchange-rate instability is one of the major barriers to

Table 2.3 The state and the regulation of tourism

General	Tourism specific
Relations with the global economy Passport and visa controls, customs (important for border-trading tourism), foreign exchange controls and exchange rates	Tourist visas and tourist exchange controls
Influencing the movement of international capital – inbound and outbound Absolute and conditional controls on the amounts and locations of investment, and levels of profit remittances	Particular incentives or controls on capital movements
Provision of legal framework to regulate production Health and safety laws, company reporting requirements, competition law, environmental protection, consumer protection	Particular laws and regulations for travel agents, tour operators, airlines, etc., dealing with issues such as guarantees against failure, travel safety, and food hygiene
Macro-economic policies Public spending and taxation policies have a particularly strong impact on tourism because of its status as a luxury/basic good	There is no 'one fit' macro economic policy which suits all economic sectors, and tourism – in common with other sectors – seeks to lobby governments to influence its direction. Some countries have social tourism policies which support tourism consumption by disadvantaged sections of society
Intervention in particular regions or localities National and local states may intervene where the local economy faces difficulties, and tourism may be one of, or the lead sector in, any regeneration strategy	Intervention to restructure the economies of tourism resorts in crisis
Reproduction of the labour force Education and training, health and housing, teaching of language and other skills at schools, regulation of wages and working conditions	Training courses in tourism at all levels, housing provision in resorts (important given high land and house prices, and relatively low wages)
Social investment State provision in response to perceived investment failures by private capital e.g. in roads or water supply	Direct state investment in and ownership of facilities such as airports, airlines, and regenerated waterfronts
Climate of security and stability International and national security and stability as an essential ingredient in the removal of uncertainty, which is a major obstacle to trade and investment	Security particularly important given the volatility of tourism demand in face of uncertainty or risk

international tourism, because of the element of risk it introduces into travel planning. In the post-1945 period, the Bretton Woods agreement, whereby exchange-rate values were fixed in relation to the dollar and the pound, facilitated currency stability, and this was favourable to the growth of international tourism. The system proved untenable in the longer term, and from the late 1960s there was transition to a system of flexible exchange rates. Subsequently, the cost-competitiveness of national tourism industries has been strongly influenced by international currency movements and, for example, this contributed to significant demand fluctuations in Spain in the 1990s (Valenzuela 1998). The introduction of the Euro has reduced such currency fluctuations in Europe.

Second, the national state *influences the movement of international capital*, both inbound and outbound. This operates in different ways: there may be general controls on, and conditions applied to, international investment, and the remittance of profits, but these have generally been reduced in the neo-liberal climate of recent decades. Additionally, there may be specific controls applied to particular tourism sectors, most notably the championing of national airlines by governments opposed to their foreign ownership.

Third, the state *provides a legal framework for production and consumption*, which includes health and safety laws, requirements for company reporting, the application of competition law, environmental protection, and consumer protection. There is some international convergence in many of these areas, facilitated by the role of the World Trade Organization in mediating barriers to international trade, and by the pressures for standardized international accountancy systems. But there remain significant regulatory differences, both among DCs and between these and the LDCs. There are also differences among countries in their attitudes to the sex industry, which has implications for sex tourism (see Box 2.2) and child labour.

Fourth, national *macro-economic policies*, including government expenditure and taxation, shape production and consumption. The Keynesian/welfarist approach of governments in the 1950s–1970s played a key role in maintaining employment and disposable income, within a relatively stable framework, and this encouraged consumption, which facilitated a golden age in the growth of tourism, especially international tourism, in the more advanced capitalist economies. These policies have mostly been economy-wide, but there are exceptions – such as social tourism, or taxation on various forms of travel – that have specifically targeted tourism.

Fifth, although there has been some withdrawal of the national state from economic intervention in recent years, *local and regional states continue to be economically active*, and perhaps have become increasingly so in response to the diminished national presence in policy (Chapter 8). There are a number of underlying reasons for this, including legitimation, economic rationality (making the most effective use of resources) and responding to local and regional political pressures and social needs. Such interventions may use tourism to restructure economies in crisis, or may be in response to crises in the tourism industry in particular localities.

Box 2.2   Sex tourism in South-East Asia

Sex tourism is a form of commodification of the body. The evolution of sex tourism in South-East Asia demonstrates changes in the organization of the work of prostitutes, and in the mode of regulation, not only of the recipient countries but also in the countries of origin of the tourists.

There have been four main stages in this evolution (Hall 1994):

1. First, there has been a long-established indigenous prostitution industry, mainly orientated to the domestic market, including domestic tourism, in which pimps have played a local intermediary role.

2. The second phase is characterized by economic colonialism and militarization. Of particular importance in this was the 'rest and recreation' provision of sex services for US military personnel on leave from Vietnam. The travel and accommodation of the military personnel was organized by the public sector (armed forces) in collaboration with private establishments.

3. The third phase has been marked by the growth of fully commercially organized international sex tourism. This was partly in response to regulatory controls over prostitution in most developed countries. International tour companies played an increasingly important role, providing access to foreign markets, either explicitly or implicitly (overtly selling general holiday packages rather than marketing sex tourism directly).

4. Most countries in South-East Asia have now acquired Newly Industrialized Country status, and their economies have become subject to increasing regulation. There has, however, been enduring international sex tourism because prostitution is deeply embedded in gender relations in strongly patriarchal societies.

The regime of accumulation has changed over time, as the scale of prostitution has increased, and as international tour companies have played an increasingly important role in sex tourism. In terms of the mode of regulation, there has usually been increasing state regulation of the tourism industry – at least formally – but prostitution is institutionalized in many of these societies. As Taylor states:

> ... there is a long history of sexual exploitation of women under colonial rule and Western men have long projected racist fantasies onto the 'primitive'/natural 'Other' ... But the long haul tourist industry is turning this kind of lived colonial fantasy into an item of mass consumption. (2000: 42)

Sixth, the national state helps to ensure the *reproduction of the labour force*. This involves a number of general interventions in respect of collective consumption – education, health and housing, for example. The provision of language training is of particular importance for tourism, as is the delivery of specialized tourism courses. In terms of consumption, the health treatment provided in spas is a significant form of tourism in many central European countries. The regulation of housing (whether through social ownership, or intervention in the private sector) is especially important in tourism resorts, given the prevalence of relatively low wages alongside high land and house prices (inflated not only by the tourism industry, but also by retirement migration).

Seventh, national states undertake *social investment* in response to the perceived incapacity of private capital to ensure its own reproduction: in other words, state investment in the face of market failure to effect particular investments. Sinclair and Stabler (1997) outline some of the reasons for this. Many tourism products are public goods, for which direct user fees cannot be charged. Public goods are non-excludable (their use can not be limited to those who pay) and non-exclusive (their use by one individual does not exclude use by others). Even if one private company did invest in such facilities, it would be unable to exclude 'free riding' by other companies and individuals. Examples of public goods include countryside views and the use of seafront promenades. Access to these cannot be gated in order to charge fees, unlike commercial theme parks, or hotels. There are two consequences of this. First, individual owners of capital will be unwilling to invest in such public goods, and they will be subject to overuse given lack of control over access. Tourism (given the tendencies of spatial and temporal polarization) is especially prone to over-use and depletion of public goods. The state may therefore invest to produce or reproduce these, although this is a contested rather than an automatic function.

Eighth, national states play a critical role in *providing a climate of security and stability* for tourism. In different ways, the coup in Fiji and the conflicts in the former Yugoslavia in the 1990s demonstrate just how volatile tourism demand can be in the face of risk and uncertainty, even for well-established tourism destinations. The collapse in international travel after the terrorist attack on New York on 11 September illustrates the critical role of the state in providing security, but at the same time its limitations.

In summary, tourism is closely intertwined with the interests and functions of the state, and – in common with all economic sectors – is affected by shifts in the mode of regulation. However, regulation encompasses far more than the role of the state. As noted earlier, regulation is the 'historical collection of structural forms or institutional arrangements within which individual and collective behaviour unfolds' (Dunford 1990: 306). In terms of consumption, for example, this includes the deeply institutionalized custom in developed countries of taking holidays. While the state plays a role in this – for example, through legislation guaranteeing minimum entitlement to paid holidays for workers, and social tourism provision for the disadvantaged – it is also based on the value that is attached to tourism, as a positional good (defining social status), and to deeply ingrained social routines. There are two important points here. First, the growth of free time and holiday entitlements in developed countries is not an automatic outcome of state development, but has resulted through worker and political struggles, as part of the wider class struggle. This partly accounts for differences between states. Second, while in

Western societies most workers take their full holiday entitlement, this is a socially constructed practice. In Japan, in contrast, workers take only a small proportion of their holiday entitlement – little more than one half, on average. There are also cultural explanations for international differences in whether tourist attractions and shops are open on public holidays or holy days.

There are also socially defined expectations of how tourists will behave on holiday – the result being often a conflict between liberation from, and continuity of, the norms of every day life (Chapter 6). Host–guest relations are mediated by the resulting routinized practices, which have to be seen in context of general mechanisms for maintaining social control and stability.

Not only is tourism shaped by the mode of regulation, but it also contributes to this. Holidays play an important role in the reproduction of the labour force (in the sense of maintaining a fit and healthy workforce). This was particularly evident in the former state socialist economies (Box 2.3), but is also significant in capitalist economies. There is the long tradition of 'work

---

Box 2.3   The mode of regulation: tourism in state socialist societies

Tourism under state socialism was shaped by an ideological legacy, rooted in the Marxist theory of production. Only the production of material goods was considered to be 'real' production, while only those service activities which directly supported production – such as transport – were considered to be productive.

Tourism was considered 'unproductive', and its main role in central and eastern Europe, during the 1950s and 1960s, was to help to regenerate the labour force. In the Soviet Union this was expressed in terms of 'to build healthy bodies and to create good citizens' (Burns 1998: 557). By the 1980s, increasing attention was being paid to other tourism policy objectives, such as generating foreign exchange, and 'promoting friendship between nations', but the prime role was still considered to the reproduction of the labour force.

The state therefore took an active role in promoting and subsidizing holidays for workers. State-subsidized holidays were organized around the principles of a 'socialist way of life', including worker solidarity; they were collectively organized for workers from the same production units. This was facilitated by the development of trade-union and company-owned hotels and other centres of recreation, some of which were located abroad. For example, in Czechoslovakia, by 1985, there were 3,147 collectively owned holiday establishments, with almost 62,000 beds, and several companies and trade unions also owned facilities in Hungary (Lake Balaton) and Bulgaria (Black Sea coast). In addition, there were numerous spas, owned by state insurance companies and the Ministry of Health, which provided a combination of rest and health care.

This extensive system of subsidized collective tourism was consistent with the so-called 'goulash socialism' strategy of the 1970s and 1980s. Improved consumption (including low-cost tourism) was an instrument for legitimating a tough political regime and restrictions on human rights, including the right to free travel.

Source: based on Williams and Balaz (2001)

outings', dating back to the origins of modern working-class tourism in the second half of the nineteenth century, when rest, reward and 'team-building' were combined in workplace-based day trips to the seaside (Walton 1997). Tourism is also evident in the incentive travel schemes used by many companies to reward individual workers, or teams of workers, for outstanding contributions to firm performance (Shaw and Williams 2002: 37–8). Holiday entitlement is also one of the ways in which 'core' workers may be differentiated from 'peripheral' workers in firms: usually through the different entitlements of full-time versus part-time or casual workers (see Chapter 3). There has been surprisingly little research on these, and many other aspects, of how tourism contributes to, and is shaped by, the mode of regulation.

This section has reviewed how regulation theory provides a framework for analysing some of the broader features of tourism. It is not, however, a rigid theoretical framework to be applied mechanically in concrete analyses. Rather, it provides level of abstraction for what must necessarily be place- and time-specific studies.

GLOBALIZATION AND THE LIMITS OF NATIONAL REGULATION

For Held, human communities have become 'enmeshed in networks and systems of interchange – a new era of regional and global movement of people, goods, information and microbes' (2000: 1). Similarly, globalization can be conceptualized as an 'ever tightening network of connections which cut across national boundaries, integrating communities in new "space–time" combinations' (Mowforth and Munt 1998:12). These interconnections operate in different ways: economic interconnections (global flows of capital, and transnational activities, resulting in increased competition); global consumerism, leading to increased homogeneity (around an increasingly hegemonic American model); and global mobility of people, whether for tourism or migration.

Globalization challenges regulation theory, which is based on notions of territoriality, with national states being key sites in the mode of regulation, and therefore having differentiated systems of production, consumption and circulation. The question here is whether globalization heralds the insignificance of the national, and therefore undermines regulation theory? This can best be considered in terms of both the regime of accumulation and the mode of regulation in relation to tourism.

Globalization and accumulation

Consumption, including tourism, has been subject to globalization (the key relationships are set out in Figure 2.2). There has been globalization of media images, aided by satellite television and the internet. This has increased the power of promotion, and the capacity to create international markets for

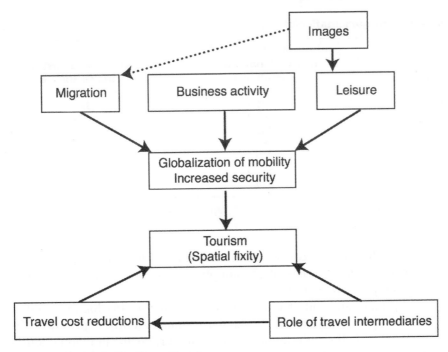

Figure 2.2 **The globalization of tourism**

destinations, whether individually or generically (for example, 'the Alps', or the 'Mediterranean' experiences). Active place promotion by particular localities, together with place and travel promotion by international tour companies, airlines, etc., have also shaped international travel (Chapter 7). Other sources of demand for international travel emanate from the globalization of business activities, and of migration, with the latter generating international VFR tourism (Williams and Hall 2002).

The extent to which latent demand becomes effective demand for international tourism partly depends on the costs of international tourism, especially in relation to intra-national travel (but also to all other goods and services). These have been sharply reduced, partly in response to de-regulation of air travel, which is discussed later. Round-the-world tickets can be bought for significantly less than £1000 (or say $1500, or Euro 1500), the Atlantic can be flown across for about £200 (or $300 or Euro 300), and low-cost carriers, such as Easyjet and Ryanair, have substantial numbers of flights between European countries for far less than £100 ($150, or Euro 150). The power of tour companies in negotiating reduced prices for hotels, car hire, etc. – at least in mass tourism destinations – has also contributed to the globalization of tourism travel.

The globalization of tourism destinations is evident from even the most cursory reading of World Tourism Organization statistics (Table 2.4). These data demonstrate globalization in the sense of intensified interchanges between places across national boundaries. However, they do not necessarily

Table 2.4 **The globalization of tourism**

Year	Numbers of international tourists (millions)	Percentage of arrivals outside Europe
1950	25.3	33.6
1960	69.3	27.4
1970	165.8	31.8
1980	286.0	34.4
1990	457.3	38.2
1995	552.3	41.2
2000	696.7	42.2

Source: World Tourism Organization (1994); World Tourism Organization website, August 2002.

mean globalization of consumption in terms of tourism practices at the destinations. Global arenas of consumption have been created, but greater homogeneity does not automatically follow.

On the one hand, international mass tourism remains highly significant while tour operators contribute to standardization of tourist experiences. They reduce unit costs through standardization, and bridge the individual tourist's conflicting needs for both security and novelty (Britton 1989). However, there has also been the growth of many other forms of international tourism, including VFR tourism, heritage tourism, nature tourism, and urban and cultural tourism. Differentiation is central to these, and – unlike mass tourism – most celebrate place differences. Urry's (1990) discourse on the growth of postmodernist tourism adds further weight to this argument. There is, of course, evidence for 'McDonaldization': international hotel chains, the spread of Disney-like theme parks, and of international restaurant styles and fast food chains, all contribute to this. They make tourism experiences 'homogeneous, calculable and safe experiences' (Ritzer and Liska 1997). This exaggerates the extent of McDonaldization, though, which captures only some aspects of the tourism experiences of some tourists. The extent to which consumption has been globalized remains contested (see Chapter 5).

At one level, McDonaldization is, of course, an expression of the globalization of production and of the organization of the labour process (see Chapter 3). But internationalization of capital is the usual measure of the globalization of production. There is striking evidence that this is a dominant feature of the world economy. In this context, the global growth rates of foreign direct investment (FDI) are three times greater than those for exports, and four times greater than output. Tourism occupies a significant place in these international flows of capital, even in relatively advanced and diversified economies. For example, tourism accounts for about 10–14% of all FDI in Australia (Dwyer and Forsyth 1994). More generally, there are major transnational corporations in particular sectors, especially air travel and hotels. In terms of the latter, there were four companies in 2000 which had hotels in more than 80 countries, while 15 companies operated in more than 20 countries (Hotels 2000).

The globalization of production is closely linked with consumption. For example, there has been significant Korean investment in tour companies, retail shops, and restaurants in Australia's Gold Coast, and in Rotorua in New Zealand (Cooper 2002), which mirrors similar Japanese and Taiwanese investments (Prideaux 2000). In all these cases, the driving force is the growth of Korean, Taiwanese and Japanese tourism overseas, creating market niches for same-nationality owned businesses in the destinations, which provide a form of culturally-specific economic mediation between the tourists and the host communities.

Perhaps the most elegant attempt to theorize international production has been Dunning and McQueen's (1982) eclectic theory of multinationals (Box 2.4). Their application of this theory to understanding (business) hotels is reasonably convincing, although the evidence of the extent to which all sectors of tourism have become globalized is ambiguous (see Chapter 3).

Two qualifications should be noted about globalization of production. First, the extent to which companies are truly globalized, as opposed to internationalized, is questionable. Sklair (1995), for example, considers that the criteria of globalization are the scope of FDI (number of countries operated in, and the lack of reliance on any single market), benchmarking of business practices against the best world rather than national practices, global

---

Box 2.4 The eclectic theory of multinationals

In 'the eclectic theory of multinationals', Dunning and McQueen posit three main reasons for multinational activity, and they argue that these apply to the hotel sector:

1. 'ownership' – of a recognized brand name, providing strong market access;

2. location – customer expectations that hotel chains will be found in particular locations, e.g. major city centres; and

3. market internalization – direct ownership provides companies with greater control and reduces uncertainty, compared with sub-contracting to locally owned companies.

While their theory works relatively well for international hotel chains in the business tourism sector, it has questionable application to other market segments. For example, Williams (1995) argues that local hotel ownership may be preferred by tour companies in the international mass-leisure tourism market. Sub-contracting under these conditions provides greater opportunities for cost reductions, given that quality control is a lesser concern. Sub-contracting also offers greater flexibility to tour operators, who may wish to shift their holiday packages among resorts, or even countries, in response to cost and market-led changes.

Source: based on Dunning and McQueen (1982) and Williams (1995) corporate citizenship behaviour (best responsible practices wherever operating), and global vision in company strategies. Very few companies – even the major airlines and hotel chains – match these stringent criteria. As Knowles et al. comment, 'the global tourism company is a caricature, where the reality is locally derived' (2001: 5). Second, although some sectors may have significant transnational presence, most tourism production remains nationally organized and owned (see Chapter 3). There is, however, a counterargument to this: even small, locally owned firms face increasing international competition, given the globalization of consumption. While this has to be acknowledged, their organization of production, their markets and their inter-firm linkages remain strongly rooted in national spaces. Therefore, while there may be globalization tendencies in the regime of accumulation of tourism, the national remains the key level of analysis.

Globalization and the limits to national tourism regulation

National states have played a major role in the growth of tourism, especially its internationalization in the second half of the twentieth century. State intervention was discussed earlier (Table 2.3), as was the existence of different models of regulation. Some of these differences are outlined in Jeffries' (2001) discussion of national tourism policies, which provides detailed case studies of how these have evolved in response to conflicting interest-group politics.

Despite these national differences, the key role of the national state has, until recently, been unquestioned. Within an overall growth paradigm, there were shifts in this role over time in the developed countries (OECD 1974): the emphasis shifted from removing barriers in 1945–55, to promotion in 1955–70, to greater involvement with infrastructure and regional policy 1970–85. The ethos of these decades was reflected in Iuoto's dictum that:

> ... it is necessary to centralise the policy-making powers in the hands of the state so that it can take appropriate measures for creating a suitable framework for the promotion and development of tourism by the various sectors concerned. (1974: 71)

Recently, however, the neo-liberal agenda and globalization have challenged the role of the national state in tourism regulation. The fiscal crisis of the state, combined with the pressures of global competition, have led to a 'rolling back of the frontiers of the state', which is linked to a parallel debate concerning the 'hollowing out of the state' (Jessop 1994). In essence, this is an argument that the locus of power has moved away from the national state: upwards to global finance bodies, the EU, the World Trade Organization, etc., as well as downwards to local and regional bodies, and outwards into civil society (to NGOs, etc.), as part of the shift from government to governance (see Chapter 8). The role of the state is changing, but we hold to the view that it remains a key site for regulation of the economy. In addition, as Hudson (2001: 71) argues, 'hollowing out' is to some extent a Eurocentric concept, reflecting the

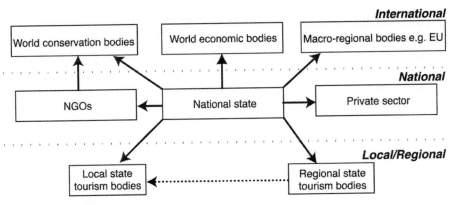

Figure 2.3 **Tourism and the hollowing out of the state**

particular strength of regional movements in Europe and the unique role of the EU as a supra-national body. In any case, arguments about the death of the state are overstatements. Rather, the state exists in new and more complex relations – including partnerships, and multi-level governance – with other tiers of state regulation and with other bodies. In short, there is more evidence for traditionalist rather than globalist theories of globalization (Box 1.1).

The same arguments apply to tourism. There have been tendencies towards the hollowing out of the state with respect to the regulation of tourism (Figure 2.3). Examples of state withdrawal from direct involvement in tourism include the ending of direct state subsidies (Section 4 grants) to tourism in England, and commodification in Spain, where the state-owned hotel chain, the *paradores*, has been privatized. In Canada, there have been pressures for the tourism industry to be more self-sufficient in marketing, leading to Tourism Canada being reorganized along corporate lines, with industry being more involved in its funding and strategic direction (Hall 1998: 207). The private sector is also vociferous in promoting self-regulation, that is, regulation by capital (Mowforth and Munt 1998: 117). Despite these shifts, the national state remains the key site for the regulation of tourism, as was evident in the earlier review of its role (Table 2.2). The example of direct foreign investment in Vietnam also bears out the importance of national regulation (Sadi and Henderson 2001). There was strong foreign investment in Vietnam during 1988–96, including the construction of a Sheraton Hotel and other major tourism facilities. Investment, however, has been constrained by the difficulties of the operating environment. Property rights are opaque, and there are high levels of perceived risk, management difficulties, weak capital markets and poorly developed banking services.

One of the most interesting debates has been about whether there has been the emergence of a system of transnational regulation, which challenges or at least co-exists with the national. This is a complex issue, not least because it is difficult to pin down in concrete analyses exactly what constitutes regulation. Hall (2000: 104) indicates some of the problems, when emphasizing that international law can be considered to be either 'hard' or 'soft', that is firm and binding laws as opposed to statements about regulatory conduct, such as declarations by international conferences, which are non-enforceable norms.

The blurred edges of tourism mean that it is the object of a wide range of international bodies. The most influential of these should be the *World Tourism Organization,* a voluntary body to which most of the world's national tourism organizations belong (Vellas and Bécherel 1995: 260–1). It plays an important role in education, consultation and data gathering, and also organizes international conferences, which seek to influence national and international regulation. For example, the 1980 Manila Declaration was an important declaration about tourism being a basic human need, which should be promoted while respecting tourism resources, improving employment, etc. However, this and other pronouncements – such as the Tourism Bill of Rights 1985, and the Bali Declaration on Tourism 1995 – are examples of non-binding 'soft' international laws.

There are, however, some 'harder' examples of international regulation impacting on tourism, particularly in the conservation field. For example, the signatories to the *World Heritage Convention,* at a UNESCO conference in 1972, committed themselves to identifying and conserving World Heritage properties. The best-known outcome of this is the World Heritage List, which provides recognition, and some protection under international law, of the designated sites (Hall 2000: 116–24). Tourism is also influenced by some of the activities of supra-national economic agencies, especially the *General Agreement on Trade and Services (GATS).* After many years of neglect, the Uruguay Round of GATS finally identified 'tourism and travel-related services' as an important sector for trade liberalization in the 1990s. There were few immediate results, but it did create the framework for incremental growth of negotiated agreements in respect of 'market access', 'national treatment' (i.e. treating foreign service providers in the same way as domestic ones), permitted maximum foreign capital participation, rights of establishment for foreign service providers, and the movement of workers between countries. While potentially a significant step in the shift of power upwards from national states, the power to negotiate the extent and form of any such agreements remained with the latter.

The other focus of the upwards shift of power has been to *macro-regional bodies.* There is considerable macro-regional co-operation in tourism, but these bodies have mostly been concerned with joint marketing, and the coordination of activities. Typical in this respect is the Inter-Sectoral Unit on Tourism of the Organization of American States, the OAS (Table 2.5). It supports national and regional bodies mostly through disseminating information and providing technical support. As such, its activities fall into the category of 'soft law', although it does influence national policies.

The most significant macro-regional body is the *European Union* (EU), which is unique in its decision-making powers, being a cross between intergovernmentalism and cooperative federalism (Kirchner 1992; see also Williams 1994: 200–8). Tourism policy has a low priority within the EU, being caught

Table 2.5 **Macro-regional tourism regulation: two contrasting examples**

The role of the Inter-Sectoral Unit on Tourism of the OAS*	The role of the European Union†
Supporting the Inter-American Travel Congress in policy formulation	Guaranteeing freedom of movement (tourists and also property purchases)
Providing support to sustainable tourism	Transport – general coordination role, but also instrumental in deregulation of air travel
Providing support to other sectors of the General Secretariat in respect of sustainable tourism	Protection of consumer rights (e.g. the package holiday directive)
Supporting hemispheric conferences	Employment and freedom of movement of labour (although practices diverge from regulation)
Formulating, undertaking and evaluating various forms of technical cooperation	
Facilitating exchange of information on tourism	Rights of establishment of firms
Conducting research on tourism	Environmental regulation
Promoting cooperation with different organizations	Culture and heritage promotion and conservation
	Regional and local development policies

*Based on Hall (2000: 129–30).
†Williams and Shaw (1998a).

in an unholy triangle: its inability to compete with other policy concerns; its submergence under other policy directorates, e.g. the Environment and the Structural Funds; and member-state resistance to loss of sovereignty . The evolution of EU interests in tourism is discussed elsewhere (Williams and Shaw 1994, 1998a), and here we only note the extent of EU tourism policy intervention (Table 2.5). It plays a direct role in the international regulation of air travel, it legislates on consumer rights, firm establishment, and freedom of movement, and it funds environmental and local development programmes, which often include tourism components. However, even though its powers far surpass those of the OAS Inter-Sectoral Unit on Tourism, and all other macro-regional bodies, it still remains highly circumscribed compared with national states. The latter decide taxation policies, employment policies, standards applied to hospitality establishments, company statutes, the public–private divide in heritage management, and many other aspects of tourism activities.

The complexities of national versus international regulation are probably best exemplified by *air transport*. There is a long history of international regulation of this, dating back to the 1929 Warsaw Convention, but of particular importance is the 1944 Chicago Convention, which established 'the five freedoms of the air':

- to cross countries without landing

- to land for purposes other than picking up passengers or cargo

- to off-load passengers and freight from an airline of the country from which those passengers originated

- to load passengers onto the airline of the country to which they are destined

- to be able to carry goods and passengers between third countries

Only the first two – and the least challenging to national carriers – were agreed by all the signatories initially. The other freedoms have since been negotiated largely on the basis of bilateral agreements, and there were few of these until relatively recently. As Wheatcroft states, 'The whole structure of international aviation regulation under the 1944 Chicago Convention was based on concepts of national sovereignty over airspace and the "ownership" of rights to carry traffic' (1998: 164).

There has, however, been gradual but generalized deregulation in more recent years. The USA led the way with the 1978 Airline Deregulation Act, which became effective by the mid-1980s. The effects were dramatic, with an expansion in the number of competing airlines, and a reduction of fares on most routes. This had significant implications for company operations, and for the labour process (see Chapter 3). As Holloway comments, 'growth was achieved at the expense of profitability, forcing airlines to cut costs in order to survive. New conditions of work and lower wage agreements were negotiated, some airlines abandoning union recognition altogether' (1998: 93).

There were also high failure rates, both among the new entrants (two thirds of which had ceased operating within a decade), and among major companies, with PanAmerican failing spectacularly. At the same time, there was the growth of a handful of mega-carriers such as Delta, American and United, whose operations were organized around classic hub and spokes routes designed to capture market share. However, these were challenged by low-cost carriers, such as South West Airlines, which competed by linking spokes to spokes, by operating out of medium-sized airports, and by minimizing operating costs.

Deregulation in Europe was later and slower. Protectionism and state subsidies persisted well into the 1990s, with major carriers such as Air France and Iberia being kept afloat by repeated state subsidies. However, the EU did liberalize between 1987 and 1997. From April 1997, in principle all European airlines could compete on any routes, including domestic ones, within the EU, with no capacity or price restrictions. Even the restrictions on cabotage (the collecting and delivering of passengers) within and between third countries was removed in 1997 within the EU (Knowles et al. 2001: 185–6). As in the USA, liberalization led to a spate of new entrants to the industry, but with equally spectacular failure rates. Of the 80 new airlines to appear following liberalization, only 20 were still operating in 1997 (Holloway 1998). There have also been dramatic failures among the major airlines, notably Sabena and Swiss Air in 2001–2, as well as the entrance of highly competitive low-cost carriers, such as Ryanair, Go and Easyjet (see Chapter 3). Liberalization, and reduced state expenditures, has also led to a public–private shift in owner-ship. The proportion of the industry in private ownership increased from 10% in 1988 to 59% in 1994 (Vellas and Bécherel 1995: 153).

The airline industry, perhaps more than any other sector, displays evidence of deregulation and of a shift of power from national states. However, progress remains highly uneven and national states remain key players even in this industry. They still have a role in licensing carriers, in allocating landing and take-off slots (which still frustrates genuine international compe-tition, at least in the major airports), and they continue to subsidize national airlines, even if at lower levels than historically. Nationally differentiated regulation applies even more to most other sectors of tourism.

SUMMARY: PRODUCTION AND REGULATION

Tourism has a number of distinctive features. It involves: the production and consumption of a bundle of goods, services and experiences; complex property rights; distinctive temporality and spatiality; a production that is incorporated into the tourism experience; an entanglement with a socially constructed Nature; and the significance of tourist–host relationships. These, in turn, give distinctive shape to the political economy of tourism:

- Symbolic values are especially significant in tourism commodification.

- Tourism experiences may be directly, indirectly, partially or non-com-modified. These are simultaneously present in capitalist societies, and their precise combination significantly shapes places and tourism relationships. Souvenir production and consumption illustrate the complexities of commodification.

- The dominant mode of tourism production is capitalist, but the examples of the less developed countries (LDCs) and the emerging market econo-mies demonstrate how this takes different, and changing, forms.

- Regulation theory conceptualizes the economy as constituted of a regime of accumulation and a mode of regulation. There has been a contested shift from Fordism and Keynesianism to post-Fordism and neo-liberalism. In tourism, the two forms tend to coexist, alongside pre-Fordist elements of production.

- Mass tourism represent Fordist mass production and consumption. The growth of new forms of tourism can be seen to herald the replacement of Fordism by post-Fordism, but we argue that this is exaggerated.

- The state plays a number of roles in relation to tourism: it mediates global relationships, influences international capital movements, provides a framework for production and consumption, implements macro-economic policies, involves local and regional policies, contributes to reproduction of the labour force, undertakes investment in infrastructure and pursues other forms of social investment, and provides a climate of security and stability. Tourism also contributes to the mode of regulation of society at large.

- Globalization challenges regulation theory, which, to a considerable extent, is premised on the primacy of the national state. Production and consumption are becoming globalized, although there are still limits to the global reach and organization of companies. And there has been a hollowing-out of the state, with power shifting down to the local/regional level, and up to bodies such as the EU and the World Trade Organization. This trend is illustrated by changes in air-travel regulation.

3 Tourism Firms and the Organization of Production

This chapter focuses on the organization of production within tourism firms, especially the labour process. It begins by considering differences in firms, through an exploration of some of the basic features of two polar types – micro-firms and transnationals. It then proceeds to examine what is understood by the labour process, and some of its dominant features in capitalist societies. The chapter then discusses some of the distinctive features of the labour process in tourism, focusing on capital–labour substitution, work orientation and occupational communities, flexibility, occupational segregation, and the central importance of quality and performance in many jobs. Before proceeding to these themes, we first note some issues relating to the fragmentation of the tourism industry.

Knowles et al. consider that the challenges of production in tourism stem from its constitution as well as being what Porter (1980) terms, a 'fragmented' industry:

> ... an industry in which no firm has a significant market share, can strongly influence the industry's outcome, and essentially involves undifferentiated products. Furthermore, the industry appears to represent what could be classified as a hostile environment. That is an environment where overall market growth is slow and erratic, there is a significant upward pressure on operating costs and there is intense competition resulting in high market concentration. Clearly, the tourism and hospitality industry possesses many of the characteristics that would classify it as fragmented with a low market share set within the context of a hostile environment. (2001: 128)

To what extent can tourism be considered a 'fragmented industry'? The key features are considered below:

- It is largely true that no single firm in tourism has a significant market share, certainly in comparison with the dominance of, say, soft drinks by Pepsi and Coca Cola, or computer software by Microsoft. The question is, however, scale-specific. While there may no dominant global players, markets in some of the smaller national tourism economies may be dominated by one or two companies. This is even more likely at the local and regional scales. However, as there is competition among, as well as within, tourism places, the extent to which a company can establish a 'significant market share' is limited.

- Tourism is not necessarily an undifferentiated product. This is, of course, highly variable. Mass tourism is founded on the lack of differentiation inherent in mass production and consumption, but there is also considerable variation in respect of the many specialized forms of niche-market tourism. Significantly, Knowles et al. (2000) make no claim that this specific condition applies to tourism.

- Market growth is not necessarily slow and erratic. In overall terms, and taking the long view, there has been sustained growth of most forms of tourism since the 1950s. Some sectors, such as spa tourism, and some individual resorts (for example North Sea coastal resorts) may have declined. There are also unpredictable annual fluctuations, notably caused by inclement weather or the impacts of political or terrorist events. But, compared with most economic sectors, tourism has experienced several decades of growth.

- Most tourism sub-sectors are characterized by relative ease of entry, which in combination with a high level of competition means that operators often face a hostile environment, in terms of competitive pressures.

In addition to these general conditions, the space–time fixity of tourism consumption makes specific demand on tourism firms in that they have to respond to highly polarized demand conditions, while also satisfying contrasting tourist expectations. The labour process is the fulcrum of how firms respond to such pressures. The labour process is understood in terms of the organization of production, in terms of employment and work practices, and the way in which labour is combined with capital in production. This is the means by which tourism firms seek to control labour costs – usually the most important element in total production costs – while also meeting tourist expectations of the service encounter.

As will be seen later in this chapter, multiple options are available to firms and these can be employed separately or in combination. The strategies adopted may be driven by human agency (the particular drive and goals of individual firm owners), but they also systematically reflect the external operating environment (especially the regulatory framework and the nature of local labour markets), and the broader characteristics of firm organization. The latter is the focus of the next section.

MODELS OF PRODUCTION: MICRO-FIRMS AND TRANSNATIONALS

There are many different models of production in tourism, as in all other economic sectors, which constantly shift during the course of economic development. Moreover, within development studies there is a debate as to whether, during modernization, 'traditional' modes of production are swept away and replaced by capitalist ones. Most commentators, such as Dahles

(1999b), consider that, rather than disappearing, traditional modes coexist with new ones in a 'dual system'.

The coexistence of different models of production is an issue not only in developing economies. The debate about the Fordist-to-post-Fordist shift in advanced capitalist economies does not, of course, envisage a simple, linear and universal transition. Instead, there is coexistence of different forms of organization at any one time, and they may even be mutually dependent: for example, large Fordist-style firms may focus on the major markets, leaving market niches for newer, smaller and more flexibly organized firms. In turn, these larger firms may sub-contract to smaller firms to ease recurrent crises in production. This applies particularly to tourism, because 'in a sector as amorphous as the travel industry, with so many permeable boundaries and so many diverse linkage arrangements to exploit, a polyglot of coexisting multiple incarnations has evolved, displaying varying traits of flexibility' (Ioannides and Debbage 1998: 108).

This 'polyglot' is constituted differently in different societies at different times, being defined by the relationships both between firms (see Chapter 4) and within them. The latter part of this chapter considers relationships within firms in terms of the labour process. As a preface to this, we consider two polar models of firm organization: the micro-firm and the transnational company. These have to be understood as being located on a continuum which runs from the smallest part-time one-person enterprises, through small family firms, independent small and medium sized firms, to large nationally based companies, and transnationals with global reach. However, within the space available to us here, we focus on just two types.

Micro-firms

There are a number of reasons for the existence of micro-firms in the tourism sector. First, tourism markets are contestable, with low entry barriers, so that it is possible for micro-firms to be established and compete in most market segments. Second, post-Fordism and the vertical disintegration of production (see Chapter 2) have also created opportunities for micro-firms, whether serving consumer or intermediate (i.e. other firms) markets. Third, the return for owners of micro-firms may be as much social (independence, status or lifestyle) as material. This contributes not only to high levels of start-ups but also to the survival of firms, sometimes on very thin profit margins. Fourth, the intersection of spatial fixity with small-scale niche or localized markets also favours micro-firms. For example, a particular beach may attract sufficient tourists to justify the establishment of only a single small café. Finally, micro-firms also benefit from the 'small is beautiful' ideology in economic policy, whereby small firms are seen as the 'seed corn' for the growth of larger companies, and as sources of innovation.

The definition of micro-firms is as problematic as the definition of small and medium-sized enterprises (SMEs). The latter is based mostly on employment, although other criteria include management/ownership, market share and turnover (Thomas 2000). Cressy and Cowling (1996) provide a typical definition:

> ... the business has no power to control prices of the products it buys and sells and the credit it gives and receives ... The business is managed by its owners who also control the business. A small business will most likely be a sole trader or a partnership but may also be a limited company. It will typically have fewer than twenty employees, but may have as few as one (the owner-manager) or as many as 500. (1996: 53–69)

Returning to micro-firms, they are even more likely to have these market control, ownership, and legal-status characteristics, while employment is restricted to the owners, their families and few, if any, additional workers. In most analyses these definitions are reduced to a set of, often arbitrary, criteria (e.g. fewer than ten employees). But these have different meanings among countries (with distinctive national regulation systems) and over time, as the conditions of production change. This is especially problematic in respect of a diverse industry such as tourism, constituted by many different forms of production. Persevering with simple definitions of 'small' or 'micro' firms leads to the trap of chaotic conceptualization. There is, therefore, a need for a more grounded approach. Here, we focus more on the characteristics and relationships of micro-firms than on any abstract definitions.

Figure 3.1 summarizes some of the key structural features and relationships of micro-firms, generalizing on the findings of the growing literature on small firms but emphasizing both similarities and differences (Williams et al. 1989; Snepenger et al. 1995; Shaw and Williams 1998; Thomas 1998; Dahles and Bras 1999; Gartner 1999; Morrison et al. 1999; Page et al. 1999; Ateljevic and Doorne 2000; Getz and Carlsen 2000). Both similarities and differences are identified.

As noted previously, there are relatively *low barriers to entry* and high rates of firm start-ups. Only limited amounts of capital, skills and experience are required to establish many types of micro-firms, such as small guesthouses, or kiosks selling food or souvenirs (Williams et al. 1989; Page et al. 1999). However, experience can be critically important in identifying market niches. For example, in New Zealand many outdoor activities enthusiasts have used their expertise to establish micro-firms successfully (Ateljevic and Doorne 2000). Similar examples can be found in the UK: in Newquay, for example, there is dynamic growth of specialist shops and accommodation establishments for surfers, mainly owned and managed by surfing enthusiasts. Sources of capital include savings from previous employment, and inter-generational transfers.

Micro-firm owners may have both *materialistic and non-materialistic goals* and aspirations, and individuals may be located on a continuum between these two poles. The characteristics of particular places will also affect the continuum; for example, non-material goals may be prominent in places with attractive natural environments or lifestyles. The non-material goals include lifestyle, social status, and independence objectives.

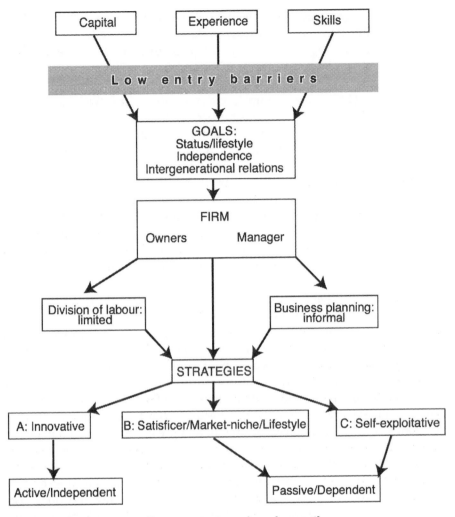

Figure 3.1 **Tourism micro-firms: context, goals and operations**

Many, if not most, micro-firms are *family businesses,* an attribute which has important implications in terms of ownership and control, and the involvement of family members in business operations (Sharma et al. 1996). Indeed, one of the aims of establishing such businesses may be to create jobs for family (and friends). In Ghana this is associated with 'Chop Economics', a system of mutual social obligations, which requires the support of friends and family, and includes the provision of jobs. The implications of this are contested, but Gartner (1999: 172) considers that Chop Economics can be detrimental to efficient business operations, as it may result in the creation of an extended family welfare net. Another aspect of family businesses is that inter-generational transfers are important, both as a goal and as a means of facilitating the supply of new business owners. For example, this was significant in Dalmatia (Croatia), where informal, privately owned tourism businesses emerged only in the 1960s (Ateljevic and Doorne, 2003). Similarly in the Margaret River area in Australia, keeping property in the family or keeping the family together were important motivations for more than one half of the tourism business owners (Getz and Carlsen 2000).

Care must be taken not to over-generalize about family businesses. Singer and Donahu (1992), for example, consider there are two main types: family-centred businesses and business-centred families. Whereas the first represent a distinctive way of life, the latter represent a distinctive means of livelihood. This raises the question of the extent to which micro-firms are 'domestically embedded'. As Sanghera notes, this can be a 'double-edged sword' (2002: 245). The family can be a resource (of capital and labour), although this has to be negotiated in the face of competing demands. At times, this can be 'dysfunctional to the business of making cash, and so threaten its survival and development'. Moreover, the priorities of different household members change through the life-cycle course, which 'highlights the temporal and contingent nature of domestic embedding' (Sanghera 2002: 247).

Property rights in, and control of, micro-firms are typically vested in owner-managers. In terms of Goffee and Scase's (1983) definition of the organizational characteristics of small firms, this links micro-firms to the self-employed and small-employer types. The former use only family labour, and have low levels of capital and weakly developed management skills, while the latter share many of these characteristics, but are more likely to employ non-family labour. Both are significantly different from Goffee and Scase's other two types, owner-controllers and owner-directors, who have more formal management systems, higher levels of capital, and greater reliance on non-family labour. This highlights one of the key features of micro-firms: there is little *division of labour* within the firm; not only are owners also the managers, but they undertake most or all management functions. There are therefore no specialists responsible for marketing, personnel, and other functions. The *weakness of business planning* is further reinforced by the reliance of many micro-firms on personal, family or other informal sources of capital; consequently, they may never have been required to produce a business plan.

Depending on their goals and operating conditions, micro-firms may have very different growth or development strategies, in terms of innovation. This takes us to the heart of the debate about the nature of entrepreneurship, which is usually considered to have three defining characteristics (McMullan and Long 1990): creativity/innovation, risk-taking, and coordination. Some commentators also emphasize that vision and leadership are essential to entrepreneurship. And, in tourism, entrepreneurs play a key role as 'brokers' between host communities and tourists/major tour companies (Jafari 1989; Shaw and Williams 2002). This is not only culturally important, but it also influences the embeddedness of tourism in local economies (Chapter 8).

The empirical evidence on the strategies of micro-tourism firms and their innovative capacity is mixed, and in Figure 3.1 we identify three main types. Studies of small firms in rural/coastal Britain and in New Zealand have found strong evidence of lifestyle-satisficer goals among their owners (Williams et al. 1989; Page et al. 1989). They are risk-averse, and may be content with relatively low returns. They may survive (and earn relatively higher returns) if they occupy specialized niche markets with little direct competition. Otherwise, they may survive through self-exploitation (working long hours for low returns) or be prepared to survive on low material earnings in exchange for non-material lifestyle rewards (type C). In contrast, Ateljevic and Doorne (2000) found that the owners of small firms in rural New Zealand were innovative where niche markets were discovered and cultivated by outdoor enthusiasts, e.g. in black-water rafting and river sledging. Although innovative, they also note that 'conscious efforts by some entrepreneurs to limit the scale and scope of their operations have captured niche market opportunities, hence simultaneously succeeding in striking a balance between economic performance and the sustainability of sociocultural and environmental values' (p. 379). These findings are broadly in line with Storey's (1994) general conclusions about small firms: that they generally spend less on research and development but are more responsive to emerging niche markets. These mixed conclusions are consistent with Shaw and Williams' (1998) identification of models of non-entrepreneurship (type B, in Figure 3.1) and constrained entrepreneurship (those who would be type A, but typically lack sufficient capital for this purpose).

Innovative firms are, of course, more likely than the other two types to be active rather than passive, and also independent, for example by avoiding dependence on intermediary tour companies. The theme of dependency is particularly strong in the literature on tourism micro-firms in the LDCs (see Shaw and Williams 1998). Britton (1989) emphasizes the diversity of small firms in enclave resorts in LDCs: some provide services outside the commercial interests of the dominant tour/hotel firms, e.g. taxis; some replicate services offered by dominant firms, but survive through lower operating costs, e.g. budget hotels and cheap restaurants; and some offer services that are complementary to those of the dominant firms, e.g. entertainment in hotels, guides, etc. Each of these is in a different dependency relationship to the major tourism companies. This issue has become entangled with that relating to the nature of the informal sector, as Dahles (1999a, 1999b) illustrates in Indonesia (Box 3.1). The distinction between 'patrons' and 'brokers' further reinforces the diversity of micro-firms and the need to make our analyses place-specific.

Transnational companies

It is as difficult to generalize about transnational companies as it is for micro-firms, but Figure 3.2 identifies some key relationships. Not the least of which is that transnational companies are engaged in many different forms of international activity. For example, several *functional or relational differences* can be identified, which may be:

Box 3.1   Small firms and the informal sector

Studies of small tourism firms in LDCs tend to associate them with operating on the margins of the tourism industry. But they are 'neither representatives of a traditional, informal, bazaar-style economy, nor do they fit definitions of the completely modern, formal, capitalist sector. They participate in both economies.' (p. 33.)

In Indonesia, there are two main types of small-firm owners. There are patrons (owners of private property, such as small restaurants) and brokers (taxi drivers, guides, etc.) who have privileged access to tourists. Brokers operate more at the informal end of the market, while patrons are more likely to be found at the formal end. In practice, they are mutually dependent for access to tourists, to earn commissions, etc. 'Patronage and brokerage actually constitute a safety belt that allows small entrepreneurs to operate in a rather flexible manner' (p. 33). They rely on networks of family, friends and business contacts, with religious and ethnic bonds also being important. The trust required for the efficient functioning of this system is embedded within these networks.

It is difficult to assess whether such firms are innovative. At one level, they seem to be enterprising and inventive in the exploitation of new market niches, as well as gaps in the regulatory framework. But, while innovative in the use of information, their products are very uniform, not least because good ideas are shared among other small firms. Therefore, 'although they do not react passively to external forces, they are not independent and self-sufficient actors either. They act within certain parameters defined by the entrepreneurial culture they depend on.' (p. 34.)

Source: Dahles (1999a)

- Trade-based: these firms engage only in sending tourists abroad, and sub-contract the provision of all services outside the national space to foreign intermediaries. For example, travel agents may purchase services from foreign hotels and transport carriers to sell in their national markets.

- Production-based: these firms have a presence in international destinations, providing tourism services for customers of any nationality. Examples include hotel chains, and international theme parks such as Disney's. These services can be provided through a number of formats:
  - (a) franchising (for example, McDonald's);
  - (b) partial production (e.g. owning hotels, but not the other services used by tourists);
  - (c) virtually complete ownership of the production of tourism services (e.g. holiday centres such as Club Méditerranée)

- Market/marketing-based: horizontal expansion into selling tourism services in other national markets, such as the take-over of British tour operator Thomson by Preussag of Germany.

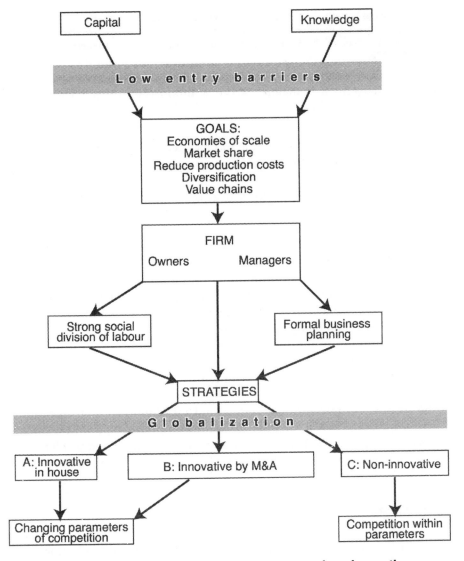

Figure 3.2 **Tourism transnational companies: context, goals and operations**

Transnational companies have a number of comparative advantages in internationalizing their activities. They have favoured access to capital, either from their own reserves, or in financial markets. In entering a new international market, they may be able to draw on sector-specific knowledge acquired elsewhere. This is important in the internationalization of business hotels, because the client group (international business people) tend to have standardized expectations and to share previous experiences. Go and Pine write that major transnational hotels:

> . . . draw on their critical mass to compete on a worldwide basis. In particular, they build on the knowledge and expertise gained over the years in many

countries to benefit the entire network. The 'global' hotel players also build on their integrated worldwide business system through resource transfer – in particular, through the transfer of technology, capital, and managerial staff. (1995: 272)

Transnational tourism companies have a number of *goals*, including the following:

- the search for economies of scale to offset technology development costs (e.g. airlines, or some forms of IT)

- the offsetting of marketing costs (e.g. the advantages of internationally known brands)

- pre-emptive strikes to secure market shares in newly emerging markets (particularly important for airlines)

- the reduction of production costs through seeking out lower labour costs (e.g. tour companies in the mass market seek out new and lower-cost resorts to maintain competitiveness)

- the diversification of their presence in different markets, given the high levels of uncertainty in international tourism markets

- the securing of international vertical linkages (between tour companies, airlines, hotels, etc.) in order to provide information flows, reduce uncertainty and provide inputs at known prices (Sinclair and Stabler 1997). This can also be represented as more effective extraction of value from the value chain

- responding to the internationalization of demand.

These goals may be contradictory and, for example, diversification (into different branches of activity) may have to be traded off against the economies of scale that can be gained from focusing on a single activity.

Ownership-management in international tourism companies can be highly personalized, as some major companies are family-owned and managed. However, they are more likely to have a formal division between ownership and management. The former often involves some form of public share ownership, although the shareholders may be major financial institutions or individuals. The ownership of larger companies is likely to be complex and internationalized and their ties to their country of origin may be weak. This highlights the difficulties of attributing nationality to many forms of increasingly globalized capital. Management involves various forms of centralized-versus-decentralized approaches to coordinating complex international activities. There will be strong social divisions of labour within such a company, with specialized departments devoted to marketing, public relations, financial control, design, etc. Business planning and reporting will be highly formalized and extend over different time-frames.

These companies have diverse *globalization strategies* and there are differences in the extent to which their operations are internationalized or globalized. For example, in a general review, Taylor and Thrift (1986) identify three types of international activity. First, global corporations, such as ITT/Sheraton, operate on a genuinely global basis in terms of their geographical range and lack of reliance on any one market. Second, multinational companies, such as Accor, tend to be more macro-regional rather than global in scope. Third, there are companies that are mainly national in orientation, but with some international interests. Go and Pine (1995) identify a broadly similar distinction in the hotel industry, based on the national–international orientation of companies (see Table 3.1). As we argued previously, there are few global or geocentric companies in tourism, although some international airlines (through strategic alliances – see Chapter 4) and international hotel chains aspire to become global corporations. One of the barriers to this stems from the spatial fixity of tourism, so that delivery of tourism services has to occur *in situ*, requiring conformity with different systems of national regulation (see Chapter 2). Globalization and internationalization also differ among the different sectors of the tourism industry, as evident with air travel, hotels and tour operators (Box 3.2).

It does not automatically follow that all transnational companies are innovative. Instead, we suggest three possible *innovation and competitiveness* statuses. Some firms may be non-innovative, and therefore may be facing increasing survival crises. Other companies may have strong internal processes of innovation, combining both formal research and development capacity, and the ability to harness the creative energies of their workers. The third status involves importing innovation capacity, through mergers and acquisitions. In the first example, there is repetitive competition (Schumpeter 1919; 1939) or competition within existing parameters, while there is strong competition, or disruptive competition, in the second two examples (see Chapter 4).

Table 3.1 **Typology of transnational companies by international orientation**

Type	Characteristics
Ethnocentric:	Home-country orientated; operations abroad mostly limited to foreign markets similar to those in home country
Polycentric:	Host-country orientated; foreign subsidiaries managed by nationals, and firms are administered on diverse national bases
Regioncentric:	Macro world-region orientated; based on markets with similar economies and cultures, and offering significant economies of scale
Geocentric:	Globally orientated; standardized products sold widely across diverse world markets

Source: Go and Pine (1995: 4–5).

Box 3.2 The extent of tourism globalization: air travel, hotels and tour operators

(a) Air travel

Incentives for globalization:

- High costs of technology

- Development of international hubs and spokes to capture market share

- Branding advantages (reinforced by frequent-flier programmes)

- Internalization of linkages, both horizontally (acquisition of other airlines as part of the development of international hubs and spokes) and vertically (purchase of hotels)

Barriers to globalization

- National sovereignty (priority given to 'flag carriers')

- National regulation (control over take-off and landing rights)

(b) Hotels

Incentives for globalization:

- Ownership or proprietary rights over differentiated and branded product.

- Internalization of transactions through local ownership, rather than relying on sub-contracting

- Relatively homogenized business-tourism market segment

- Multinational business corporations' favouring of preferential trading rights with single hotel groups

- Pre-emptive investments in key sites (e.g. city centres, or near airports) to secure market share

Barriers to globalization

- Some forms of mass tourism are cost-driven, and ownership considerations may not be relevant in these markets.

- Individualization provides competitive advantages (while this does not preclude global chains, it provides a market niche for hotels which can claim they are independent, and owner-managed)

- National regulations (especially in respect of accountancy, building and environmental standards)

Box 3.2 *Continued*

(c) Tour operators

Incentives for globalization:

- Brand loyalty to tour companies, in the face of perceived risks in international travel

- Globalization of demand requires tour operators to have presence in increasing numbers of destinations

- There are some economies of scale in IT, brochure production costs, and in negotiating costs with local suppliers. This favours presence in both increasing numbers of international markets and destinations.

Barriers to globalization

- National differences in tourism consumption practices

- National regulations (about flights, rights of consumers, etc.)

THE TOURISM LABOUR PROCESS

The labour process is central to the competitiveness of micro-firms and transnational companies, and indeed of all firms. This is understood as the organization of the labour force, in terms of work practices/workplace organization and, therefore, implicitly includes the way in which labour is combined with capital (and technology). The labour process is not 'given' in any economic system, but is negotiated among capital, labour and (through its regulatory role) the national state. This does not mean that it is necessarily the focus of either cooperation or conflict between capital and labour. Instead, it should be conceptualized as a 'tension', which requires management (Burawoy 1979).

The emphasis on management of the labour process is significant. There are examples – such as slave labour or non-registered migrants forced to work in sex tourism – where workers are coerced through fear and violence. But these are exceptions and, as Baldacchino argues, usually 'workers cannot be forced to work without a modicum of consent on their part; nor do workers agree to sell an exact quantity of labour' (1997: 92). The amount of work done, and the manner in which it is done, requires consent and active worker input. This applies particularly to 'frontline' service employees, who have to respond (perform) to the emotional needs and expectations of clients, as well as to the requirements of managers and, at a more abstract level, of capital. Therefore, workers are not passive participants in the labour process.

One of the key features of the labour process in tourism is that, in most sub-sectors, it represents a relatively high proportion of total production costs.

For example, labour costs accounted for 32% of US hotel revenues in 1989 (Pannell, Kerr, Forster Worldwide survey reported in Go and Pine 1995: 116). Labour costs are less dominant in other sub-sectors, for example airlines, but still of central importance. There is considerable downwards pressure on wages, as confirmed by Riley et al.'s review of the empirical evidence on labour costs in tourism (2002: 40–1). They note, first, that over time – between 1980 and 1997 – the ratio of labour costs to total sales has been held constant. Second, that average tourism earnings have consistently been lower than average earnings in all non-managerial, manual non-agricultural jobs. Figure 3.3 updates and extends their analysis. In most countries, average tourism wages are lower than average wages as a whole. In more than half the countries for which data are available, they are less than 80% of average wages, falling to less than 60% in seven countries. At the other extreme, in only 11 countries do they exceed average wages and, with the unusual exception of Hong Kong, these are all less developed countries (LDCs) or early emerging market economies. Therefore, Riley et al. present evidence of not only the effective downwards pressure on tourism wages, but also the critical need for management of the 'tension' in the labour process.

Riley et al (2002: 59–69) provide a seminal review of the downwards pressures on tourism wages (Figure 3.4). They identify three main sets of factors. First, there are the job attributes of attractiveness, a desire to learn, acquisition of transferable skills and ease of learning. These contribute to a large potential labour market, high levels of mobility and detachment of productivity from skills level. As a result, managers take a short-term view of employment whilst seniority is also poorly rewarded through higher wages. Second, there are industrial-structure and economic factors. Fluctuations in demand lead to greater emphasis on employment flexibility, while the small scale of most tourism enterprises means there are only limited opportunities for advancement within firms, and weak occupational hierarchies. Third, there are psychological issues. Employees obtain non-material job satisfaction from employment as well as wages, so are more tolerant of low pay. There is also a tendency to use other tourism jobs (rather than similar skilled jobs in other industries) as the referents for pay. Realistic low expectations of pay and advancement also contribute to lower recruitment wage rates.

There are many examples of how tourism firms exert downwards pressure on wages. For example, Vellas and Bécherel (1995: 135–6) report that airlines facing financial crises in the 1990s responded mainly by reducing wage costs. This is not surprising, given that the cost of crew per ton/km was seven times greater in the most expensive airline (Pan Am) compared with the lowest-cost operator (Korean Airlines). British Airways, which made large operating losses before privatization, illustrates how restructuring the labour process sharply reduces operating costs (Vellas and Bécherel 1995: 154). The labour force was reduced by 15,400 employees between 1981 and 1983, while an employee bonus scheme linked to company profitability was implemented. The combination of changes in recruitment, employment levels and working practices resulted in a productivity gain of 40% between 1981 and 1986. The

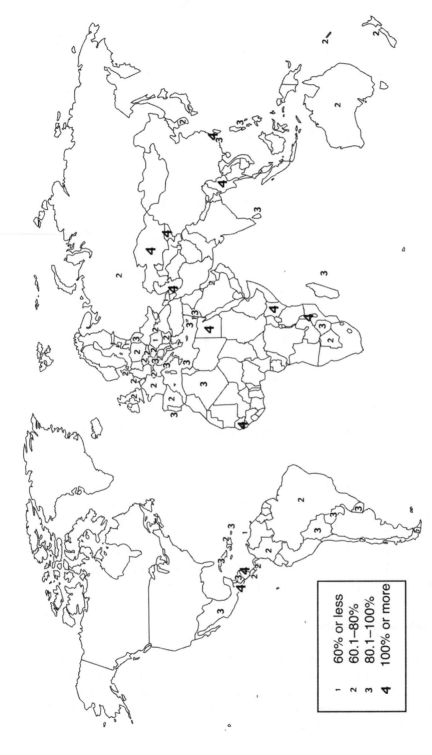

1	60% or less
2	60.1–80%
3	80.1–100%
4	100% or more

Figure 3.3 **Average earnings in tourism: percentage of average non-managerial earnings, 1999 (source: based on *ILO Yearbook* 2001)**

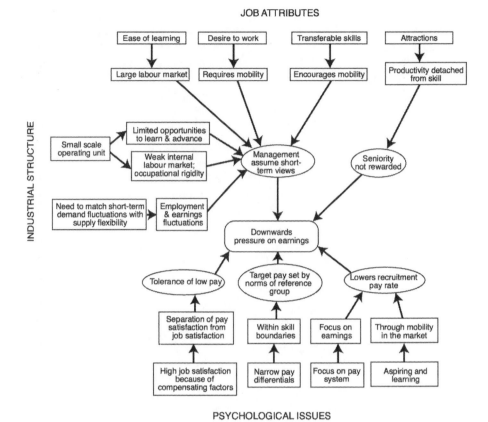

Figure 3.4 **Downwards pressures on tourism wages (source: after Riley et al. 2002: 58–9)**

management of the labour process should not, however, be regarded as some ruthless and incessant body substituting capital for labour, increasing the rhythm of working practices, and eroding working conditions and pay. The provision of tourism services is too complex for this view, as the discussion below indicates. There is a need to consider the subtleties of the labour process within tourism firms, as well as how these interact with the external environment, especially that of changing labour markets.

Capital–labour relationships

The deployment of capital and labour is strongly intertwined with employment and working practices in tourism. Milne and Pohlmann (1998) stress that even the hotel industry – considered one of the most labour-intensive sectors – is both labour- and capital-intensive. Technological change leads to a rebalancing of capital and labour inputs in the provision of most services: 'Front-office and back-office duties, cleaning and room preparation, security, maintenance, laundry, and restaurant/bar services are all experiencing major technological change' (p. 188). For example, mini-bars and kettles may be provided in rooms in place of room service, check-out facilities may be provided via PCs at reception or via televisions in clients' bedrooms. This has the advantage for management of releasing frontline staff to actually focus on the clients.

It should not be assumed that there is an inexorable substitution of labour by capital in hotels, or in any other sector. Management decisions take into account the relative costs of capital and labour. The costs of labour in relation to unit costs of production can fall, even if capital and technology remain constant. Wages may fall in response to changes in the external labour market (e.g. increased unemployment or increased in-migration) or because of changes in working practices within the firm. Given these conditions, costs can be reduced by employing more labour relative to capital. Management may also decide not to increase the capital–labour ratio in response to technological developments because the company strategy prioritizes the quality of labour, such as, highly attentive and individualized service. There are also limits, within current and foreseeable technology, to the substitution of capital for some types of workers: for example, ski and other sports instructors.

The substitution of capital for labour has implications for the recruitment and retention of workers, as well as for working practices, but the outcome is contested, not least because it is place- and sector-contingent. On the one hand, Milne and Pohlmann argue that:

> ... technology seems likely, therefore, to raise the overall skill profile of the hotel work-force by reducing, in absolute terms, the number of unskilled or repetitive background functions, and broadening the customer contact functions and technological skill requirement of front-line workers. (1998: 188)

On the other hand, technology facilitates the replacement of craft production by Fordist production, where work is routinized, repetitive, de-skilled and specialized; for example, the de-skilling of cooking in McDonalds and other fast-food chains. However, capital–labour shifts in the restaurant business can also lead to removal of unskilled and skilled jobs, as Bagguley (1987) demonstrates in respect of the introduction of automatic dishwashers and cook-chill technology. In fact, the argument about de-skilling is more complex than is suggested by direct substitution effects. Given the pressures for more flexible working practices in tourism (discussed below), the emphasis may be on multiskilling rather than de-skilling, and the former may constitute a form of re-skilling. For example, in many budget hotels a single worker is responsible for, and needs the appropriate skills to deal with, customer accounts, check-in, and providing bar and breakfast buffet services.

Work orientation and occupational communities

Tourism employment is characterized by distinctive psychological features, that are considered here in terms of work orientation and occupational

Table 3.2 **Tourism-work orientation**

Orientation	Attributes
Instrumental utility orientation:	Tourism perceived as a means for economic advancement
Positive commitment to tourism:	Tourism favoured for the intrinsic values of the jobs it offers, for example their image, pleasant surroundings, and variety of tasks
Refugee orientation:	Tourism as offering an escape route from a declining industry, an unpleasant job or unemployment. For some, tourism is the 'least worst' option, whereas for others it is an opportunity for improvement. Tourism is seen as a contingency or convenience
Entrepreneurial orientation:	Tourism appreciated for its suitability for private business activity, or at least seen as a potential route to entrepreneurship

Source: Szivas and Riley (2002).

communities (see Figure 3.4). Goldthorpe et al. (1968) developed the use of the term 'work orientation', emphasizing that there are 'holistic' (i.e. not simply materialistic) attitudes towards work, derived from the way people live in society. Recently, Szivas and Riley (2002) extended this research to tourism employment, and suggested a fourfold classification of 'work orientation' (Table 3.2). Material considerations are foremost among those with instrumental utility orientation, while refugee orientation is also largely materially driven, as is entrepreneurial orientation. In contrast, positive commitment orientation emphasizes perceived and valued non-material concerns.

There are a number of aspects of non-material rewards. Some are related to place association – tourism jobs happen to be in attractive locations. This is particularly important in attracting migrant workers (Box 3.3). Other attractions may include the variety of tasks to be undertaken in a flexible environment, and opportunities for host–guest interactions (including work as performance – discussed later). Job satisfaction also stems from the notion of occupational communities (Lee-Ross 1999), where the spheres of work and non-work are blurred. This typically may involve small businesses in local communities, or businesses with regular return visitors, where there are possibilities for close social interaction with family, friends who are co-workers, or customers who are friends (Marshall 1986). The psychological rewards from such occupational communities may offset considerable self- and family exploitation by those working in these establishments. In other settings, the 'isolation' of many tourism workers through working long and unsociable hours also reinforces the identities of occupational communities (Riley 1984). These features make the industry more attractive to employees than would otherwise be the case, and more willing to accept lower wages or greater intensification of the labour process.

Box 3.3 Work orientation, tourism and migration

Uriely (2001) conceptualizes migrant workers in terms of their engagement in tourism, and their tourism oriented motivations:

- *Travelling professional workers*: mainly work-related, and engage in tourism activities as a byproduct of travelling

- *Migrant tourism workers*: travel in order to make a living, but only among tourism places, given their pleasure orientation

- *Non-institutionalized working tourists*: work while travelling to support their trip

- *Working-holiday tourists*: work is part of their tourism experience, e.g. volunteer conservation workers

The first type are migrants, with dominantly material motives, who work in tourism mainly because of the employment opportunities available. The second type, migrant tourism workers, have mixed economic and tourist motivations. They are attracted to particular tourism destinations because of their tourism attractions, and they work in order to support their visits (often seasonally). For the third type, the primary motivation is the experience of travelling abroad, and for some this may be a form of adventure tourism, with elements of self-discovery. Work (in any sector) is instrumental in supporting their tourism objectives. Finally, there are those for whom work is part of their tourist experience, notably those working on conservation projects in attractive or challenging locations.

Uriely's work mirrors that of a number of other commentators (see the overviews in Williams and Hall 2000, 2002). Bianchi (2000), for example, uses the concept of 'migrant tourist-worker'. This refers to a category of mobile resort workers (ski and surf instructors, cooks, etc.) who have, to various degrees, abandoned their places of origin and habitual residence, and have opted to seek adventure, work and self-fulfilment through working in and around tourism resorts. In another example, Mason (2002) comments on the young New Zealanders travelling around Europe as part of the 'Big OE' (overseas experience). This constitutes a 'rite of passage', and when they work to support their travel, they represent 'non-institutionalised working tourists', following Uriely's terminology.

Flexibility and temporality

The spatiality and temporality of tourism means that most sectors face uneven and unpredictable demand from tourists, and therefore uneven need for tourism labour. In family businesses, the required labour adjustments 'have been internalized; family members, and even the owners, may work in the tourism business for only part of the year, or part of the week, while providing a labour reserve to counter erratic fluctuations in demand' (Agarwal et al. 2000: 249). Non-family businesses adopt sophisticated strategies to meet such demand fluctuations.

Atkinson (1984) set out what has become the classic conceptualization of flexibility, in his work on manufacturing. He distinguishes between numerical and functional flexibility: the first implies changes in employment levels in response to demand fluctuations, while the second suggests the movement of workers between tasks in response to changing demand at different points within the establishment. This gives rise to the notion of core and peripheral workers in the labour force: the former have greater security of tenure, while the latter are employed on various forms of short-term, part-time or casual contracts. Shaw and Williams (1994) extended this conceptualization to the tourism industry, when they classified the variety of employment conditions in tourism – casualization, temporary, seasonal, part-time, homeworking under contract, etc. – in terms of four axes: regularity of working hours, functional–numerical flexibility, employment security, and availability of material and fringe benefits.

Atkinson's conceptualization has been subject to considerable scrutiny. Urry (1990) raised two significant questions. First, he argued that while core and peripheral workers are observable in terms of employment conditions, there is no simple association with numerical versus functional flexibility strategies. Second, he argued that permanent and temporary workers often do exactly the same jobs within a company. Lockwood and Guerrier (1989) have also critiqued the model, based on their survey of major UK hotels, in which they observed relatively little functional flexibility in hotels, with tight job demarcations prevalent. There was evidence of numerical flexibility strategies, but the wages and benefits of part-time workers were not significantly different to those of 'core' workers. In part, this is because – contrary to the implications in Atkinson's dichotomy – employers value casual and part-time workers as much as 'core' workers. Similarly, Milne and Pohlmann (1998: 188) found numerical flexibility was common in Montreal hotels, with just over one third of them reducing casual staff by at least 50% in the low season, and another third making lesser cuts. Both studies were restricted to larger hotels, and this leaves open to question whether their findings are applicable to smaller hotels or to other tourism sectors. However, in the most comprehensive review to date of the evidence in the hospitality sector, Wood (1997: 172–3) concludes that there is extensive use of numerical flexibility, but less evidence of functional flexibility; not least, departmentalism within larger organizations militates against the latter.

Although Atkinson's work has received most attention, it is not the only conceptualization of flexibility. A related but broader perspective is provided by Rimmer and Zappala (1988), who identify five main types of employment flexibility (see Figure 3.5). In addition to numerical and functional flexibility, they also identify working-time flexibility (incorporated by Atkinson into the numerical category), wage flexibility, and procedural flexibility. The latter refers to the framework for consultation within the firm and, in a sense therefore, is germane to all other types of flexibility. Numerical flexibility is of course directly dependent on relationships to the local labour market, but the latter also conditions other forms of flexibility, e.g. downwards wage

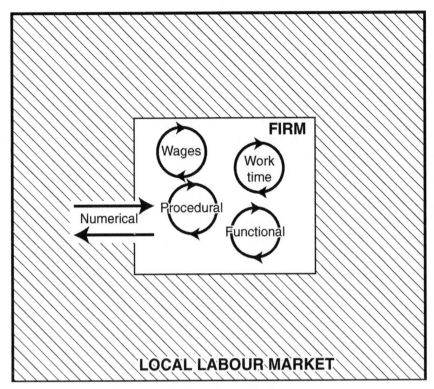

Figure 3.5 **Sources of labour flexibility in tourism (source: after Rimmer and Zappala 1988)**

flexibility is less likely in tight labour markets. This model has been applied to registered clubs in New South Wales, Australia (Buultjens and Howard 2001), where they found that employers valued work-time flexibility most highly, followed by functional and numerical flexibility. Wage flexibility was considered least important, perhaps reflecting the difficulties of adjusting nominal wages downwards in most advanced economies for institutional and regulatory reasons (trade union resistance, customary firm behaviour and worker expectations, minimum or national wage-setting, etc.).

There is evidence, then, of some forms of flexibility being used to reduce costs, through de-skilling and re-skilling. However, it would be misleading to see this as all-pervasive. Instead, as Wood argues, 'the suspicion is that managerial strategies towards deskilling ... are less a matter of unified intent than uncoordinated stumblings towards some hazy ideal of efficiency and administrative improvement' (1997: 181).

Labour-market segmentation

Labour-market segmentation, drawing on wider social cleavages, offers employers possibilities to depress labour costs. The segmentation of workers (by race, age, gender, etc.) provides a basis for paying lower wages to some (usually more weakly organized or vulnerable) social groups. The nub of this is the social construction of job content, linked to the system of remuneration. Some jobs are constructed as 'unskilled work', because they are undertaken by particular social groups, and this is used to justify paying lower wages, irrespective of the real skill content of these jobs. Employers can use social divisions as a systematic way of simultaneously unifying and dividing workers. This is not to say that workers should be viewed as passive in the labour process, since they contribute to creating such divisions, as well as contesting them (either individually through the courts, or working through unions and other labour organizations). However, the role of human agency should not be exaggerated, for the social divisions that people bring with them into the labour force are structurally and collectively created (Hudson 2001: 200). These ideas are explored further through consideration of two major dimensions of labour-market segmentation: gender and migration.

Women and men are segregated both vertically and horizontally in tourism employment. There is a growing number of studies which document the different ways in which labour market segmentation is expressed, including:

- Milne and Pohlmann (1998). In Montreal, 50% of hotels claimed that more than half their management were women but they tended to be in the lower and middle grades, trapped beneath a 'glass ceiling'.

- Jordan (1997). Women occupy only 4% of middle and 1% of top management posts in the British hospitality industry.

- Hennessy (1994). In the English county of Cornwall, most tourism jobs are part-time and two thirds are for less than 40 weeks. Two thirds of the part-time jobs are taken by women.

- Mackun (1998). Two thirds of tourism employees in Emilia Romagna, Italy, are women. There is strong occupational segregation, with women performing 60% of room servicing and waiting on tables, and men performing 67% of bar-tending work. There are also different patterns of seasonal working, with only 19% of women, but 42% of men, working more than four months a year.

- Bird et al. (2002). A study of hotel front offices in Ireland and the UK found few gender differences in terms of employment as core or peripheral workers, and in fact found that men were more likely to be casually employed. However, there was a 'worrying lack of progression of women into the managerial level' (p. 115).

Gender is one of the most fundamental social divides, and it strongly informs tourism labour-market segmentation in a number of ways (Box 3.4). First, there are contingently gendered jobs, which are done by women either

---

Box 3.4   Gender and tourism labour-market segmentation

Gender is significant in three main ways in determining women's employment in tourism:

Labour price

That is in contingently gendered jobs, for which the demand for labour is gender-neutral, but which happen to be done by women because of the low wages paid, and perhaps because of women's willingness/desire/need to work part-time, or seasonally.

Sex

Where sexuality and sex-related attributes are explicit or implicit in the job description. Women (and men) are expected to package and present themselves in a particular way, where their sexuality is implicit or even explicit. At one extreme, this includes the sex industry, but more routinely it is found in the way that individuals are selected for jobs such as air hostesses, receptionists, bar-tenders, etc., on the basis of what are considered to be masculine or feminine characteristics.

Gender

Where some jobs are patriarchally prescribed. These usually involve the transfer of patriarchally defined women's household roles to waged labour-market roles as housekeepers, cleaners, etc.

Source: after Purcell 1997

---

because they are low waged, or because their part-time nature fits with women's combination of dual roles in waged labour and in the home. Jordan, for example, quotes one senior manager in the UK hospitality industry as stating that 'Tourism is a very low margin business and the major cost is payroll. Women are more likely to accept the low pay and conditions than men' (1997: 532).

Second, there are jobs where sex attributes are used by employers, either openly in job descriptions, or in some of their covert recruitment practices. For example, Adkins' study of the 'Globe Hotel' contrasts the job specifications for waitresses and barmen:

> Waitress: 'attractive, average weight and height, must have enthusiastic attitude'
>
> Barman: 'strong, average weight and height, very smart, able to communicate well with general public, enthusiastic and helpful manner' (1995: 10)

Jordan (1997) expresses this as the commodification of perceived female characteristics to project images of glamour and sex appeal. Some of the more obvious examples included the commodification of female glamour or sex appeal in occupations such as air hostesses and receptionists.

Third, the domestic division of labour, where women's role has been constructed around notions of care, and domestic duties, is carried over into the labour market. Tourism is engaged in 'offering the amenities of the private or household sphere for sale in the public market' (Crompton and Sanderson 1990: 135). Typically, jobs are socially classified as 'naturally' women's work (making beds, etc.), This is often accompanied by socially constructed de-skilling, reduced promotion possibilities, and generally poorer working conditions. While these three dimensions have been presented separately here, they often act in combination.

Gender is intertwined with social class in labour-market segmentation. For example, jobs such as bed-making and cleaning in hotels are not so much done by women, as by working-class women. Sinclair (1997) summarizes the often complex intermingling of Marxist and feminist theories, identifying three main types, according to whether they emphasize class differences (Marxist), patriarchy and men's control over women's access to jobs (feminist), or dual systems (combined). The dual-systems theoretical approach is particularly appealing:

> Dual systems analysis posits that capitalism creates a hierarchical structure in the paid labour force but is indifferent as to whether men or women occupy specific positions within it. Access to occupations is, instead, determined by patriarchal relations which involve men's control over women's labour, resulting in women's employment in low wage jobs, continued dependence upon men and greater unpaid work within the household (Sinclair 1997: 7)

In this perspective, capitalism and patriarchy are separate but related systems. This is perhaps too simplistic, for it ignores the interweaving of class and gender structures. A further theoretical variant, 'patriarchal capitalism', argues that the very nature of gender divisions in society shapes how the labour process is structured, and how jobs are created. In other words, 'capitalism takes advantage of prevailing gender definitions' in actually creating particular hierarchies of jobs (Sinclair 1997: 7). While these contested theoretical perspectives are acknowledged, it is also important to remember, when moving from abstraction to concrete analyses, that the class–gender structuring of tourism employment is worked out differently in different places.

Migration also tends to be highly segmented, and any one place is likely to be at the confluence of a number of interrelated flows. The links between these different flows of tourists, consumption (or lifestyle) migrants, and labour migrants evolve over time, as suggested by Williams and Hall (2002) (see Figure 3.6). Tourism may generate subsequent flows of seasonal or permanent lifestyle or retirement migrants. If the labour needed to produce the services they require cannot be met locally, inmigration may follow. Some migration will be in response to generalized labour-market opportunities in the destination, while others may be nationally or culturally specific, serving particular sub-markets e.g. Korean speakers to serve Korean tourists in Australia (Cooper 2002), or English speakers to serve the British community in Spain (Eaton 1995). Salvà-Tomàs (2002) provides a concrete example of some of these flows in the Balearic Islands (Box 3.5).

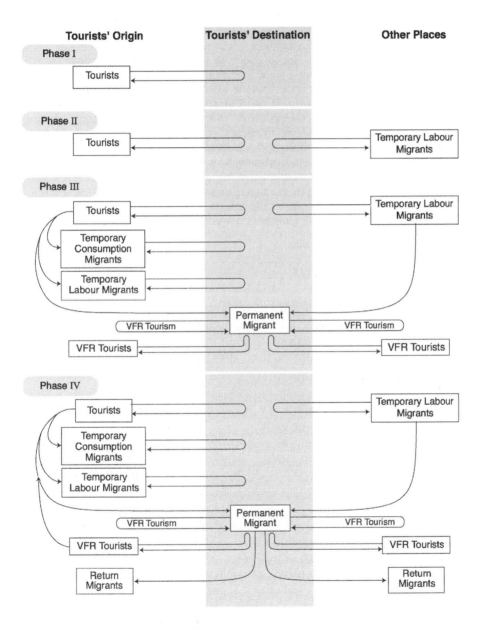

Figure 3.6 **Stages in the evolution of tourism–migration relationships: an idealized model (source: Williams and Hall 2002)**

Box 3.5 Tourism and migration in the Balearic Islands

The development of tourism in the Balearic Islands has led to labour shortages, not only in the hospitality industry but also in other sectors, such as agriculture (unable to compete for reduced labour reserves) and construction (benefiting directly and indirectly from tourism demand). This has contributed to labour inmigration from Africa, Latin America and Asia, totalling more than 32,000 registered and non-registered migrants.

Immigrants from less developed countries (LDCs) have been strongly segmented in the Balearic labour market. Moroccans tend to work in agriculture, the Senegalese work mainly as street traders, while Philippine, Dominican Republic, Peruvian and Colombian immigrants work mostly in domestic service. This labour segmentation is, however, disappearing, partly because of the strong labour demand in construction, which, in recent years, has attracted most of the immigrant workers from Morocco and sub-Saharan Africa.

There has also been more recent immigration from Northern Europe, especially Germany and the UK. While most of these migrants are consumption led – retired, partly retired, or teleworking for northern European companies – there is also an important stream of entrepreneurial migrants, to provide services targeted at niche national markets among both tourists and residents.

Source: after Salvà-Tomàs (2002)

There are entrepreneurial and labour flows that aim to serve general, and particular national, markets – whether residents or tourists. In part, labour migration is responding to absolute shortages of workers, but international tourism across language and cultural divides creates specific demands for particular types of workers: those who speak the languages, and understand the values and motivations of the tourists. This is particularly important given the nature of the tourism experience, which is partly constituted of a sequence of interactions with service providers.

'Foreign skills' (really, linguistic and cultural skills) are useful not only to entrepreneurs, but also to prospective employers. Dawkins et al. (1995) found that the 'foreign skills' most valued by employers were foreign-language proficiency, experience of contacts with foreign people, having lived or worked in a foreign country, specific cultural knowledge, knowledge of foreign business ethics and practice, and formal study of a foreign country. This is broadly confirmed by Aitken and Hall's (2000) findings that in New Zealand the most important 'foreign skills' were considered to be specific cultural knowledge, followed by extensive contacts with foreign people, and knowledge of foreign business practices and ethics. These skills are likely to be particularly important in nationally segmented niche markets. For example, many Koreans and Japanese are employed in hotels and restaurants in Australia and New Zealand, where there are significant numbers of Korean or Japanese tourists.

Migration status is another important source of labour-market segmentation, particularly the divide between registered and non-registered migrants.

The precise nature and form of this depends on national regulatory frameworks. Unregistered migrants are more likely to be condemned to more marginal jobs, whether in the formal or the informal economy. But they also probably constitute a significant component of the labour force in many destinations, making a substantial contribution to reducing absolute and relative labour costs (accepting lower wages and, through their presence, reducing labour shortages and therefore depressing overall wage levels). Arguably, this constitutes a vital component of labour-cost regulation and firm management and competitiveness in some places.

Labour quality, performance and real costs

While labour costs are a primary concern of enterprise managers, the role of labour is more complex than this, particularly in tourism. Managers may be more concerned to obtain satisfying tourist–worker encounters than reduce labour costs, and may seek to increase rather than reduce labour inputs per tourist. This is counterbalanced by expectations of being able to raise the price charged for the service, or ensure increased total demand (e.g. through increased repeat visits, or more positive recommendations to other potential customers).

It is therefore more accurate to argue that capital seeks to minimize real rather than absolute labour costs. The real costs take into account the full costs to capital of hiring labour, including retraining costs if workers leave due to low wages, and the motivation and skills required for delivering quality services, etc. Employers who take anything other than a very short-term view must seek to balance two goals. On the one hand, as argued throughout this chapter, they attempt to shift risk and uncertainty to their employees (i.e. employ them flexibly and preferably on low wages). But they also have to try and retain or attract workers with experience and skills, who can provide continuity in work practices and quality. In other words, as Hudson argues in a broader context, 'in the final analysis . . . companies are concerned about unit production costs, not nominal wages per se' (2001: 109). It is true that, by and large, formal skills and training are relatively low in much of the tourism industry, as Riley et al. (2002) argue, although there are exceptions, such as airline pilots, top chefs, and white-water rafting instructors. However, there are also less formal skills, such as personal interaction or language skills, or close familiarity with, and knowledge of, the needs and tastes of a regular client group. In such cases, knowledge and skills are embedded in individuals, and in many tourist jobs are not easily transferable. Once such workers are lost, they will not easily be replaced. One of the particular characteristics of these embodied skills is 'performance', which is discussed below.

Tourism is part of the experience economy, and one of the more telling comments by Pine and Gilmore is that 'while the *work* of the experience stager perishes upon its performance . . . the *value* of the experience lingers in the memory of any individual who was engaged by the event' (1999: 12–13). This applies particularly to frontline workers, who are in direct contact with tourists (Drucker 1992). The service provided by frontline workers is part of the tourism experience. Urry (1990) expresses this differently, arguing that in every transaction there is 'a moment of truth', when satisfaction/dissatisfaction is realized. The emotional content of these transactions between tourists and tourism employees is as important as their manifest function, that is, the sensory feelings engendered by the encounter with the hotel receptionist may be at least as important as the fact that he or she served you efficiently.

'Total quality management' is one strategy whereby employers seek to enhance these encounters, or moments of truth. But they can also be culturally interpreted. For example, Crang argues that there is a need to look at 'socio-spatial relationships between producers and consumers and their organisational geographies of display' (1997: 139). McDowell writes incisively on the embodiment of attributes involved in labour performance. Although writing specifically about financial services, her comments have resonance for tourism:

> Workers with specific social attributes ... produce an embodied performance that conforms to idealised notions of the appropriate 'servicer'. Firms or organisations have explicit and implicit rules of conduct and these inform 'the desirable embodied attributes of workers'. (1997: 121)

The embodiment or performance can be seen in tourism in the examples of airline stewards, beach attendants, and receptionists in hotels, among others. Employers have expectations of how these employees should look (in terms of gender, age, clothes, etc.) and also how they act (in terms of their voices, bodily movements, etc.). This performance adds to the tourist encounter, because it is part of the expectation of many customers; indeed, many tourists may be prepared to contribute to the overall performance. Work as performance applies not only to employees, but also to owners of micro-firms; for example, the small-hotel or restaurant owner who performs the role of a warm, slightly eccentric but 'characterful' host.

Crang's (1994) study of waiters in themed restaurants provides one of the most detailed case studies of performance in the hospitality industry. Customers expect certain performances from waiters and waitresses, and contribute to these performances. More generally, Crang argues that tourism products are experiential, interactional (involving employees and tourists), and involve the 'temporal co-presence of producers and consumers in tourism production processes' (1997: 139). Moreover, he argues that 'Tourist places are not just imagined places, they are also performed places; and tourism employees are not just actors on a stage, they have to act out that stage' (1997: 147). This means not only that the labour process cannot be predetermined by managers, but also that managers need to attract and retain the staff who have the valued embodied skills.

These arguments are important in highlighting the significance of performance in the labour process. But caution is required in drawing conclusions about tourism production. Even if all activities are construed as performance, employers will have very different expectations of, for example, a 'greasy spoon' café as opposed to a themed restaurant, or of the kitchens versus the front of house in a restaurant. In other words, there are many different ways of extracting surplus value from labour even within a single establishment. Also, even if employers cannot entirely control the performance, they will try to manage both its quality and its costs. For example, they may try to routinize the encounter; employees will be trained to use a number of stock phrases, such as 'have a nice day'. This is particularly important for chains of hotels or restaurants, which require the 'ability to market a guaranteed experience to consumers across a geographically dispersed set of production/consumption sites' (Crang 1997: 140). Finally, Wood (1992) reminds us not to romanticize tourism employment by exaggerating the performance element, so that we lose sight of the drudgery that is the reality of the labour process for many workers.

SUMMARY: ORGANIZING PRODUCTION WITHIN THE FIRM

This chapter has focused on the internal organization of firms, although it is recognized that any such analysis cannot be divorced from the operating environment. In general, this is one of the more neglected aspects of tourism research, but within the constraints of the available literature we emphasize the following.

- The definition of micro-firms is problematic, but there is a need to focus on their actual relationships and organizational forms rather than on abstract definitions. These include low barriers to entry, the combination of materialistic and non-materialistic goals, family association with the firm's activities, and the vesting of property rights in owner-managers.

- Micro-firms are not homogenous, and we identify three main types, distinguished by their approaches to growth, risk and innovation.

- Transnational companies have strongly internationalized activities, and are characterized by the separation of ownership and management, and the pursuit of globalization strategies.

- Transnational companies are not homogenous, and we identify three different innovation and competitiveness situations.

- The labour process is understood as the organization of the labour force, in terms of work practices/workplace organization. While there are downwards pressures on wages within the labour process in all industries, these are particularly strong in tourism, because of the nature of job attributes, industrial structure and psychological issues.

- Firms are faced with various strategies for combining capital and labour in order to reduce total production costs.

- Non-material rewards, including membership of occupational communities, play an important role in tourism labour practices.

- The strong temporal rhythms in tourism demand reinforce the need for flexible labour practices. Flexibility can be conceptualized in different ways.

- Labour-market segmentation, linked to broader social divisions, provides one means whereby tourism wages can be reduced. Lower wages are paid for some types of work, which have been socially constructed as 'low skilled women's or migrants' work', irrespective of their actual skill content.

The importance of the 'service encounter' in tourism gives particular weight to issues of quality and performance. Employers therefore have to focus on their real labour costs, rather than just on nominal wages.

4 *Inter-company Cooperation and Competition*

One of the most important dimensions of globalization is the intensification of competition (see Chapter 1). Firms respond to competitive pressures in a number of ways. The starting point for this analysis is Schumpeter's (1919, 1939) classic distinction between 'weak' or 'repetitive' and 'strong' or 'disruptive' competition. The former occurs within existing parameters, while the latter challenges and changes these. Whether there exists predominantly strong or weak competition depends on the time period, place or sector under review. However, it is not a matter of a dichotomy, with firms engaging in either weak or strong competition; and there are many different strategies within each of these generic types. Individual firms are likely to adopt multiple strategies in responding to competitive pressures, as illustrated by the tour operator Thomson (see Box 4.1).

Although firms may adopt diverse strategies, these all aim to increase the ability of the firm to extract surplus value or profits from producing tourism services and experiences. Some strategies centre on increasing the competitiveness of the firm's existing position in the value chain, while others explicitly seek to reposition it, perhaps though investment to develop new products, through alliances with other companies, or via mergers and acquisitions. In other words, some of these strategies involve the reshaping of company connections, and the relocation of what are already essentially the blurred boundaries of the firm.

Although the responses to competition in tourism have many features in common with other industries, they do possess some distinctive features: tourism experiences are dependent on the production and consumption of multiple elements, property rights are often difficult to establish and protect, and production is characterized by temporality and spatiality (Chapter 2). Moreover, tourism makes particular demands on host communities and local environmental systems, which necessarily mediate how firms respond to competition challenges. For example, 'cultural brokerage' (Crick 1989) is often required between host communities and tourists, and among the many different constituent groups of these, who possess vastly different levels of cultural capital in terms of linguistic skills and cultural knowledge. Not only does this factor mediate the competitive strategies of firms, but firms can also use their role as cultural brokers as part of their competitive strategies. There is a need therefore to rethink the dichotomy between tourism production processes and their impacts, given that the latter can be intrinsic components of the former.

Box 4.1   Competition and collaboration: Thomson

Thomson, the UK-based tour operator, has responded in a number of ways to competition:

Competition within existing paradigms

The company introduced price cutting in order to increase market share. For example, in 1987 prices were reduced by up to 20%, compared with the previous year. Although market share increased, profits fell. This, and persistent rounds of price cutting, and late price discounting, contributed to very low profit rates, which barely rose above 1–2% in the early 1990s.

Competition and changing paradigms

Information technology
Thomson was a leader in the introduction of IT-based systems of providing information on, and selling, holiday products. For example, its direct-sell subsidiary, Portland, increased the number of holidays sold per employee by 21% in 1985, largely through the introduction of an advanced booking system.

Products
Thomson, in common with most other major tour operators, has diversified its product range over time, adding new short-haul destinations (such as Turkey) to its core southern European destinations, as well as long-haul holidays.

Collaboration

Supply chain
The company formed supply strategies with hotels, coach companies, and car-hire companies in order to deliver its core 'package holidays' and optional services to its clients.

Mergers and acquisitions
The history of Thomson is dominated by mergers and acquisitions, which contributed to product and market diversification, as well as vertical integration. In 1965 the initial tour operator business was acquired, along with Britannia Airways, by the Thomson Corporation (Canada). Thomson Skytours became Thomson Holidays in 1972, after merging with Riviera Holidays, Gaytours, and Luxitours. In 1972 it internalized some of its linkages to travel agencies by acquiring, and then investing heavily in, the Lunn Poly travel agency. In 1988 it acquired Horizon Holidays (including the Wings and Orion charter brands). Country Cottages (an agency for renting self-catering houses and apartments) was acquired in the late 1990s, as part of a diversification strategy. In 2000, the Thomson Corporation sold its interests in Thomson to Preussag of Germany.

This chapter focuses on two central themes: competition within existing paradigms and competition within changing paradigms. The latter commences with a review of how an individual firm can change competition paradigms via innovations in technology, products, markets, processes, knowledge, etc. The focus then shifts to a firm's connections, and the development of various forms of collaboration, alliances, interdependencies and mergers/acquisitions. In different ways, all these strategies seek to blur or redefine the boundary of the firm, and to create or recreate networks of firms. The general structure of this chapter, and several of the ideas developed herein, are based on an interpretation of Hudson (2001, chapters 5 and 6).

COMPETITION WITHIN EXISTING PARADIGMS, OR REPETITIVE COMPETITION

Competition within existing parameters is variable among the different sub-sectors of tourism. The extent of competition depends on the ease of entry to a particular sector, and this reflects both the mode of regulation (competition laws, etc.) and the regime of accumulation along with the availability of capital, economies of scale, etc. New firms face only minor barriers to entry to some markets, such as agritourism in Tuscany, and once in the market they face little in the way of competition through changing parameters, for this is a relatively stable product, with mature markets.

Sinclair and Stabler (1997: Chapter 4) review the extent to which markets are contestable, or are characterized by various forms of monopoly and oligopoly. They conclude that in the sectors of hotel accommodation and travel intermediaries 'there are elements of contestability ... alongside the dominant market forms of monopolistic competition and oligopoly' (p. 93). But they also emphasize the heterogeneity of these sectors, and of transport, and the existence of distinct sub-markets. Each of the three main sectors is reviewed below.

- *Hotels.* In business centres, and many major holiday resorts, large hotels experience oligopolistic conditions; but elsewhere – for example, in small towns or smaller resorts – local market structures are closer to monopoly. This broad picture is, however, obscured by two considerations. First, the extent to which markets are localized is questionable. A tourist seeking a large luxury hotel in a particular rural area may find there is only one such hotel. However, the real choice facing many leisure tourists may be between several such hotels in different areas, because they are selecting a product type rather than a place-specific product. Second, if the tourist is tied to a particular place, he/she still has intra-sectoral choice: they can choose to use a smaller hotel, or rent a cottage/villa. In contrast, there are large swathes of the hotel industry, such as clusters of guesthouses and small hotels in resorts, which operate in strongly competitive markets.

- *Travel intermediaries*. Tour operating is a highly competitive sector, despite a degree of oligopoly in some major markets, such as the UK and Germany. In both of these markets two or three companies tend to account for about one half of total sales (Shaw and Williams 2002: 127–32). Nevertheless, these markets remain highly contestable; as Sinclair and Stabler argue:

> A number of factors such as ease of entry and exit, the number of tour operators, fierce price competition, low margins and often significant losses all point to contestable if not highly competitive market conditions in the UK and many other countries. (1997: 75)

The keys here are the relative ease of entry for new tour operators, combined with a consumer culture of comparative shopping on the basis of price, and expectations of last-minute price discounts. Travel agents are in an even more competitive market in many countries. Even a relatively small city such as Exeter (UK), with little more than 100,000 inhabitants, has 15 travel agencies. Travel agents are squeezed between an imperfectly competitive market (the domination of total sales by a small number of companies) and the need to sell in a highly contestable consumer market (Sinclair and Stabler 1997: 78).

- *Transport*. Competition in the transport sector is highly variable. Over certain distances, there is inter-modal competition: for example, between flying or taking the train from London to Paris, or between flying and taking a coach from the Netherlands to the Alpine ski resorts. But the regulatory framework also heavily influences competition in transport sub-markets. Air transport, and the effects of deregulation in the USA and the EU, were discussed in Chapter 2. Many of the high-volume air routes within the USA and Europe, but also in the long-haul market, have become highly contestable. There is rarely direct competition within the train sector (although some exists in the privatized UK market), but trains usually face strong competition from other modes of transport, including the car, depending on the distance travelled. The same applies to coaches.

Tourism markets, in particular, tend to be contestable, although there are exceptions: there are long-haul routes served by only a single carrier, there are destinations with 'unique' tourist attractions, and there are market economies dominated by just one or two tour operators. But these are exceptions, and in most markets firms face considerable competition. If they compete within existing paradigms, a number of options are available. The first, and most obvious, is price competition. This has been the dominant strategy followed in the UK tour-company sector. For example, there is a common belief that, in some markets, a 15–20% price cut can double market potential. Few companies engage in across the board price-cutting because there are differentiated returns by product or sub-market. Hence, there is likely to be differential pricing, e.g. the Saturday night rule has long been used as a way to demarcate between lower- and higher-price tickets for the same journey, in effect distinguishing between business and leisure travellers. Price cuts – such as bargain fares – are also likely to be selectively targeted at particular routes or hotels, notably those with the greatest capacity to generate volume gains, or where there is already significant over-capacity.

Firms also compete by holding prices constant but reducing their production costs, thereby enhancing the extraction of surplus value, or profits. There are numerous ways of achieving this. Fixed capital costs can be reduced by using cheaper buses or furniture, by not heating the hotel swimming pool in the shoulder season, or by increasing the length of the redecorating and renovating cycle for decorations and furnishings. But the most significant strategy available to management is the reduction of labour costs. Most tourist facilities are spatially fixed, although there are exceptions such as caravans. Managers therefore have to reduce labour costs *in situ*, through either recruitment practices or the intensification of the labour process (see Chapter 3).

There are examples of what can be described as 'raw competition', usually combining price cutting with reductions in operating costs. Box 4.2 offers an example of two European budget carriers that have combined competition within existing and changing paradigms. However, few firms compete on price alone, as this example illustrates. Instead, price–cost–quality ratios are the focus of competition in many tourism sub-sectors. This is consistent with the view, explored in Chapter 3, that what matters is not nominal but real labour costs. The latter incorporates the elements of skill, including perform-ance skills, and the maintenance of quality standards. As Hudson (2001) argues, what matters are not nominal wages *per se* but unit production costs.

There are, of course, other ways of competing within existing parameters. For example, Arbel and Woods (1990) report that in the 1970s and 1980s the American hotel industry was able to borrow capital at negative or zero real interest rates in the USA because of a combination of special factors: using real estate as collateral, and having longer-term debt than most industries, which was advantageous in the face of relatively strong inflation. These benefits were, however, potentially available to all hotels, and so did not constitute changing paradigms, which is the theme of the next section.

COMPETITION WITHIN CHANGING PARADIGMS, OR DISRUPTIVE COMPETITION

Competition within changing paradigms is disruptive of existing markets, and tends to be based on innovation. Firms aim to secure a competitive advantage and, if possible, one that is not easily imitated by competitors. Companies seek to protect such competitive advantages in different ways: via patents excluding other users over a number of years, distinctive branding (e.g. as used by Disney) or, perhaps, the sheer scale of investment required to

---

Box 4.2 Competition within existing and changing paradigms: budget airlines

The US company Southwestern Airlines has been a model that many European scheduled airlines have sought to emulate. Two contrasting examples are provided by Ryanair (based in Ireland, but with a strong presence in the UK) and Easy Jet (based in the UK, but founded on Greek capital). As can be seen from Figure 4.1, they both have significant cost advantages over the major scheduled carriers, even the relatively efficient British Airways. Their costs are broadly in line with those of the charter airlines, but they tend to operate over much shorter distances. However, these statistics also demonstrate that the two companies have significantly different operating costs, and this reflects differences in their competition strategies.

- Easy Jet competes in terms of relative costs, compared with the scheduled airlines. Production is based, as with Ryanair, on direct selling, stripping out services such as free drinks and meals, the rapid turnaround of aircraft, and reduced staffing. Marketing focuses on the benefits the airline brings to consumers in contrast to the major airlines.

- Ryanair has an 'extreme' focus on low costs, which means that 'it is really competing with other uses of passengers' time and money'. It competes on absolute costs. Production is based on the same changes in the labour process as are employed by Easy Jet, plus using secondary and less convenient airports (with lower landing charges). Marketing is based on advertising very low-cost fares (less than £10 for some seats on some flights). Selling the bulk of tickets through its websites means that sales costs are about 4% of ticket prices, compared with 12% for British Airways.

The changes introduced by these airlines constitute both competition within existing paradigms (price cutting) and competition within changing parameters, via process innovation. As in most industries, firms use mixed strategies in their response to competition.

Source: Martin (2002)

---

compete in a particular market. In general, the capacity to create a competitive advantage through changing paradigms of competition is dependent on firm size, but this is not invariable. For example, relatively small firms can establish competitive advantages in niche markets – at least over the short to medium term. A.J. Hackett's bungy jumping at Kawarau Suspension Bridge (New Zealand) illustrates how a small company established a competitive edge through product innovation; but the large number of later competitors demonstrates how difficult it is to protect an advantage over the longer term in a contestable market.

Innovation is the key to changing paradigms. David and Foray even suggest that 'the need to innovate is growing stronger as innovation comes closer to being the sole means to survive and prosper in highly competitive and globalised economies' (2002: 11). Baumol (2002) similarly argues that compe-

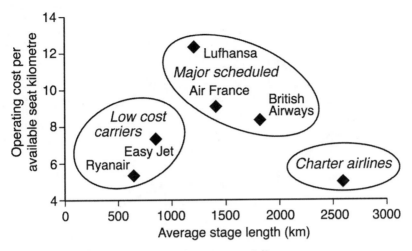

Figure 4.1 **Operating costs of selected European airlines**

tition forces companies to invest in innovation, or risk losing out to their competitors. Moreover, his study of 46 major product innovations (across all areas of the economy) reveals that the time-span in which innovations offer a competitive edge – measured as the gap before competitors enter the market – has fallen from 33 to three years, between the late nineteenth and the early twenty-first centuries. Firms protect their interests in this increasingly competitive environment partly through technology licensing, but this does not obviate the central importance of innovation in market economies.

The increasing rate of innovation makes for what Schumpeter (1919) termed a 'turbulent' environment for competition. Innovation is intrinsically linked with uncertainty, risks, and instability. Hence, two key features of entrepreneurship are the abilities to innovate and to take (or manage) risk (see Chapter 3). Firms try to destabilize existing markets – usually at some risk to their own position, or their capital – in order to increase their market share or total sales. Although it is possible to conceptualize innovation, it is more difficult to pin it down in concrete analysis. For example, when does the introduction of new technology, or an attempt to break into new markets, represent innovation as opposed to an extension to existing firm practices? If a firm has been marketing in six of the UK tourist-board regions, and decides to enter a seventh region, can this be considered innovation, and how risky is it? Similarly, at what point does upgrading or amending a firm's website represent innovation?

One response to this dilemma is to recognize that there is an innovation continuum, which, for illustrative purposes, can be turned into a typology. For example, Chan et al. (1988) consider there are three types of innovation:

- *Incremental.* This does not require a major breakthrough in either markets or technology. For example, speeding up the turnaround of passengers on rides in a theme park, or upgrading hotel bedroom furnishings.

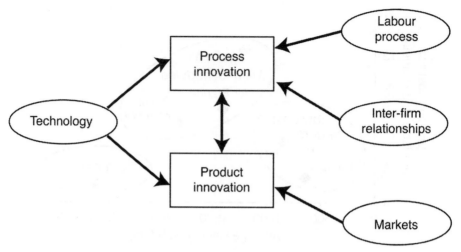

Figure 4.2 **Competition within changing paradigms (disruptive competition)**

- *Distinctive*. This usually demands adaptation of consumer behaviour, and possibly of company organization. For example, the addition of pay-as-you-use video equipment in hotel bedrooms, or decentralized management structures to run locally themed restaurants within a major chain.

- *Breakthrough*. This involves a new approach in consumer behaviour, system organization, or new technology. Examples include direct internet sales of airline tickets, the first 360-degree theme-park rides, or the introduction of budget hotels.

Each of these three innovation types involves differing levels of creating and using new knowledge, investment, uncertainty and risk. Breakthrough innovation usually promises greater returns for the firm, but also greater risk. As indicated above, there are different strategies for changing the paradigms of competition. However, these are all essentially either product- or process-based innovations (Figure 4.2). Process-innovation is about changing the way in which products (new or existing) are produced and distributed. While they are treated separately here, in reality they are closely interrelated. New products often require new processes if they are to be realized and effectively marketed and sold. In both cases – process and product innovation – most of the innovation that drives economies occurs within existing companies, and in the more successful companies it becomes routinized behaviour (Baumol 2002).

Product innovation

Firms' strategies in tourism are broadly similar to those that can be observed in other sectors. Firms seek to maximize their revenues relative to costs, and there are a number of ways to achieve this. In this context, Go and Pine identify the following strategies: expanding market size, increasing local market share, decreasing costs, and adjusting the sales mix. Inevitably these are linked, and often involve product innovation (1995: 28–9). For example, the first US tour operator to offer 'package holidays' to China may have expanded the total market (attracting tourists who would not otherwise have considered a 'package' holiday), increased its local market share (of 'Asian' holidays), decreased travel costs to China (compared with independent tourism), and adjusted company sales mix by adding a new product.

Product innovation involves product differentiation in relation to competing companies. The essence of product differentiation is:

> ... where the firm offers perceived added value over the competition at a similar or somewhat higher price. In offering what the customer believes is a better product at the same price or enhancing margins by slightly higher pricing, it is possible to achieve higher market share and therefore higher volume. (Knowles et al. 2001: 133)

Product differentiation need not involve significant innovation – for example, the differentiation of leisure from business air travel, through the imposition of the Saturday night rule as a qualifying condition. Even where product differentiation involves product innovation, it is variable, as Chan et al. (1988) indicated. This theme is explored below, through the example of mass international tourism, illustrating the complex ways in which technology, marketing, and place commodification are often interrelated in product innovation.

Tourism is an experience, with enormous symbolic values, reflecting its constitution as a positional good and an instrument of cultural capital (Chapter 5). Simply taking holidays abroad, rather than domestically, has significance, but there are also differences in the cultural capital attached to holidays in Tuscany, say, compared with the Costa del Sol. The opening up of mass tourism products depends, therefore, as much on marketing and advertising, to reinforce the signifiers of particular places, as on investment in new products or technological changes. The growth of package holidays from northern to southern Europe depended partly on enabling technological changes in jet engines and the development of air-inclusive tours (process innovations). But it also depended on investment to create new products: for example, hotels, beach facilities, or evening entertainments. In this sense, places can be tourism products, created as part of the changing competition paradigm in the industry. Air-inclusive package holidays certainly increased the total tourism market, but their growth was also partly at the expense of domestic tourism destinations in northern Europe. While the early growth of tourism resorts, such as Torremolinos or Benidorm, may be viewed as 'breakthrough' innovations, the subsequent growth of new tourism resorts in Spain – and in other Mediterranean countries – probably represents 'distinctive' or even 'incremental' innovations. This type of place-based product innovation need not occur only on greenfield sites, as the rather different example of the Rick Stein phenomenon indicates (Box 4.3).

Box 4.3 Rick Stein's Padstow: the re-creation of tourism place

Padstow is a historic fishing village on the north coast of Cornwall (UK) where tourism benefited from the arrival of the railway in the nineteenth century. Its growth was relatively modest, however, and it retained much of its 'traditional' architectural character. The modestly successful tourist destination has, however, been transformed since the 1990s, and human agency has played a key role in this: the activities of restaurant owner, media star and chef, Rick Stein.

Stein opened the Seafood Restaurant in Padstow in the mid-1970s, and gradually established its reputation as one of the UK's leading restaurants. In the 1990s he started to appear regularly on television and, although some of his programmes took him around the UK, and indeed around the world, Padstow was always in the background. The recurrent themes in his presentations were the quality of food, the sea, and Cornwall, and he effectively changed the image of Padstow, turning it into a major tourist attraction. Not untypical is the 2002 website description of his flagship Seafood Restaurant:

> The Seafood Restaurant has established a national reputation for imaginative cooking of the very freshest fish and shellfish. The restaurant is situated just across the quay from where the lobster boats and trawlers tie up and most of the fish comes, literally, straight off the boats and in the kitchen door.

While this description does not specifically mention Padstow, it does not need to. Repeat television series, and a host of successful cookery books, such as *Rick Stein's Taste of the Sea* and *Rick Stein's Seafood*, have imprinted images of the harbourside restaurant on the consciousness of many people in the UK.

Building on a high public-relations profile, Stein has invested in a number of related businesses in Padstow (Figure 4.3): hotels, a café, a delicatessen, a gift shop and a cookery school. These are reinforced by online shopping facilities and, hinting at global reach, there is also a web link to his partly owned vineyard, hotel and restaurant in Australia's Hunter Valley.

Tourist numbers have soared in the wake of the increased promotion of Padstow, and the Seafood Restaurant has become a tourist attraction in its own right – those who cannot, or choose not to, pay the high prices in the restaurant, come to gaze at this icon of middle-class cultural capital, and perhaps to spend some money in some of his other businesses. Other local businesses have also benefited from the media exposure, so that the harbour front has a scatter of new or revamped restaurants, and the steep hillsides of the town are dotted with investments in small hotels and guest houses. This is, unquestionably, higher-order product and place innovation.

Source: based on www.rickstein.co.uk (May 2002)

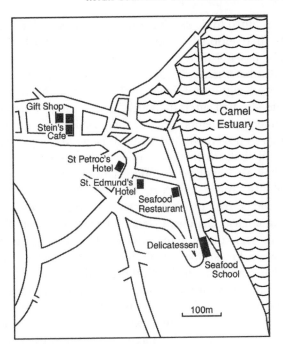

Figure 4.3 **Rick Stein's enterprises in Padstow**

Process innovation

At the macro-scale, the shift to Fordist production, and latterly to post-Fordist production, provide examples of process innovation in response to systemic crises in the system of production (see Chapter 2). But there are also many other forms of process innovation that do not represent such a radical change in production. We have already touched upon one of these in respect of budget airlines in Europe (Box 4.2). They use a combination of marketing and sales innovation (direct sales to the public, including internet sales), with changes in the labour process (reduced service levels, no pre-take-off allocation of seats, increased turnaround of aircraft), to reduce costs and shift the competition paradigm.

Process innovation can be realized in a number of ways; but in the general literature – although less so in tourism – most attention has been given to technology. Changes in technology can be generic, being linked for example to the revolution in information technology (IT), or specific to a particular sector, e.g. the introduction of jet technology. Usually, however, the generic and specific sectoral innovations are linked, as in the adaptation of e-commerce technology to travel agents' sales of tour operator products.

There are a wide range of process innovations based on technological developments. We have already referred in Chapter 3 to the manner in which the introduction of automatic dishwashers and chill-cook technologies revolutionized the labour process in restaurant kitchens (Bagguley 1987), leading to cost reductions allied to de-skilling and changing social divisions of labour. The firms that first adopted these technologies established a competitive advantage on the basis of their sharply reduced costs.

There have also been major breakthroughs in transport technology. In air transport, the most significant have probably been the introduction of jet engines and wide-bodied aircraft, which, respectively, sharply changed the cost and volume parameters of air travel. Similarly, marine transport innovations, such as hovercraft and hydrofoils, gave competitive advantages, in terms of travel time, to the companies that adopted these technologies. Coach design has also evolved through a number of process innovations, enhancing visibility, speed, comfort, and the provision of entertainment, although their effect has probably been less paradigm-shifting than the impacts of technological changes in air and marine travel. There have also been changes in technology which have affected cycling and walking/ rambling: mountain bikes have revolutionized the former, while the use of new materials has transformed the clothing and footwear used in the latter. In fact, mountain bikes are an example of both product and process innovation, depending on whether the bicycle is itself considered to be the object of consumption, or the cycling experience that it gives access to.

Theme parks and heritage sites have also been subject to technological changes (Chapter 10). This is particularly evident in the intense competition between theme parks on the basis of introducing new, more thrilling rides. At a smaller scale, there are also many examples of new multi-media technology, which is changing the presentation of heritage. For example, the Jorvik centre, which presents York's Viking heritage, used new audio-visual technologies to establish a competitive edge over many of its rivals in the heritage tourism market.

While there are many examples of technological developments shifting the competition paradigm, IT innovations have probably had the most profound impact on tourism. The series of 'breakthrough' innovations that constitute the IT revolution have changed the paradigm of the entire tourism sector. Buhalis (1998: 410) summarizes the extent of these shifts:

> ITs reshape the nature of competition in most economic activities, whilst they link consumers and suppliers, adding value to organizations' products. Hence ITs change the competitive frame for almost all organizations, regardless of the industry they operate in, their location or size. In particular, technology affects competitive advantage as it determines the relative cost position or differentiation of organizations.

IT offers firms a number of competitive advantages. According to Buhalis, these include establishing entry barriers (because of the cost of hardware/ software, or of skilled IT workers), affecting switching costs, differentiating products and services, limiting access to distribution channels, ensuring competitive pricing, decreasing supply costs and easing supply constraints, and increasing cost efficiency. IT can also become a product in itself, as in virtual tourism. While IT costs have generally been reduced over time, innovation usually requires significant capital inputs, either in terms of equipment or in labour (re)training. Therefore, there are risks associated with IT innovation, despite the seemingly inexorable march of technological development.

IT has had a major impact on both 'backstage' processing and distribution channels, that is e-commerce. Developments in backstage processing are more or less common to all sectors of tourism, whether hotels, tour operators, or airlines. IT revolutionizes the way information is stored, recalled and transmitted. This has had immense implications in terms of the speed of data processing, labour requirements, and the range of services available. For example, a guest's reservation information (and preferences known from previous stays) can be quickly recalled at hotel reception, allowing rapid registration. By the time the guest reaches his or her room, the ubiquitous television will already display a personalized message, with information about services available and automatic check-out procedures. IT can have a major impact on backstage employment in hotels (and to some extent the frontstage, although the personalized nature of tourist–tourism worker encounters constrain the latter). Milne and Gill provide an example of the employment consequences, reporting that 50% of Via Rail's (Canada) marketing staff were made redundant after the company adopted the Sabre IT system (1998: 134).

The impact of IT on distribution channels has been even more dramatic, which is not surprising given that 'information is the life blood of the travel industry' (Sheldon 1994). IT does not simply change the range and volume of information transfer, but rather it 'is altering long-standing relationships in the channel system in tourism, and creating new forms of competition – sometimes overnight' (Go and Pine 1995: 310).

The development of computer reservations systems (CRSs) from the 1970s typifies the impacts of IT on distribution. There are substantial development costs associated with CRSs, which have to be met through realizing large volume sales. This has contributed to strong concentration so that, by the early 1980s, five major CRSs dominated the North American travel industry. One of these – Sabre – still accounted for 40% of all air-travel bookings via US travel agents in 1995 (Milne and Gill 1998). Although the initial innovations were located in North America, these became globalized, and by 1994 three Global Distribution Systems (GDSs) (Galileo, Sabre and Amadeus) accounted for 78% of all hotel, car and airline reservations.

A number of structural implications follow from this shift in competition. First, as Poon argues, it contributes to 'the transformation of travel and tourism from its mass, standardized and rigidly packaged nature into a more flexible, individual oriented, sustainable and diagonally integrated industry' (1993: 13). It provides new ways to coordinate production, creating new methods for firms to gain competitive advantage from more flexible products and processes. Indeed, IT is often considered to be one of the facilitators of the shift to post-Fordist production. It also challenges the traditional organization of the value chain; in particular, tour operators, hotels, airlines and other producers can reduce their reliance on travel agents by selling directly to customers via the internet. In other words, this can lead to 'disintermediation' (Macdonald-Wallace 1999).

IT is also not power-neutral. Vellas and Bécherel (1995: 190) raise several issues relating to GDS, but these have resonance for other areas of IT development. There can be uneven rights of access to GDS systems, and in the distribution of costs among system vendors, carriers and suppliers of tourism services. Small firms are particularly disadvantaged, and the commission they pay can equal 20–30% of the total room-rate they receive (Milne and Gill 1998: 131). Small firms may also find it difficult to adopt many forms of IT because they lack the necessary management and technical skills (Buhalis 1993). The sheer costs of GDS development may also lead to reduced market contestability. For example, developing countries may find it difficult to develop complementary, let alone competing, systems.

The issue of disintermediation and, in particular, the implications for the future of travel agencies have been subject to considerable debate. For example, Morrell (1998) has discussed the impact of three technological developments – GDS, ticketless air travel, and the internet – on the future role of travel agents in the sale of airline tickets. Not surprisingly, the paper concludes that disintermediation tendencies will be mediated by several considerations: customer ignorance and lack of confidence; weak individual purchasing power, compared with travel agencies, in securing price discounts; and consumer immobility (a reluctance to leave home to collect information and buy tickets).

Despite these reservations, we generally concur with Knowles et al. that 'technology will shape the future of marketing programmes, product design and corporate strategies' (2001: 131). This is increasingly the driving force behind many of the competition paradigm shifts. Of course, technology is not an autonomous force for change, and it is given meaning only by the way it interacts with the labour process, inter-firm relationships and markets (see Figure 4.2). The ability to compete depends not only on the development and implementation of new technologies, but also on the capacity of the firm to learn, and to adapt to change. That, in turn, raises a further issue – competition is not simply about how firms compete against one another, but also about how they cooperate with other firms in order to increase competitiveness.

COLLABORATION AND FIRM INTERDEPENDENCIES

There are many situations in which firms consider that cooperation is preferable to competition. They therefore may pursue collaboration strategies in order to improve their competitive position. In extreme form, if there are only a small number of companies, and there are difficult entry barriers obstructing new firm formation, companies can form a cabal to fix prices and eliminate direct competition in a particular market or sub-market. Examples, at very different scales, would be two beach-chair hirers agreeing to charge the same price, or all airlines on a particular route charging the same air fare. The extent to which any such collaboration may occur is dependent on the regulatory framework.

Other than reinforcing the need to study the mode of regulation and the regime of accumulation, this brief discussion emphasizes the need to see firms not as free-standing economic entities, but in context of their networks. The key points are that the borders of firms are fuzzy and they should be conceptualized as centres of strategic decision-making, rather than discrete economic units (Dicken and Thrift 1992). This links with Porter's (1985) seminal work on value chains, which are constituted by firms' linkages to other firms and/or the final products they provide. These interdependencies are critical to competitive advantage in tourism, as in most sectors. Indeed, given the fragmented nature of production, and the fact that tourism experiences involve multiple contributions from firms (Chapter 2), they are of particular note in the tourism sector. Changing these relationships therefore constitutes one of the keys to competitiveness

There are many different forms of networks. Hudson comments that 'networks are constituted in varied and often complex ways, with different degrees of closure, with different structures of power relationships within them, and with varying degrees of formality and informality' (2001: 192). Moreover, as indicated earlier, their precise form depends on national regulatory differences, especially competition law. This is evident in the extent to which differing degrees of liberalization pervade inter-firm networks in the USA compared with, say, Japan. Three main forms of networks are recognized here, as discussed later in this section.

Informal inter-firm relationships have been an important focus of research in economic geography in recent years. Storper (1995) wrote compellingly of the need to study firms' 'untraded interdependencies', that is the practices, routines and agreements that effectively create mutual expectations. Through these, firms build up mutual trust and acquire shared values, which are critical for the effective operation of markets. Storper's research represents an extension of the work of Granovetter (1985) and others on the need to look at interpersonal relationships and therefore at social networks. These are the networks through which firms acquire information, seek partners or sub-contractors, as well as clients. They also underpin market allocation mechanisms.

Not surprisingly, such networks may be place-based, or centred, although there is a debate about the importance of proximity in maintaining such relationships. We return to this theme later in the chapter in context of the notion of 'industrial districts'. Alternatively, they may be ethnically based. The Chinese diaspora arguably provides one such network, wherein individuals and firms collaborate in a number of ways, including the provision of capital and knowledge (Thrift and Olds 1996). Sometimes place and the ethnicity of networks are strongly entwined, as in the case of the Finnish community in Lake Worth, USA (Box 4.4).

Box 4.4 Informal business networks: the Finnish community in Lake Worth

There is a substantial Finnish business community in the Lake Worth area of the USA, which has grown up around the demands generated initially by tourists and retirement migrants of Finnish extraction (either permanent residents or first or second generation immigrants). According to the American-Finnish business directory, by 1997 there were more than 120 businesses in the Lake Worth-Lantana area which served the Finnish community, although they also produced goods and services for the non-Finnish population. Tourists, winter residents and permanent residents all frequent these 'Finnish' businesses, which include bakeries and restaurants, and professional and personal services, such as doctors, hairdressers, and insurance agents.

These businesses, which are utilized by residents and tourists alike, create a network of services that fulfil almost every need among the Finnish population, from cradle to grave:

> In Lake Worth, a little Finn can be delivered by a Finnish-speaking doctor, be baptized in a Finnish church, live in a Finnish-built home, work in an English-Finnish office, and spend his last days in a rest home with Finnish orderlies, nurses and fellow Finnish senior citizens (*Daily Gazette* 1982: 10).

Language is the key here, because many Finns do not speak English very well; but this is only part of a much wider informal network of relationships among these businesses. They serve a highly segmented market, give preference to relationships with other Finnish firms, and have developed an operating environment of trust and collaboration.

Source: based on Timothy (2002)

Informal networks are no less – and often more – important in less developed countries (LDCs), perhaps by virtue of the weakness of formal mechanisms for regulating inter-firm connections. For example, Dahles comments on their role in the Indonesian economy:

> The creation of complex networking relations among entrepreneurs appears to be the central strategy in the development and operation of small enterprises. Networks are used to develop not only business contacts but also to raise social standing and enhance political influence, which in turn contribute to economic success. Networks – a source of 'social capital' – are essential not only for successful business dealings and the enhancement of prestige, but also as insurance against an uncertain future. (1999b: 9)

In other words, business networks perform three main functions. They lubricate business relationships, they provide the element of trust necessary to reduce harmful uncertainty, and they provide a basis for wider social networks, especially in relation to political power. In turn, these social networks can be utilized for the benefit of the firm.

While much of the recent debate in economic analysis has been on such networks and informal relationships, they also constitute the background against which a series of overlapping formal relationships are created and re-created. For example, Mackun's (1998) study of tourism firms in central Italy has noted the increasing interest of small-hotel owners in business networking, through joining business and community organizations that fulfil a number of functions: they promote local tourism, conduct market research, and communicate the collective needs of their members to government. Moreover, these are effective, precisely because they are built on dense informal networks: 'The close-knit nature of the communities (extended families and neighbors) has increased the strength of these groups' (p. 268).

In the remainder of this chapter we consider three kinds of inter-firm cooperation: supply strategies, long-term strategic alliances, and mergers and acquisitions. These imply increasingly strong and more formalized cooperation among firms. Alternatively this can be expressed in terms of an increased blurring of the boundaries between firms, leading – in the case of mergers – to dissolution of the boundary between two or more firms, or at least its weakening. Which of the three strategies is adopted is contingent on the internal organization and assets of the companies involved (Knowles et al. 2001: 138), the competitive and regulatory environment, and their long- and short-term goals.

Supply and marketing strategies

Firms normally face a basic decision: whether to internalize part of the value chain, or purchase these goods and services from external firms. If they rely on inter-firm relationships, then they have a choice between relatively short-term market relationships or some form of relational contracting. The latter is the object of this section.

Tour companies illustrate some of the choices faced by companies. Their operations are characterized by a high degree of vertical integration. Their main role is to link the buyers and sellers of tourism services, that is, effectively creating or expanding markets. This requires linkages to travel agents, which sell most of their holiday products (although some tour operators sell directly to consumers). They also have linkages to the hotels and transport companies that provide core services. The tour operators' role is to assemble packages of services, provided by airlines, accommodation suppliers, local transport firms and, in some cases, firms providing local excursions and car-hire companies. The tour operators then invest massively in promoting these packages.

The tour operators have to balance risk and uncertainty, on the one hand, against internalizing the profits that can be made from selling individual components of the holiday package, and securing quality control over these. They take on the risk that they will sell sufficient 'packages' to cover costs and make a profit. This benefits other service providers who do not have to invest

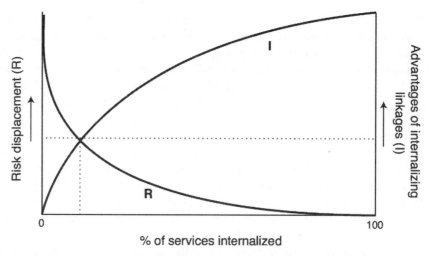

Figure 4.4 **Supply linkages: the trade-off between risk and internalizing linkages**

in promotion and secure large volume sales via the tour operator. However, they have to sell their services at discounted prices, and do not manage to transfer all the risk to the tour operator. Instead, many tour operators have a system of block reservation, whereby they release unwanted beds in hotels at agreed cut-off points. American Express Vacations uses 7 or 14 day cut-offs (Ioannides 1998; Sheldon 1986), which give the hotels little opportunity to find alternative markets for their services for those particular nights. If the tour operator decides to internalize these links, through a programme of acquisitions or mergers, to create a vertically integrated company, the effects are mixed. They now absorb the risk carried by the hotels and airlines, but they can also reduce costs through preferential purchasing from suppliers (e.g. hotels) that they own. They can also internalize the profits generated – in large part – by their promotional activities. Figure 4.4 presents the trade-off faced by large tour operators: small companies will probably be excluded from this trade-off because of lack of resources to develop internalization strategies.

Tour companies adopt different strategies in the face of such choices. Thomson, for example, established its own travel agency (Lunn Poly) and airline (Britannia), while not investing significantly in hotels and other services in destinations. In contrast, German tour operators are more likely to own a significant proportion of the holiday accommodation they use (Williams 1995). Smaller tour operators, in both countries, are unlikely to own any of these facilities/services, and instead rely entirely on inter-firm relationships. Each of these different strategies has important implications, not only for the companies involved, but also for the destinations. As Ioannides (1998) observes, tour operators are effectively gatekeepers in resort evolution. Buhalis (2000) explores some of these issues in respect of the role of tour operators in Mediterranean destinations (Box 4.5).

Box 4.5   Tour operators in the Aegean: consequences of supplier strategies

Supplier relations between tour operators and local resort hotels are characterized by asymmetrical power relationships. The tour operators deal not only with large numbers of SMEs, but also, potentially, have a wide choice of destinations. Large tour operators can provide large client volumes, but individual hotels have to balance their occupancy rates against the average room yield.

A number of uneven relationships stem from this fundamental asymmetry:

- Hotels have to accept low prices, which would lead to operating losses were it not for the use of unpaid family labour.

- The smallest firms are most disadvantaged, because they lack marketing budgets for promotion, and are less autonomous.

- Human agency is important: the more-experienced and better-qualified managers are better able to negotiate with the tour companies.

- Tour operator bankruptcies have a double negative effect: not only do hotels suffer creditor losses, but they come under even greater pressure to accept lower prices from other operators.

- Tour operators use their cancellation rights, especially outside the peak season, and pass on risk to the hotels.

- Payment delays: the norm is two weeks after departure, but can be up to one year.

Overall this pattern of relationships increases the competitiveness of individual tour operators, while trapping many small hotels in weak competitive positions.

Source: based on Buhalis (2000)

The supplier decisions taken by tour operators may represent a special case, but similar decisions are faced by all tourism businesses. The major question is whether it is in the interests of the purchaser to have a fluctuating system of sub-contracting, or whether to build closer relationships with selected firms. In car manufacturing, for example, some larger companies have preferred to develop links with 'system integrators', which effectively manage the work of a number of sub-sub-contractors. They may also be drawn into working closely with the lead company in the design stage, that is, enter into some form of informal partnership with the principal company. This is not very different to some of the supplier relationships seen in the tourism industry. For example, hotels act as integrators for tour operators: they assemble the catering, accommodation, and entertainment services provided to the tourists.

Moreover, those firms that take at least a medium-term view are likely to be selective in their suppliers. This is consistent with Gertler's (1997) comment that, over time, relationships between firms may come to be more reliant on non-market than market interactions. There is considerable economic value grounded in trust, knowledge of suppliers, and shared goals in respect of production. Hence, most major tour operators are selective in the travel agents they work through. Holloway comments that 'few operators now deal indiscriminately with all retail agents. As with most products sold, some 80 per cent of package holidays are actually sold through 20 per cent of the retail agents' (1998: 223). Typical of the larger UK tour operators is Thomson, which, in the second half of 1990s, decided to target its sales, other than via Lunn Poly, through the largest independents (more than 1000 sales per year) and the top 20 'miniples' (small chains) with more than 3000 bookings each. These are provided with additional sales support, such as sales materials, daily late-availability updates, educational study trips, and staff incentives. This represents a strategy of developing closer relationships with suppliers, without entering into formal alliances with them.

While the advantages of fostering relationships of trust are clear, such strategies may face a number of obstacles. It is difficult to unravel a background of lack of trust, and an unwillingness to share information. There may also be high levels of volatility in the ownership of local firms, so that it is difficult for the principal firms to maintain continuity among suppliers. This is compounded by the lack of quality concerns and technical knowledge among many suppliers (Jenkins 1982). Yet, even in the most challenging of circumstances, it is possible to foster links with local businesses, as Telfer and Wall demonstrate in the example of the Sheraton Hotel in Lombok (Indonesia) (Box 4.6).

The Indonesian example takes us into the question of the extent to which location or proximity is a precondition for the development of inter-firm relationships, and whether there are geographical clusters of firms as a result of agglomeration and external economies. There is considerable debate on this subject in the industrial literature, centring on the existence of Marshallian districts. Even the literature on manufacturing is far from conclusive as to whether organizational features (e.g. centralized buying in larger corporations) or spatial proximity are more important in determining firm linkages. One of the key points in this literature is the existence of different types of industrial districts (Storper 1997). On the one hand, there are 'nonprogressive' industrial districts, based on competitive advantages from cost savings resulting from agglomeration. And on the other hand there are 'progressive' technology districts, based on competitive advantages derived from product-based technological learning. Hudson comments that 'an innovative industrial district is a dynamic constellation for mutually adjusting firms, responding to new challenges and opportunities via continuous redefinition of interfirm relations and the external boundaries of the district' (2001: 204). The research on industrial districts has been extended in the 1990s to an emphasis on innovative capacity and the concept of 'learning regions', but we return to this theme in the conclusions. Is there any evidence for any of these types of districts in tourism?

There is obviously geographical clustering in tourism – this spatial feature is inherent in the social construction of tourism, and the nature of tourist attractions. Firms in resorts are mutually interdependent, for the tourism experience has multiple components (Chapter 2). Museum districts also

---

Box 4.6   The Sheraton Hotel, Lombok: supplier relationships

Telfer and Wall address the question of whether large, externally owned hotels are more or less likely than small locally owned hotels to have local linkages to suppliers. Centralized purchasing in large corporations supposedly reduces local linkages, but the authors' study of the Sheraton Hotel on the Indonesian island of Lombok demonstrates the complexity of supplier connections.

First, the Sheraton Hotel was exceptional in that the Canadian chef had a passion for locally sourced quality ingredients. He therefore instigated two special projects to encourage this against a background of weak marketing of island produce:

- *Sheraton Fish Programme:* The Sheraton provided ice tanks to a local fisherman, who was contracted to visit and buy in local fish markets. Strong quality control was imposed on the produce by the hotel. This reduced imports, provided cheaper and fresher produce for the hotel, and gave them local products to publicize on their menu.

- *Sheraton Vegetable and Herb Programme.* The Sheraton provided seeds to a local farmer who contracted to supply produce exclusively to the hotel. This eventually failed, because of seasonal variations in occupancy rates and demand, the need for constant supervision by the hotel, and the lack of other hotels showing interest in the project (hence, a lack of economies of scale).

Both experiments failed, eventually, for two reasons: first, the lack of knowledge and expertise among local producers, and a failure by the state to invest in these; and second, because human agency was the key to the project's success, when the chef moved on the projects collapsed.

Even when the special case of the Sheraton was put aside, the authors found a complex picture. Large and medium-sized hotels sourced about 70% of their produce locally, while small hotels sourced virtually all of their supplies locally. However, the view that the small guest house is likely to use local suppliers, and benefit local agriculture, is problematic because:

- The Sheraton purchases 45 times more supplies than the average small hotel.

- The Sheraton sells far more food per tourist than the small guest house.

- The large hotels are usually more regular and reliable payers of suppliers than are small hotels.

- Some of the materials sold in local markets are imported.

Source: based on Telfer and Wall (1996; 2000)

Box 4.7 The museum district

Museum districts are usually the product of public policy. According to Santagata they have the following positive externalities (i.e. common benefits available to individual firms, without direct costs):

- Networking externalities: proximity provides 'cultural connections' to other museums for tourists as well as curators and historians.

- Consumption externalities: increased utility is enjoyed by customers as a result of cultural connections. The increased number of visitors also reinforces the signifiers that individual museums within the district are important tourist destinations.

- Externalities of time: temporary exhibits are important magnets of attention, with visits to these being combined with time spent in the permanent exhibition.

- Economies of scale and scope: these reduce unit costs, and increase product variety.

Museum districts can offer potential critical mass, so that significant numbers of additional visitors are attracted to the area, bringing net benefits to all the individual museums. However – in contrast to Santagata's view – these advantages do not come about automatically. Proximity does not necessarily result in all these forms of positive externalities. Instead, the realization of such externalities is partly dependent on the social relationships between the key managers of the museums, and on complementarities in terms of markets and the production of tourism services.

Source: based on Santagata (2002)

---

present a specialized form of industrial district in tourism (Box 4.7). But to what extent are there 'progressive' tourism districts? The most comprehensive attempt to address this question is Hjalager's (2000) Danish study. She stresses that such districts should have five main characteristics:

- *Interdependence of firms: horizontal, vertical and diagonal systems of contracting and sub-contracting to achieve economies of scale.* In tourism, strategic alliances often tend to be non-local, e.g. between external tour operators and local hotels, or involve marketing consortia, such as Best Western hotels, which deliberately aim at wide, rather than local, geographical coverage.

- *Flexible firm boundaries: firms tend to be temporally and functionally flexible.* Tourism firms are certainly temporally flexible, responding to seasonal and other variations in demand. But the sharp discontinuities in employment, and reliance on seasonal labour migrants, mean that it is difficult to build up local pools of 'solid knowledge repositories', i.e. a skilled human capital base in the locality.

- *The coexistence of cooperation and competition*. Firms both compete and cooperate, to their mutual advantage. Free-riding makes this difficult in tourism, so that most collaboration tends to be formal and imposed top down, rather than emerging organically as long-term strategic collaboration among firms.

- *Trust in sustained collaboration*. Mutual expectations of trust are nurtured by repeated face-to-face contacts. This is problematic in tourism because of the rapid turnover of firms (deaths and births) and free-riding problems.

- *A 'community culture' with supportive public policies*. While this is variable, tourism public policies have tended to focus more on marketing rather than on knowledge transfer.

It is, of course, difficult to generalize about such districts, and there is a need to consider the contingencies of time and place. Over what scale do agglomeration economies operate? In other words, is spatial concentration a requirement, or is it possible to secure agglomeration economies if firms are distributed across a wider area, e.g. through several linked small clusters of tourism firms? The evidence is not conclusive. The balance of advantages and disadvantages may change over time. And the effectiveness of clusters is also dependent on the capacity for building effective coalitions and partnerships. While further research is necessary on all these issues, Hjalager's work does at least sensitize us to the difficulties of creating 'progressive' districts in tourism. In particular, she highlights the importance of non-localized strategic alliances, and this is the theme of the next section.

Longer-term strategic alliances

Firms may decide it is advantageous to formalize their relationships with suppliers through a strategic alliance. A strategic alliance is defined by Johanson et al. (1991) as an inter-organizational relationship, in which partners invest time, effort and resources while collaborating to achieve both individual and shared goals. Such strategic alliances typically come about when there is a high degree of mutual reliance among companies in the value chain, leading to a need to 'exert control over other suppliers through transaction arrangements' other than commissions or supply contracts (Hall and Page 1999: 94). In the economy as a whole, strategic alliances are most common in those sectors characterized by high risk, high technology costs, globalization and economies of scale (Dicken 1998).

Strategic alliances in tourism take a number of different forms. They can be with individual competitors or with groups of these, while their objectives are also varied: to improve market access (mutual, shared distribution of costs and benefits), market development, especially in face of risk (e.g. in emerging market economies), sharing the costs of R&D (as in some GDS and CRS systems), or economies of scale in production. The alliances may be based on shared development costs, reducing competition in key markets (such cartels may be illegal under some regulatory systems), and they may be either for production or distribution, especially marketing. Marketing alliances are relatively common in the accommodation sector. For example, the Logis de France consortium markets 4500 establishments, which account for over 60% of France's total room supply. The largest single such consortium, however, is Best Western, with more than 3000 hotels and more than 250,000 rooms (Vellas and Bécherel 1995: 103). Franchising is a special case of strategic alliances, and is relatively common in catering. In this format, the parent company provides production and marketing knowledge, monitors quality, and provides its brand name, while the franchisee provides the capital and local organization. The McDonalds fast-food chain is one of the best-known examples of franchising.

Connections among firms in the airline sector are also characterized by strategic alliances (Table 4.1). These mostly involve horizontal integration with other air carriers, although there are examples of vertical integration, as in the alliances they have formed with hotels, tour operators, or car-rental companies. There is a long history of such alliances in the air-travel sector, dating back to at least the regulatory framework provided by the 1944 Chicago Convention (Evans 2001). Under the terms of the Convention, firms made alliances to cooperate on the transfer of passengers, baggage, etc., which extended to pooling arrangements for revenues on shared routes. In recent years, there has been a shift to broader forms of cooperation, including code-sharing, so that the connecting flights of different airlines appear as a single airline when purchasing the ticket. There has been exponential growth in such alliances: from 280 alliances in 1994 to 513 in 1999 (Evans 2001: 231). The major airlines are also seeking global alliances. For example, British Airways acquired British Caledonian in 1988, took a minority share in Qantas in 1992 and a 44% share in US Air in 1993 (it later sought an alliance, instead, with American Airlines) (Holloway 1998: 97–8).

Table 4.1 **Principal objectives of airlines' strategic partnerships**

1. To merge commercial activities: sales, reservations and passenger services.

2. To organize flight hubs: arrangements with feeder airlines.

3. To establish joint management agreements for setting up ground handling at airports.

4. To create commercial representation agreements in order to capture market share.

5. To operate joint investment and operating expenditure agreements: for example, block purchases of aircraft, shared maintenance workshops etc.

6. To set-up holding groups for strategic planning, marketing etc.

7. To merge reservation services, including code-sharing.

Source: based on Vellas and Bécherel (1995: 147–50).

Airlines have a number of objectives in establishing such strategic alliances, and Evans (2001) classifies these as external and internal drivers (Box 4.8). These are clearly compelling, as evidenced in the growth of alliances in recent years. However, there is surprisingly little research that actually establishes the effectiveness of such alliances. One of the few studies to attempt this – at least conceptually – is Dundjerovic's (1999) assessment of the economic advantages of various forms of consortia in the hotel sector, compared with outright ownership of chains of hotels (Table 4.2). He identifies a number of potential scale advantages, including: indivisibilities (i.e. fixed costs), the economies of increased dimensions (i.e. costs of capital equipment such as water storage tanks), specialization or managerial economies, massed re-sources (reduced stockholding), purchasing economies (discounts for volume sales), external finance economies (better credit terms), lower unit advertising costs, and – arguably – lower average wages (based on Dundjerovic's questionable assumption that larger firms offer more training and promotion prospects). He concludes that hotel chains have advantages over full strategic alliances in terms of increased dimensions, specialization and external finance.

Box 4.8   Airlines' strategic alliances: internal and external drivers

External drivers

- The information revolution: CRS systems allow companies to manage and control passenger flows more intensively and flexibly, but strategic partner-ships are required to develop and operate these.

- Economic restructuring: liberalization and privatization are intensifying compe-tition, and strategic alliances provide a means to reduce market entry and competition.

- Global competition: there is a need to develop global reach in context of globalization pressures.

Internal drivers

- Risk sharing.

- Securing economies of scale, scope and learning, e.g. through using each other's hub and spokes, and sharing the costs of handling, desk space and marketing.

- Accessing assets (especially landing slots at airports), resources and compet-encies of other companies.

- Shaping competition, e.g. making allies out of potential competitors.

Source: based on Evans (2001)

Table 4.2 **Comparison of scale economies achievable by hotel chains and different types of consortia**

Types of economies of scale	Hotel chains	Full consortia	Marketing consortia	Other consortia
Real economies of scale				
Indivisibilities	No	No	No	No
Increased dimensions	Yes	Partial	Partial	Partial
Specialization	Yes	Partial	Partial	Partial
Massed resources	No	No	No	No
Pecuniary economies of scale				
Supplier discounts (consumables)	Yes	Yes	No	Maybe
Supplier discounts (advertising)	Yes	Yes	Yes	No
External finance	Yes	Partial	No	No
Advertising	Yes	Yes	Yes	No
Lower wage levels	Yes	Not known	No	No

Source: Dundjerovic (1999).

The findings about the limitations of consortia are echoed by Crotts et al. (2000), who identified a number of problems rooted in the tension between balancing competition and cooperation. Individual firms will have concerns about becoming too dependent on their partner, may fear that the alliance excludes them from potentially more rewarding partnerships with innovative partners, and may be reluctant to share commercially sensitive information. This is why many commentators argue that strategic alliances are:

> ... inherently unstable and transitory forms of organisation, a 'second-best' solution that is disturbingly likely to break up under commercial pressure. It can be argued that the benefits of alliances can probably be achieved more completely and effectively through mergers and thus alliances are only generally a stopping off point on the way towards full mergers if the lifting of regulatory and legal restrictions were to make them possible. (Evans 2001: 239)

The next section considers acquisitions and mergers as an alternative to strategic alliances.

Acquisitions and mergers

The 'internalization' of firms' links was one of the three key factors that Dunning and McQueen (1982) used to explain the growth of transnational companies (see Chapter 3). Sinclair and Stabler (1997) have elaborated on the advantages that result from both vertical and horizontal mergers and acquisitions. In this context, the advantages of vertical integration (based on Sinclair and Stabler 1997: 134–6) include:

● reducing transaction costs

- improving synchronization of transport, accommodation provision and entertainment

- facilitating information collection

- providing inputs at known prices

- reducing uncertainty about future demand

- increasing market power

Similarly, the advantages of horizontal integration are:

- scale economies: linked to increased market power they may enable firms to raise prices and profitability, increase market share, raise barriers to entry, and obtain easier access to finance

- acquiring market share, in order to develop greater market control e.g. through creating an oligopoly

- improving access to technology

- diversifying into new markets

- adding a qualitative difference to the firm's activities and therefore strengthening its competition within changing paradigms

Mergers and acquisitions provide quick routes to increasing market share, developing market control, gaining access to a rival's technology, or diversifying products and markets. Above all, it is the speed of change that is critical, compared with new investment.

In each case, mergers and acquisitions are a means to relocate the firm in respect of inter-firm competition. Horizontal mergers to increase market share are mostly a form of weak competition within paradigms. This can be across international boundaries, as evidenced in recent tour-operator takeovers in Europe. Such mergers can also secure cost reductions, which is another expression of competition within parameters. For example, in August 2001, Preussag – which had taken over Thomson – was seeking to achieve 'synergies' and 'economies of scale' through cooperation in fuel purchasing and maintenance costs for its Britannia and Hapag-Lloyd charter operations (*Financial Times* 28 August 2001). But mergers and acquisitions may also be used to change the paradigms of competition. For example, large firms often take over innovative small firms, as Thomson did with Country Cottages.

The extent of mergers and acquisitions is dependent on market and production conditions, as well as on the prevailing global and national regulatory context. Therefore, there are significant sectoral variations, as we note below in respect of airlines, hotels and travel agencies. There have been both vertical and horizontal mergers and acquisitions in the airline sector. Airlines have taken over other airlines – for instance, British Airways acquired Dan Air wholly in 1992 and has part ownership of Deutsche BA, Air Russia, and TAT, the French regional carrier. Additionally, in the early 1970s some airlines saw potential in owning hotels as they increasingly moved into using larger jumbo aircraft. However, by the late 1970s and 1980s, mounting losses in the face of stiff competition led to de-mergers in a number of cases, and renewed focus on their core businesses, aided by the introduction of CRS systems (Holloway 1998: 74).

Conditions in the hotel sector have also been variable. It is notable, for example, that the ratio of concentration of ownership in the hotel sector is 15 times higher in the USA than in the Mediterranean region (Sinclair and Stabler 1997: 72). Over time, there has been a trend to greater merger and acquisition activity. Go and Pine comment that:

> Starting with the take-over of Inter-Continental by Grand Metropolitan in August 1981 and Hilton International by Ladbroke in September 1987, mergers and acquisitions have increased in frequency and magnitude to the point where they are perhaps the most crucial trends with the largest impact on the structure of the international hotel industry. (1995: 9)

In general, however, this remains a polarized sector, with increasing concentration being balanced by a continuing high degree of fragmentation. This leads to our conclusion that, in the face of increasingly globalized competition, firms are faced with various strategies. These involve different forms of competition within existing and changing paradigms. There is no inevitability in the outcome. Instead, it depends on the confluence of structural industrial conditions and the contingencies of place, as well as an element of human agency.

SUMMARY: INNOVATION AND FORMS OF COMPETITION

Globalization contributes to the intensification of competition, and firms can respond to this in a number of ways. Here, these are characterized in terms of 'competition within existing or changing paradigms' – that is, whether firms can innovate in such a way as to fundamentally challenge the basis of inter-firm competition. In addition, firms need to both compete and collaborate with other firms, and we consider different forms of collaboration: alliances, interdependencies and mergers/acquisitions. The main points to emerge in the course of this chapter are:

• There are differences in the degree to which there are contestable markets in different tourism sectors, and these are also temporally and spatially varied.

- Price is the most common basis of competition within existing paradigms. 'Raw competition' focuses almost entirely on prices, with intense pressures to drive down costs, especially labour costs. However, most firms actually compete on the basis of the price-to-quality relationship.

- Competition within changing paradigms involves market disruption, usually via innovations. Their impacts vary, and can be characterized as incremental, distinctive or breakthrough.

- Product and process innovations are the two main forms of innovation, and the latter is increasingly associated with IT developments.

- In some circumstances firms prefer collaboration to competition.

- Collaboration may be informal, based around networks of mutual trust and reciprocity. These can be community- or ethnicity-based.

- Collaboration may be built around supply or marketing strategies. Tour operators' supply chains provide the best-known example in tourism.

- There is a debate as to the extent to which such collaboration is territorially based, and whether there are effective 'tourism clusters'.

- Firms may also enter into long-term strategic alliances, when these are considered mutually beneficial. They are particularly evident among airlines, but also exist in other sectors.

- Mergers and acquisitions can be the logical outcome of a desire to foster collarboration. These provide rapid means to acquire market share, gain access to technology, and diversify products and markets.

5 Mapping Tourism Consumption: from Fordism to McDonaldization

PERSPECTIVES ON TOURISM CONSUMPTION

During the 1990s Urry sought to outline a sociology of consumption 'concerned with the differential purchase, use and symbolic significance of material objects' (1995:129), but more especially in relation to the consumption of tourism. This is part of a wider set of studies that recognize that people's lives are shaped not only by their occupations and the nature of production but, more importantly, also by consumer goods and services. According to some commentators, consumer processes and goods are the most significant elements of developed societies (McCracken 1990). In such societies, McCracken argues, consumption is a cultural phenomenon and without consumer goods 'these developed societies would lose key instruments for the reproduction, representation and manipulation of their culture' (1990: xi). Lee (1993) takes the argument further, stating that consumer goods take on some form of magical quality. The thrust of such perspectives is that consumption is both 'an economic and cultural touchstone' (Miles 1998: 3).

Within this emergence of a sociology of consumption, Miles (1998) has drawn attention to a number of key contributions, including the writings of Saunders and Bourdieu, which can help shed some light on aspects of tourism consumption. Saunders (1981) for example, has stressed the importance of access to consumption rather than class or relation to the means of production. Although criticized for a number of shortcomings (see Warde 1990), the work of Saunders has highlighted the fundamental division in societies between those with and those without access to different aspects of consumption. This is an important but relatively neglected issue within tourism studies, and it is one that will be discussed in the concluding part of this chapter. A more direct impact on tourism research is contained in the work of Bourdieu (1984), who stressed the social significance of consumption.

In the context of tourism consumption, it could be argued that tourists are motivated by the need to reproduce a pattern of preferences based along the lines of class demarcation. The notions of cultural and symbolic capital are important within tourism, with different classes of consumers being better equipped to 'take advantage of symbolic capital' (Miles 1998: 21). Such capital, Bourdieu argues, plays a pivotal role in the construction of lifestyles. Within tourism, it has been argued that new forms of tourism consumption can be identified with the so-called 'new middle classes' (Mowforth and Munt 1998; 2003). Such notions will be explored more fully in the middle sections of this chapter. Of course, to see consumption entirely as a sociocultural phenomenon would be misleading, as there are material relationships and strong global economic trends at work, as discussed in the early chapters of this book.

A final, and perhaps the most important, perspective on tourism consumption is the attempt to develop an overarching theory linking societal change and consumption (Urry 1990; 1995). This assumes a shift from an older form of consumption (Fordist or modernist) to new forms (post-Fordist or postmodern) and, as we argued above, the emphasis in tourism studies has been on identifying these new forms. At a broader level, Urry (1995) argued that disorganized capital is increasingly involved in dissolving tourism's specificity, when tourism as a form of consumption starts to become hegemonic and organize much of contemporary social and cultural experiences. The extension of this perspective envisages what Urry (1995) terms the 'end of tourism', describing a situation in which people become tourists for so much of the time. Certainly, in many postmodern societies, there is increasing evidence that the edges of tourism and other forms of consumption are becoming increasingly blurred (Shaw et al. 2000).

The issues surrounding the nature and development of tourism consumption are contested and complex, in part because many of the conceptual frameworks remain untested. There appear to be two main areas of debate; first, the changing patterns of tourism consumption, or what can be termed the Fordist/post-Fordist dialectic, and second, the so-called 'new' forms of tourism consumption. These contested issues will form the central part of this chapter.

THE SHIFT TO POST-FORDIST CONSUMPTION

There has been a good deal of comment on the shift to new forms of tourism consumption, away from the rigid, large-scale Fordist modes, towards more flexible forms of post-Fordist patterns. Authors such as Britton (1991), Urry (1995), Sharpley (1994), and Mowforth and Munt (1998; 2003) have, to different degrees, stressed such changes. This shift is complex and embedded in macro-level changes in both production and consumption systems, including processes of globalization (Short and Kim 1999). Indeed, it is far easier to recognize the so-called 'old' and 'new' forms of tourism consumption than it is to detail the processes themselves.

Table 5.1 **Characteristics of mass tourism**   *use!*

Collective consumption by undifferentiated tourists

Collective gaze of tourists – focused on signifiers designed to concentrate tourists' seasonally polarized consumption

Demands for familiarity by tourists

Undifferentiated product – similarity of facilities and experiences

Rigidity of production – highly standardized, large-scale, dependent on scale economies

Low prices – importance of discounting and price cutting

Large numbers of tourists related to a circuit of mass production

As a starting point, it is useful to identify the characteristics of mass tourism, which are representative of the Fordist mode of consumption. The main defining features are given in Table 5.1, but it is worth emphasizing that mass tourism is characterized by large numbers of tourists related to a circuit of mass production (Boissevain 2000). Moreover, there tends to be a rigidity of production that is highly standardized, large-scale and strongly dependent on scale economies (Ioannides and Debbage 1998). Products are, however, offered to the tourists along both cost and stage in the family life-cycle segments. On the whole, tourists want a high degree of familiarity, which is provided by a similarity of facilities and experiences. Mass tourists participate in what Urry (1995) terms a collective gaze, which focuses on recognized signifiers in the landscape, partly designed to concentrate tourists in particular destinations.

In contrast, post-Fordist tourism consumption involves the creation of more specialized, individual and niche markets, which are seemingly tailor-made to meet the changing needs of tourist demand. Such tourists have been recognized by a variety of writers, who have labelled them in different ways (Mowforth and Munt 2003, Chapter 5). For example, early work by Krippendorf (1987) in Alpine Europe described the emergence of what he termed critical consumer tourists. Others have attempted to identify green, sustainable and ecotourists (Mowforth and Munt 2003, Chapter 5; Fennell 1999; Holden 2000), while Milne (1998) talks of 'better' tourists. These all tend to be associated with wealthier, better-educated and, some would argue, more desirable tourists (Milne 1998). These tourists have, according to Poon (1993) and Urry (1995), more control, created through increased consumer purchasing power, leading to tourism products defined by tourist tastes and preferences.

The essence of post-Fordist tourism consumption and production regimes is that of flexibility (see Table 5.2). This is manifest in far less structured, more independent forms of tourism, which are the antithesis of the highly structured, rigid, mass package tour. Tourists are offered, and demand, a diverse product, which is highly differentiated, giving a wider degree of consumer choice.

Table 5.2 **Characteristics of post-Fordist tourism consumption**

Characteristics of post-Fordist consumption	Tourist examples
Consumers increasingly dominant and producers have to be much more consumer-oriented	Rejection of certain forms of mass tourism (holiday camps and cheaper packaged holidays) and increased diversity of preferences
Greater volatility of consumer preferences	Fewer repeat visits and the proliferation of alternative sights and attractions
Increased market segmentation	Multiplication of types of holiday and visitor attraction, based on lifestyle search
Growth of a consumers' movement	Much more information provided about alternative holidays and attractions through the media
Development of many new products, each of which has a shorter life	Rapid turnover of tourist sites and experiences, because of rapid changes of fashion
Increased preferences expressed for non-mass forms of production/ consumption	Growth of 'green tourism' and of forms of refreshment and accommodation individually tailored to the consumer (such as country-house hotels)
Consumption as less and less 'functional' and increasingly aestheticized	'De-differentiation' of tourism from leisure, culture, retailing, education, sport, hobbies

Source: Urry (1995, 151).

As we argued earlier in this chapter there are two key debates. The first concerns whether it is possible to recognize changing consumption patterns and, if so, just how widespread are they? Second, when did these trends occur and, more especially, what are the processes involved? Finally, and most crucially, how realistic is this perspective? Clearly, these issues overlap, although it is convenient to discuss them individually. We should also recognize that these issues are a sub-set of a wider set of debates on the nature of consumption and consumerism, of which tourism is but a part, albeit a significant one (see Gottdiener 2000). Lee (1993) argues that there was a rebirth of consumer culture, following the rise and fall of mass consumption, which involved the 'emergence' of a new diversified commodity form (Miles 1998: 9). According to Lee, this transformation occurred during the 1980s, when the aesthetics and style of consumption became more diverse in response to an increasingly sophisticated consumer market (see also Chapter 2). Reinforcing this view, Slater claims the '1980s saw one of the most powerful rediscoveries of consumerism' (1997: 10), and also witnessed the 'subordination of production to consumption' (p. 10). At this time, Fordism, the pioneer mode of mass consumption, started to give ground to a new form of consumption. In terms of change in the marketplace, there are a number of intertwined factors. These include what Urry sees as 'changes in the structuring of contemporary societies' (1990: 88), producing a significant growth of the service class, as well as the way in which such classes have disturbed or disrupted pre-existing cultural patterns (Martin 1982). There is also the increased importance of the media in structuring the tastes and fashions of tourism consumption (Urry 1990; Nielsen 2001).

The new middle classes and tourism consumption

Numerous perspectives on consumption have sought to establish the links between postmodern shifts in tourism consumption and the development of the new middle classes (Urry 1990; Munt 1994; Mowforth and Munt 2003). All draw from the work of Bourdieu (1984), who argued that different social classes are engaged in a struggle to distinguish themselves from one another through education, occupation, residence and consumption. As Munt emphasizes, the latter includes 'both objects and experiences, such as holidays' (1994: 105). Central to this is the growth of what Bourdieu terms the new petty bourgeois and what Urry identifies as the new service class. Urry, using the UK as an example, argued that 'in western societies there is both a major service class and, more generally, a substantial white collar or middle class' (1990:88).

Within the expanded new middle classes, Munt (1994), following Bourdieu, identifies important subdivisions, most notably the 'new bourgeoisie', and the 'new petit bourgeoisie'. The former is 'firmly located in the service sector' and is rich in both economic and cultural capital (Munt 1994: 107). In terms of tourism consumption, this would express itself in discriminating, possibly luxury, holidays (Bruner 1989), with an interest in, for example, certain types of eco-tourism (Mowforth and Munt 1998). The so-called 'new petit bourgeoisie', according to Bourdieu, are within occupations involving 'presentation and representation' (Munt 1994: 107). They are important taste-makers, but are also relatively low on economic capital compared with the 'new bourgeoisie'. Within this group are to be found those involved in more so-called postmodern forms of tourism, such as backpacking and other independent forms of travel. According to Bourdieu, these two groups within the expanded service class are the 'major consumers of the postmodern' (Urry 1990: 89). The boundaries between the 'new middle classes' are not clear cut, but both are important in the emergence of postmodernist, or what Urry (1990) terms post-Fordist, forms of tourism consumption.

This brings us to the second dynamic in the transformation of tourism consumption, involving the way such classes have disrupted pre-existing patterns. According to Urry, it is the significance the new middle class attaches to cultural capital that is crucial, along with its continual need to augment this. Within this context, Lash (1991) points to the increasing commodification of tourism, starting with mass package holidays and extending into the new forms of tourism consumption, certainly in terms of 'the international tourist infrastructure' (Munt 1994: 109). The notions and extent of commodification are more fully discussed in Chapters 2 and 7, while Chapter 6 discusses more fully the way holiday experiences are engineered. For postmodern tourists, increasing emphasis is also given to 'sign value' or the symbolic importance of tourism as a means of accumulating cultural capital (Munt 1994). This may take a variety of forms, including the search for the traditional, authentic or exotic, leading to a competition for uniqueness (Cohen 1989). For many tourists, the need to have unique experiences becomes of critical importance, although it is debatable whether such experiences have to be 'authentic' (see Chapter 6).

The final contribution to the shifts in consumption is provided by the impact of the media, especially television, film and, more recently, the internet. Lash and Urry (1989) and Urry (1990) argue that the media have been important in exposing all social groups to representations of different lifestyles and forms of consumption. This influence transgresses the boundaries between different social groups. Certainly, within tourism there is a strong emphasis on constructing representations of places and experiences through all types of media (Morgan and Pritchard 1999) (see also Chapters 7 and 10).

Of course, the impact of these structural changes in consumption is variable and mediated by social class and lifestyles. For example, in many Western societies, there has been an increase in the number of holidays being taken alongside more flexible types of trips. This increased flexibility in holiday-taking has also been aided by two important innovations. The first is the credit card, which has released unearned income to allow more tourism trips, and aided short-term decision-making in the context of, for example, taking short-break holidays. In a broader context, Ritzer and Liska view the credit card as a 'meta-means of tourism (and consumption)' (1997: 105). Of course, there are national variations in its use, even in Western economies, but in the UK credit cards accounted for some £6–7 billion of spending in 2001–2. The second innovation is the coming of the internet, which, along with the credit card, has made the selection, booking and payment of holidays far easier and more flexible. Indeed, recent websites, such as Lastminute.com, take the purchase of trips and other forms of consumption to a new degree of decision-making flexibility, rewarding the tourist for making quick purchase decisions with discounted holidays. In the UK, for example, the period from the mid-1980s has witnessed an increase in the number of short-break holidays (defined as 1–3 nights' stay in commercial accommodation) being taken among the social groups AB and C1, which contain the new middle classes. The volume of the short-stay market (defined differently to short breaks, as it also includes people staying with friends and relatives) increased from 45% of all holiday trips in 1986 to 67% by 1995 (Shaw et al. 1998). By 2000, the short-break market accounted for 64% of all holiday trips taken by UK residents (English Tourism Council 2000). Conversely, Mowforth and Munt (1998) show that in Britain during the early 1990s, the number of package holidays sold fell by 10%, pointing to the fall-off in some aspects of mass tourism associated with Fordist tendencies (see Chapter 2).

However, as Mowforth and Munt point out, 'data on the increased importance of new forms of tourism are difficult to come by' (1998: 98). Much of the information that does exist is from particular sites or destinations, often collected via a case-study approach, and consequently general figures are hard to assemble. Nevertheless, a number of authors (Urry 1990; Poon 1994; Sharpley 1994; Mowforth and Munt 2003; Williams 1998; Shaw and Williams 2002) have commented on the rise of new forms of tourism, identifying a number of key types:

- heritage/cultural tourism
- ecotourism
- adventure tourism
- visiting theme parks/mega-shopping malls

Some have argued that these 'new tourisms are truly contested ideas' (Mowforth and Munt 1998: 102), and in some areas there is no clear agreement on either conceptual or practical boundaries.

NEW FORMS OF TOURISM: CREATING THE TOURIST EXPERIENCE

As we discussed previously, for the new middle classes mass tourism is no longer sufficient; these new tourists (Poon 1994) want their holidays and leisure times to be full of 'worthwhile' experiences and to provide a means of increasing their cultural capital. The new forms of tourism, and the way they are presented, are seen to offer these benefits. Furthermore, there are a number of common characteristics of tourism consumption associated with these new forms. Some of these have already been identified (Table 5.2); others include the notion that these forms of tourism consumption are less functional and increasingly aestheticized, with strong emphasis on experience. In a broader, more economic, context, Pine and Gilmore (1999) argue that there is a move from a service economy to an experience-based one, where goods and services in themselves are no longer sufficient, but rather they are valued insofar as they are enhanced by the experience offered (Chapter 2). Such experiences, according to Pine and Gilmore, are based on engaging consumers by providing a memorable and personal product, appealing to their sensations.

The desire for experience gained from leisure time and holiday-taking has also changed the nature of tourist behaviour. Urry (1990), following earlier work by McCannell (1976), encompassed some of these changes in the notion of the tourist gaze. Increasingly, authors are arguing that in terms of new modes of tourism consumption, the tourist gaze, as a passive activity, is being replaced by the notion that the tourist is both the source and the object of the gaze (Richards 2001; Coleman and Crang 2002). Many of the new forms of tourism are felt to present opportunities to engage all of the senses in the creation of the tourist experience. More generally, Holt (1995), from a broader perspective on consumption practices, has suggested that people consume in four different ways:

- as experience, involving subjective and emotional reactions;

- as integration, by gaining information;

- as classification, by defining the individual or group through that which is consumed, with strong links to ideas of identity and cultural capital;

- as play, through socializing and communicating.

Holt's study, though based on a reading of the general literature, is also specific in its application to spectators at baseball games. However, all of these 'metaphors for consuming' (Holt 1995: 2) are of relevance to tourism consumption, for example, the notion of consuming as involving emotional states in the process. Such experiences are of critical importance in the acts of travel and the novelty of visiting different places and settings. These experiences are, however, very rarely newly constructed by tourists, but rather assimilated through a series of interpretative frameworks (see Chapter 6). Similarly, consuming as integration refers to the process by which consumers acquire and manipulate object meanings. For tourists, this again is important, as such integration practices allow the consumption object, say for example a certain type of holiday, to be integrated into their self-identities. It therefore becomes incorporated into symbolic capital. Related to this is the idea of consuming as classification, which describes how consumers use consumption objects to classify themselves relative to others. These ideas, as we have discussed, are especially important to consumption practices associated with tourism. The final dimension of consuming is that of consumption as play, which, as we shall see, figures importantly in some of the new forms of tourism consumption.

Of course, these four dimensions of consuming are not normally divided within the mind of the tourist, nor are they part of a set of easily identifiable processes. Rather, they are more like automatic responses. Their significance lies in part with the fact that they have the potential to provide an explanatory framework for understanding some of the characteristics associated with the new forms of tourism consumption. They are also much more than metaphors of the sociology of consumption, as some of the ideas have penetrated the world of marketing, and it can be argued that they are embedded in the manipulative processes of many commercial organizations. Pine and Gilmore's (1999) thesis of the experience economy highlights, as does Meethan (2001) in terms of post-Fordism, the range of marketing niches.

The consumption of heritage

An important characteristic of postmodern societies is the merging of different time periods (Urry 1995; Hollinshead 1997), as symbols of the past are

Table 5.3 **Characteristics of visitors to heritage attractions identified in selected British studies**

Author	Main visitor findings
Bagnell (1996)	Small sample survey of heritage sites – identified higher social classes
Light and Prentice (1994)	Survey of visitors in Wales – identified them as middle-class, well-educated and without children
Mellor (1991)	Work in Liverpool – identified age as important variable, mainly older visitors
Prentice (1993b)	Survey of Isle of Man attractions – identified importance of education to middle classes
Rowe (2002)	Survey of heritage sites in Devon – identified importance of education for middle-class visitors
Silberberg (1995)	Identified importance of cultural factors among early retired couples

reconstructed via a thriving heritage industry and represented in the present. Within the heritage industry, history becomes a commodity – a tourist spectacle. In addition, the compression of the past into the present is linked to the issue of tourist experiences and the search for authenticity (Sharpley 1994). Moreover, the concern for image and authenticity is strongly associated, as we have argued, with the new middle classes, especially the 'new bourgeoisie'. Work by Light and Prentice (1994) on heritage visitors in Wales has found that they tend to be drawn from the middle classes and are well-educated (Table 5.3). As Richards (1996) argues, within post-Fordist forms of tourism consumption there is a constant search for new experiences and sources of stimulation to help distinguish particular social groups. This is what Munt (1994) and others stress is the importance of cultural capital as a means of personal distinction. However, to view heritage tourism as merely a preserve for the new middle classes is somewhat simplistic, because heritage is a broad spectrum, which attracts a range of visitors. In this context, the past means different things to different people. To some it is part of a 'high culture' while to others the new heritage attractions sell other kinds of memories (Williams 1998; see also Chapter 6).

The rise of heritage tourism in Britain has been dramatic, and has occurred in a wide range of settings (Chapter 10). Since the mid-1970s, Britain has seen the opening of well over 1000 new registered museums, together with over 210,000 listed buildings. During 1997–8, some 32% of adults visited historic properties, with the figure rising to 39% for those in the 45–59 age group. Organizations such as the National Trust saw its membership increase from a mere 278,000 in 1971 to almost 2.8 million in 2000, generating an income of £200 million. In 1998, over 58.5 million visits were made to historic houses in England alone, alongside 12.5 million visits to gardens and 63.25 million visits to museums and galleries.

Richards (1996) concludes that the demand for heritage tourism has been stimulated by increasing levels of income, more education and the emergence of the new middle classes. To this may be added the rapid de-industrialization of many Western economies, a process that was especially pronounced in the UK in the late 1970s and 1980s. This had two key associated components. First, as certain ways of life disappeared, people showed greater interest in gazing on their past. Second, the redundant industrial sites provided 'natural' homes for the new heritage industry (Williams 1998: 185 and see Chapter 10). Some case studies have indicated that there is a strong link between the growing levels of employment in the culture industries and the choice of heritage attractions (Richards 1996). It is contested that this form of tourism consumption is especially engaged in by these workers and is part of a lifestyle in which the boundaries between the spheres of work and leisure are becoming blurred. As one would expect, there is an over-representation of short-stay or day visitors, with strong concentrations in urban areas. Within the UK, there has also been the more specific influence of the Heritage Lottery Funding, which is discussed in Chapter 10.

Forms of ecotourism consumption

Mowforth and Munt (1998, 2003) have drawn attention to the plethora of terms used to describe alternative or 'new' forms of tourism. They argue that this terminology of 'new' tourisms is indicative of the attempts to distance it from mass tourism. These 'new' tourisms tend to be grouped around eco-tourism, but in effect involve a range of forms that exhibit degrees of de-differentiation (see Table 5.2), i.e. ways in which these 'new tourism practices may no longer be about tourism *per se*, but embody other activities' (Mowforth and Munt 1998: 101). Munt conceives the new tourisms as 'more characteristic of less formalised . . . tourism, such as backpacking' (1994:108) (see Chapter 6). Page has drawn attention to the use of the term 'traveller' to emphasize 'a de-differentiated form of activity which has tenuous links with tourism' (in Page and Dowling 2002: 90). More specifically, Gordon (1991) has argued that many of the 'new' tourists are people with lifestyles strongly motivated by creativity, health, new experiences and personal growth. Consequently, environmental issues are an area of considerable importance. Cleverdon (1999) and Mackay (1994) have identified specific characteristics of these ecotourists, although attitudes and commitment to environmental issues are conditioned by a range of lifestyle variables (Shaw and Williams 2002).

As with heritage tourism, ecotourism is a diverse and somewhat contested area. Some authors claim that 'ecotourism has certainly emerged as one of the least clearly defined areas of study' (Page and Dowling 2001: 55). Certainly, there is a lack of clear agreement about how it should be defined. For some, it is a subset of Nature tourism (Beaumont 1998), while others, such as the International Ecotourism Society (formerly the Ecotourism Society), claim it is 'responsible travel to natural areas, which conserves the environment and

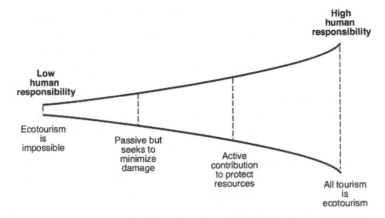

Figure 5.1 **Ecotourism as a continuum of paradigms (source: Orams 2000)**

improves the welfare of local people' (see Lindberg and Hawkins 1993). The term is contested not only among academics, but also among national tourism authorities, such as those, for example, of Australia and Canada. Against this background Orams (2000) has helpfully suggested that ecotourism can be viewed as a continuum of paradigms relating to levels of responsibilities (Figure 5.1). Similarly, Fennell (1999) has conceived of ecotourism being linked to both adventure tourism and forms of cultural tourism, emphasizing both the areas of similarity and difference (Figure 5.2). In this context, he argues that the main overlap between ecotourism and adventure tourism is similar environmental settings. However, others, such as Dyess (1997), view adventure tourism as merely a subset of ecotourism. These different views appear to be as much about the perceptions of tourist experiences (see Chapter 6) as they are about definitions (Page and Dowling 2002).

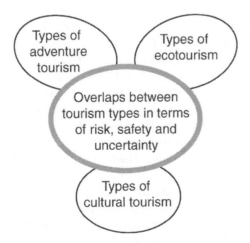

Figure 5.2 **The links between new forms of tourism (source: Fennell 1993)**

Given this background, it is not surprising that there are similar debates over the definitions and identification of ecotourists. Hvenegaard (1994) argued that while ecotourists are representative of a cross-section of society, most are well-educated, between 40 and 50 years old, and with above-average incomes (see Chapter 6). Similarly, work in Louisiana identified ecotourists as being from higher-income groups (Luzer et al. 1995) while other commentators have characterized them as being affluent older people with above-average amounts of discretionary free time (Ballantine and Eagles 1994). As we have previously pointed out, such characteristics fit with the new middle classes, a topic most clearly discussed by Mowforth and Munt (2003). They go further, distinguishing between ecotourists and what they term 'ego-tourists', both being drawn from the new middle classes, but the latter using this interest in the environment to help differentiate themselves from other social groups. Nevertheless, a profile does emerge from the literature about ecotourists and their forms of consumption, but this is in part blurred by variations among the types of ecotourists and their consumption patterns.

Existing research also highlights the global and rapidly growing nature of ecotourism as a form of consumption. As with other forms of tourism consumption, estimates of its size vary depending on how measurements are made and the definition being used. For example, Hvenegaard (1994) suggests ecotourism (including related wildlife tourism) accounts for $1 billion per year worldwide. Using different estimates, the Ecotourism Society (1998) argues that somewhere between 20% and 40% of all international tourist travel was associated with some interest in wildlife-related tourism, although this is necessarily a chaotic conceptualization.

THE 'McDISNEYFICATION' OF TOURISM

The debates on old and new forms of tourism consumption, or Fordism and post-Fordism, have, in part, been extended by Ritzer (1998). His initial thesis centred on the notion of 'McDonaldization', which emphasized efficiency, calculability and predictability. The 'McDonalidization' of society is a form of 'grand narrative', viewing the world as growing increasingly predictable and dominated by controlling technologies. The standardized production practices inherent in such businesses as McDonald's and Disney's theme parks (Ritzer and Ovadio 2000; see also Chapters 2 and 10) have been involved by Ritzer and Liska (1997) as ideas to describe some of the new meanings of consumption, especially tourism, using the terms 'McDonaldization' or 'McDisneyfication' as metaphors. They argue that consumers want their tourist experiences to be as 'McDonaldized' as their day-to-day lives. In this context, they want holidays that are predictable, efficient (in terms of value for money), calculable and controlled (containing acceptable routines).

Ritzer and Liska (1997) apply the McDisneyfication thesis to mass, package tourism, describing the uniformity of global cultures. Debates on post-Fordist consumption are also part of this thesis, with Ritzer suggesting that

McDonaldization and the growing diversity of consumer choice inherent in post-Fordism are not necessarily mutually exclusive. Within this context, the main emphasis is on the choice available to tourists and the fact that, increasingly, many tourism packages are more flexible (Chapter 2). According to Ritzer and Liska this represents not 'De-McDonaldization' but rather mass customization, presenting the tourist with greater choice and more flexible products. Under these conditions, Ioannides and Debbage (1998) prefer to talk about conditions of neo-Fordism to describe the production and consumption changes. In terms of the McDisneyfication paradigm, Ritzer and Liska (1997) illustrate the organizational principles that characterize the Disney theme parks. These are based on set prices, tourist routes that are regulated by an elaborate system of signposting and guides, and an experience which is highly predictable (see Chapter 10).

Whatever the terminology, Ritzer and Liska (1997) explain that while McDisneyfication brings satisfaction to travellers, it does so only in terms established by the general processes of McDonaldization. As they go on to argue, consumers 'not only accept them but embrace them' (p. 100). Moreover, such consumption patterns are carried over into holiday-taking, and Ritzer and Liska identify rationalization as a major force conditioning tourists.

Such ideas are, however, strongly contested, with Rojek arguing that the weakness of the thesis 'is that it tends to present rationalisation as a monolithic process' (2000: 56). As a consequence, it underestimates levels of reflexivity and resistance, along with the diversity of tourist experience. There is certainly strong evidence, as we have discussed earlier in this chapter, that many 'post-tourists' are becoming increasingly concerned about environmental issues of tourism and, as such, have sought different types of holiday. Set against this, the McDisneyfication thesis argues that, in spite of seemingly increased choice, in reality the tourist experience is becoming far more standardized. These issues will be explored further in the next chapter.

Shopping as tourism consumption

The ideas associated with Ritzer's thesis can be illustrated through the rise of the theme park and the shopping mall, both being examples of specific forms of tourism consumption. To some authors they are certainly new forms of leisure consumption, representing the 'artificial construction of post-modern tourist spaces' (Williams 1998: 180). As concepts, neither are that recent; after all, Disney opened his first theme park in 1955. Other versions had grown out of fairground-type amusement parks in the early 20th century, such as Blackpool's Pleasure Beach in north-west England (Walton 2000: see Chapters 9 and 10). Similarly, shopping malls were being constructed across North America and western Europe in the 1960s (Miles 1998). However, in both cases, the late 1980s witnessed a change in the scale, presentation and consumption of these new tourism spaces (see Chapter 10). In the USA and Canada, for example, attendance at theme parks increased by at least 24%

between 1980 and 1990 (Loverseed 1994). Of equal importance has been the global spread of theme parks, reflecting the growing universal appeal and success of the concept (Jones 1994; Williams 1998).

Theme parks, such as Disney World and EuroDisney, are clearly targeted at the family market, although their appeal is broader, encompassing middle-aged people too. Such environments are exciting places, but the tourist experience is standardized and controlled. The visitor is immersed in a fantasy world that provides 'entertainment and excitement, with reassuringly clean and attractive surroundings' (Smith 1980: 46). What is significant is not that visitors are deceived by the pseudo-realities of the theme parks, but rather, as Miles (1998: 65) argues, that 'Disney builds on such images in order to naturalise the process of consumption'. It is in this context that leisure and shopping experiences become blurred. Shopping is a key part of the Disney experience, and in Euro-Disney, Paris, for example, visitors have an extensive range of themed retail outlets to choose from. This choice, in turn, provides visitors with a series of narratives, which make consumption a central part of the experience. As Bryman explains, images within the theme park act as prompts 'to remind the visitor of his/her identity as consumer of both the corporation's products and the Disney merchandise' (1995: 154). Other commentators view the process in starker terms, as merely encouraging visitors to spend as much as possible (Ritzer and Liska 1997).

Miles (1998), along with Urry (1990) and Hollinshead (1997), argues that in the context of consumption, theme parks such as Disney are associated with post-Fordism or postmodernism. More specifically, Ritzer and Liska see these and similar developments as a different form of post-Fordist consumption, 'McDisneyfication'. Significantly, their appeal is to the elements of the new tourists, or 'post-tourists', for whom the playful consumption of 'signs' is of importance (Williams 1998). The ideas of the theme park have also been developed in other leisure environments, especially mega-shopping malls and some heritage environments (see Chapter 10). The mega-shopping malls, which grew from the late 1980s onwards, have done much to combine retail and leisure elements (Falk and Campbell 1997). According to Lehtonen and Mäenpää a trip to the modern shopping mall is, for many consumers, like visiting 'somewhere else, where the real world is challenged by the possible world' (1997: 147). Like theme parks, shopping malls are selling an experience, and one that is highly controlled. Butler (1991) argues that mega-malls, such as the West Edmonton Mall in Canada or the Mall of America in the USA, create an image of 'elsewhereness'. These malls have endowed consumerism with almost religious-like qualities, according to Miles (1998), becoming cathedrals of consumption. They provide safe, controlled environments for consumers to spend their leisure time, merging acts of consumption into notions of play. Socially, they tend to be selective, as Table 5.4 suggests, attracting the more affluent as well as increasing proportions of younger consumers.

Table 5.4 **Socio-demographic characteristics of visitors to Meadowhall Shopping Mall (Sheffield)**

Characteristic	1997 adult visitors (%)	1999 adult visitors (%)
Age		
16–24	23	26
25–44	55	46
45 +	22	28
Social class		
ABC1	62	63
C2DE	38	37

Source: modified from Outhart et al. (2000).

OLD AND NEW FORMS OF TOURISM CONSUMPTION

The perspectives on the changes in tourism consumption outlined in this chapter describe a series of broad shifts. These range from mass tourism as defined by Fordist consumption to 'new' tourism forms as encapsulated by post-Fordism, through to the ideas of neo-Fordism or McDisneyfication. Clearly, such broad descriptions contain many – perhaps too many – generalizations, and, not surprisingly, they are contested within the tourism literature. The difficulties are increased because the edges of tourism and other forms of consumption are becoming increasingly blurred (Falk and Campbell 1997; Lury 1996). It is evident, however, that for many, consumption practices are the domain 'within which people explore and define their own identities or, at least, a kind of identity that exists away from the place of work' (Gottdiener 2000: 22). Indeed, as Gottdiener argues, 'there exists a proliferation of consumer cultures' (p. 21), because of the increasing differentiation of consumption patterns. This is the case generally and also within tourism. Not surprisingly, it is possible to recognize the coexistence of old and new forms of tourism consumption. For example, Sharpley has argued that recent experience in the UK 'demonstrates that the original Fordist-type basis of the package holiday remains as popular as ever' (1994: 25). While this may be an overstatement, in that many holiday packages have become more flexible in format, it is true regarding volume, since the number of holidays by inclusive packages increased by 22% between 1986 and 1997 (British National Travel Survey 1998). What has changed is the way that packages are constructed and marketed around particular markets, experiences and destinations (Box 5.1). As we shall see in Chapter 6, it is possible to go on an adventure/backpacking-type holiday as part of an inclusive package. Rather than simply old and new forms of tourism consumption, there appears to be a range of tourisms and related experiences, many of which have some characteristics, to a greater or lesser degree, of post-Fordism.

One of the main criticisms of these broad meta-theories of consumption is that they fail to recognize that there are different types of consumers or, more

---

Box 5.1   New forms of flexible holiday packages offered in the UK

The following are types of packages available to the Mediterranean for short-stay holidays aimed at more flexible types of tourism consumption. They are based on low-cost flights, flexible deals and a variety of packages. Examples include:

- 'Villa for a weekend': villa rental companies traditionally rented only for seven nights or more, but some are now offering villas for a few days. Companies include Simply Travel, offering 2–4 nights, and International Chapters.

- 'Short-stay high life': established up-market operators, who developed their business mainly on the back of long-haul travel, are now offering short breaks to high-class hotels. Operators include Elegant Resorts, ITC Classics and Seasons in Style.

- 'Quickie charters': operators are increasingly recognizing that charter flights can be used for mini-breaks. These are based around well-known Mediterranean destinations that have busy airports, such as Majorca, Costa del Sol, Alicante, Corfu and Crete. Typical breaks are usually 3 nights, based on a long weekend. Companies include Flightline '48 hour party' – making use of cheap flights enables tourists to go clubbing for the weekend in Ibiza. These are based on short-break packages utilizing charters or links with low cost carriers. Breaks are offered by Thomson Club Freestyle, while BMI Baby has linked with Go to service the Ibiza club scene.

Source: based on *The Sunday Times*, 21 April 2002: 1–2

---

especially, different types of tourists. Ritzer and Ovadia have moved to rectify this, and they argue that a 'McDonaldising society is characterised by neither singular consumers nor singular settings' (2000: 45). They believe that types of consumers require much greater specification; but it could be argued that such classifications already exist within tourism. Poon (1993) for example, attempted to distinguish between 'old' and 'new' tourists and tourism (Figure 5.3). More generally there was work, dating largely from the 1970s, that saw the development of a range of tourist typologies. In part, these contributions could be said to fulfil Ritzer's and Ovadia's conditions of attempting to understand 'the relationships between consumers and the range of ... settings' (2000: 47). The tourist typologies have been widely reviewed (see Murphy 1985; Sharpley 1994; Shaw and Williams 2002) and it is not our intention to detail them here, but rather draw out some key features. Most of the typologies are based on the identification of significant traits of tourists and, more especially, their demands as consumers. Early work by Cohen (1972) drew attention to the fact that all tourists are seeking some element of novelty or strangeness, while at the same time most feel the need to retain a degree of familiarity. In this context, Cohen identified different types of tourists, based on a range of combinations of demand. They varied from those for whom familiarity was given priority, through to those tourists for whom

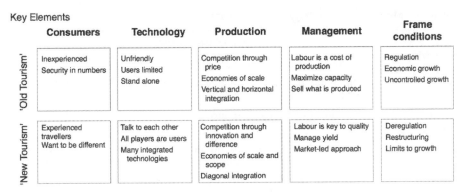

Figure 5.3   **Poon's 'new' and 'old' tourists (source: Poon 1993)**

novelty of experience was important. To a large degree, these typologies were directed at identifying the differences between mass tourism and degrees of individual travel. It could be argued that they were, themselves, a response to the dramatic rise of mass tourism in the 1960s and 1970s. Some approaches were broader in their aims and, as such, have stimulated longer term debates within tourism (Shaw and Williams 2002).

Progress on identifying different types of post-Fordist tourists has been variable. Within studies of ecotourism, a number of new tourist typologies exist, as summarized by Fennell (1999) and Page (in Page and Dowling 2002). As Table 5.5 shows, these attempts are wide-ranging and based on a variety of criteria. For example, Lindberg (1991) constructed a simple fourfold division of ecotourists based on levels of interests, while work by Boo (1990) used a measure of the importance shown towards types of protected areas to distinguish tourists. Set against these narrow perspectives are those typologies suggested by Kusler (1991) and Mowforth (1993). The latter author derived a threefold typology based on the motivations of tourists, their behaviour and the degree of organization involved in taking the holiday.

Table 5.5   **Typologies of ecotourists**

Author	Basis of classification
Boo (1990)	Degree of importance attributed to protected areas
Budowski (1976)	Identification of scientific and nature tourists
Duffus and Dearden (1990)	Degree of physical effort and level of interest; generalist and wildlife specialists identified
Kusler (1991)	Degree of organization from independent (do-it-yourself) ecotourist to organized tours
Lindberg (1991)	Degree of dedication and time spent; four types identified: hardcore, dedicated, mainstream and casual
Mowforth (1993)	Motivation, behaviour and level of organization; three types identified (see Table 5.6)

Table 5.6 **Mowforth's typology of ecotourists**

Variable	Type of ecotourist		
	Rough	**Smooth**	**Specialist**
Age	Young and middle-aged	Middle-aged and older	Young and old
Group composition	Individually or in small groups	In groups	Individually
Organization	Independent	In tour-generated trips	In independent and specialist tours
Costs	Low, basic, cheap hotels	High; 3/5-star hotels	Mid to high 3-star hotels
Type of holiday	Sport and adventure	Nature and safari	Scientific/hobby pursuit

Source: Mowforth (1993).

Using these criteria, Mowforth recognized three main types of ecotourist – the 'rough', the 'smooth' and the 'specialist', as shown in Table 5.6.

However, as Acott et al. argue, such perspectives fail to recognize that it is 'possible for individuals to be ecotourists in non-ecotourist locations or, conversely, to be non-ecotourists in an ecotourist location' (1998: 239). In other words, there is a strong tendency for tourists to be defined in terms of the activity they are engaged in. While this may inform us about ecotourism, it tells us little about the overall consumption patterns of tourists. For example, there may be many families and individuals who have adopted more environment-ally concerned lifestyles, but they may not be ecotourists as such. This has been investigated in part by Dinan (1999) in her study of notions of environmental concern held by tourists in south-west England. The research identified two main types of tourist, based on an index of environmental attitudes: 'concerned' and 'unconcerned' tourists. Significantly, Dinan was unable to distinguish any clear socio-demographic differences between the two groups.

Lifestyles and tourism consumption

This brief discussion of the attempts to identify a new form of tourism consumption, and to classify variations in consumption patterns, raises a number of broader issues. The first is that identifying consumers by one type of tourism is, perhaps, a limited way of understanding the complexity of tourism consumption patterns. Second, it may well be that some of the standard measures of socio-demographics only partially explain variations in tourism consumption. One reason is that the middle classes are now such a broad, diverse and powerful economic group. That said, it does seem that the stage of the family life-cycle is a significant distinguishing variable across a range of new forms of tourism, as is the notion of lifestyle.

It may be easier to discuss these new tourisms, of which there are many variations, in terms of their appeal to different middle-class lifestyles. Such a perspective levers tourism consumption away from the sociology of consumerism and more towards marketing theory. We would argue that there is a strong rationale for this, in that marketing has, in large part, been responsible for helping to create and sell new tourism experiences (Middleton and Clarke 2001). It is the marketing system and especially ideas associated with relational marketing that have helped target and deliver the new consumer experiences to the middle classes. This system is part of what Pine and Gilmore (1999) term 'the experience economy', which has supported the symbolic meanings and differentiation associated with new forms of tourism consumption. Thurot and Thurot (1983) gave early recognition to some of these ideas, as did Urry, who argued that the desire of consumers to be treated in a differentiated manner gave rise to 'life-style research on the part of the advertising industry' (1990: 87).

Work on tourism consumption and lifestyles is hardly new, as Gratton (1990), Cooper et al. (1998), Shaw and Williams (2002) and Schott (2002) have all reviewed or applied value and lifestyle typologies to an understanding of tourism trends. During the 1990s the importance of the links between lifestyle and consumption patterns was increasingly recognized through the construction of broader sets of typologies. Underlying these approaches has been the recognition of key changes in consumer attitudes, including:

- the increasing pursuit of individuality by consumers

- a greater emphasis on informality and spontaneity

- the employment of, and belief in, the use of all the senses for personal wellbeing

- a willingness to integrate technology into the consumption process (Gratton 1993; Horner and Swarbrooke 1996)

Some have claimed that such processes have led to the creation of an increasingly recognizable set of international consumption patterns (Gratton 1993). In terms of tourism, these have tended to be associated with cultural and 'green' tourism, but only among certain lifestyle segments. At the European level, Mazanec and Zins (1994) have attempted to identify a typology of European lifestyles and relate these to patterns of tourist behaviour. They were able to identify five broad types, based on travel motives and activities. These encompassed:

- new experiences – especially cultural/heritage-based tourism

- fun experiences – especially cultural ones (of a more popular type), sports and leisure shopping

- pleasure experiences – a variety of interests, including sport

- movement experiences – including sports for recreation

- nature experiences – including the importance of the landscape, hiking and walking for leisure

Within Mazanec and Zin's work, each of these types could be associated with a particular 'Euro-lifestyle', of which they identified 16 main clusters. Of course, there are important cultural variations, but the key point is the association of particular lifestyles with certain expressions of consumption (see also Gabriel and Lang's typology of consumers (1995)).

Such typologies have clear drawbacks in their rigid and static nature. However, they are suggestive of changing lifestyles, and may be used to characterize certain patterns of consumption. Moreover, they allow us to consider a range of consumption types that may be associated with new forms of tourism and the new tourist.

The final point we want to make regarding old and new forms of tourism consumption is that most of the debates have centred on those tourists that Poon (1993) terms 'new' tourists (Figure 5.1). In doing so, there has been a relative neglect of other forms of tourism consumption, especially those where holidays are a more marginal and constrained activity. We have already discussed elsewhere the means of improving access to tourism (Shaw and Williams 2002), which raises issues of the so-called disadvantaged tourist (Smith and Hughes 1999). Such disadvantaged tourists are in a very weak position to negotiate the new forms of consumption. For many marginal economic groups, acts of tourism consumption are restricted along fairly narrow lines, and often centre on elements of Fordist patterns of consumption.

SUMMARY: CHANGING FORMS OF TOURISM CONSUMPTION

This chapter has detailed the importance of tourism consumption, especially as a cultural process. One of the key debates concerns the changing patterns of consumption associated with the growth of the so-called new middle-classes. In this context, there has been increased importance given to the accumulation of social and cultural capital from the holiday experience. Of equal significance is the suggestion that there has been a shift in tourism consumption, characterized at its most basic as a move from Fordist patterns – with their emphasis on mass tourism – to post-Fordist ones. In the latter, attention shifts to the individual and to new forms of tourism. The debate has been taken further by Ritzer's McDonaldization thesis, which stresses the mass customization of tourism and leisure consumption.

In broad terms, three main forms of tourism consumption can thus be mapped out:

- Fordist or mass consumption – characterized by a highly standardized product, differentiated mainly by stages in the family life-cycle and costs

- post-Fordist consumption – involving the creation of apparently individual, tailor-made holidays, which are comparatively less structured and more independent, and in which tourists are offered highly differentiated products, with more choice

- McDonaldization of tourism consumption – characterized by a form of mass customization, presenting to tourists flexible products, based on efficient and calculable holidays

We have also argued that to see tourism consumption entirely in terms of these major shifts is somewhat simplistic. Set against these generalizations is the reality of a range of tourisms and related experiences.

6 *Engineering the Tourist Experience*

DIMENSIONS OF THE TOURIST EXPERIENCE

As we discussed in Chapter 5, a culture of tourism has emerged since the 1960s that is now a 'highly significant component of most metropolitan and national economies' (Rojek 2000: 53). This is characterized by its global dimensions and its increasing complexity in terms of the different forms of tourism consumption that have emerged since the 1980s. In this chapter we explore the relationship between old and new forms of tourism consumption, and the ways in which they shape different types of tourist experiences. These are reflexive in nature, as experiences and types of tourism have evolved. To understand such relationships it is necessary to make sense of the complex factors affecting the tourist experience. We start, however, with a brief discussion of various perspectives on this experience.

In essence, it is possible to identify three broad, but somewhat overlapping, dimensions of the tourist experience as reflected in the literature. Boorstin (1964), for example, defined it as a popular act of consumption, which was essentially a prefabricated experience of mass tourism. In contrast, MacCannell (1973) argued that it was more an active response to the pressures of modern living, with tourists searching for an 'authentic' experience in order to pacify the difficulties in their lives. However, as Li states, both approaches 'attempt to define the experience with the notion that it has significance for individuals and for their societies' (2000: 834). This view of the search for authenticity is part of the first of the three main perspectives, which is the role of authenticity within the tourist experience.

The second perspective, which extends this debate, begins with the work of Cohen (1979), who argued that different tourists require different experiences, which in turn hold varying meanings, all of which are mediated by their societies. Within this context, the tourist experience is seen by Cohen as the relationship between a person and a variety of 'centres'. These 'centres' relate to the individual's spiritual centre, which holds symbolic meaning. Using such ideas, Cohen (1979) constructed a typology of tourists that categorized individuals into two broad groups, depending on their demand of the tourist experience; those concerned with a 'modern pilgrimage' and those in 'search of pleasure'. Such descriptions of tourist activities, Cohen argued, also reflect patterns of motivations, which differentiate and characterize various modes of tourist activities. As Li explains, such motivations are 'linked to the "private-ly" constructed worlds of tourists and represent patterned ways of satisfying a wide range of personal needs' (2000: 834–5).

These ideas overlap with the third perspective on the tourist experience, which relates to behaviour. This is encapsulated by Ryan (2002), who sees the tourist experience as being a multifunctional leisure activity. Ryan argues that this experience 'is one that engages all the senses, not simply the visual' (1997: 25) as implied by the notion of the tourist gaze (Urry 1990). In this context of the nexus between experience and behaviour, Krippendorf identified the relationship between the two in raising the following basic questions: 'What do people do, what do they experience when they travel? How are their numerous wishes and expectations reflected in their behaviour?' (1987: 30).

We would argue that there are three main dimensions of the tourist experience relating to:

- experience and consumption, involving the nature of authenticity
- the relationship between experience, motivation and tourist types
- the experience–behaviour nexus

These will inform the key debates in this chapter, alongside the notion that, increasingly, the tourist experience is being controlled and engineered by a range of agencies.

THE TOURIST EXPERIENCE AND AUTHENTICITY

The views of authenticity in the tourist experience offered by Boorstin (1964) and MacCannell (1973) not only emphasize two different perspectives but, more importantly, are suggestive of different kinds of authenticity. Boorstin saw mass tourism producing a homogenization and standardization of the tourist experience through the commodification of culture, with the latter producing contrived tourist experiences constructed around pseudo-events. According to Boorstin, the 'tourist seldom likes the authentic', preferring 'his own provincial expectations' (1964: 106). Such a view is to be found in much of the literature on forms of mass tourism, and is to some extent echoed by Ritzer and Liska's (1997) views on the McDonaldization of tourism (see Chapter 5). More extremely, Rojek (2000) sees this as removing novelty and excitement from the tourist experience, and destabilizing tourist responses to authenticity.

Set against Boorstin's perspective, MacCannell argues that tourists search for experiences that embrace authenticity as an antidote to the 'shallowness of their [ordinary] lives' (1973: 589). Such ideas are, in part, based on the notions of 'back and front regions' in relation to authenticity in the tourist space, which we will explore in Chapter 7. In addition, MacCannell saw authenticity in two different ways, 'as feeling and as knowledge' (Selwyn 1996: 6–7). For MacCannell, the tourist experience involves the search for authenticity as feeling. In terms of authenticity as knowledge, MacCannell argued that tourists more often than not become the victims of staged authenticity (see Chapter 7).

Types of authenticity

This debate suggests that authenticity in the tourist experience is complex, and that there are different types of authenticity. Such a view nullifies the criticism of Urry that 'the search for authenticity is too simple a foundation for explaining contemporary tourism' (1995: 51). According to Wang (1999), the importance of authenticity within the tourist experience is clarified by examining it in three different ways: objective authenticity, constructive authenticity and existential authenticity.

To understand these different types of authenticity, we need to explore the basis of the term itself. Wang (1999), following Trilling's (1972) work, claims that its usage has been extended from its museum-based origins into tourism generally. Within tourism, festivals, rituals, art and other cultural artifacts may be viewed as either authentic or inauthentic. However, the complexity of authenticity exceeds this notion, involving, as it does, 'authentic experiences' and 'toured objects' (Wang 1999: 351). The former refers to experiences through which the tourists feel themselves to be 'in touch both with the real world and their real selves' (Handler and Saxton 1988: 243), while in terms of the latter, tourists perceive toured objects as authentic 'because they are engaging in non-ordinary activities', based on these objects.

Wang's three different types of authenticity within the tourist experience relate to objective, constructive and existential authenticity, as outlined in Table 6.1. In terms of objective authenticity, the emphasis is on the tourist experience gained by the recognition of toured objects as being original and authentic. The debate between Boorstin and MacCannell thus relates to objective authenticity, in that experiences are seen as either authentic or inauthentic. However, as Wang argues, 'authenticity is not a matter of black or white, but rather involves a much wider spectrum' (1999: 356). For

Table 6.1 **Types of authenticity in the tourist experience**

Object-related authenticity	Activity-related authenticity
Objective authenticity refers to the authenticity of originals (see work of MacCannell 1973)	*Existential authenticity* refers to a state of 'being' that is to be activated by tourist activities. Unlike 'objective' and 'constructive' types it does not relate to the authenticity of toured objects (see work of Wang 1998).
Constructive authenticity refers to the authenticity projected onto toured objects by tourists, in terms of images, expectations, preferences and beliefs (see work of Bruner 1994)	

Source: modified from Wang (1998).

Table 6.2 **Viewpoints concerning authenticity and the tourist experience under constructivism**

There is no absolute authenticity

Traditions are invented and constructed involving power and social constructs

Authenticity is pluralistic, depends on the viewer (tourist) and the perspective

Authenticity is a label attached to visited cultures in terms of stereotypical images and expectations held by tourists

The inauthentic or artificial can become an emergent authenticity

Source: modified from Bruner (1991, 1994) and Wang (1998).

example, while experts and academics may judge a particular experience as inauthentic, some tourists may view it as authentic.

This leads on to what Bruner (1989) and others term 'constructive' authenticity, where experiences of the authentic are socially constructed and not objectively measurable. In this perspective, as Table 6.2 indicates, authenticity is no longer 'a property inherent in an object, forever fixed in time' (Bruner 1994: 408). Rather, it is a notion that is relative to varying experiences and interpretations of authenticity of different types of tourist. In this context, Cohen (1988) argues that if mass tourists perceive certain contrived toured objects as authentic, then such perspectives are as valid as so-called expert views. These views are socially and culturally constructed via stereotypical images held by tourists, where authenticity is a projection of the visitor's expectations and beliefs, i.e. mainly 'projections of Western consciousness' (Bruner 1991: 234). In Western societies, notions of authenticity are primarily consequences of replicated interpretations, commodified for mass consumption (McIntosh and Prentice 1999). Authenticity is therefore context-bound and part of an emerging process, as illustrated by Salamone's (1997) study of two San Angel Inns serving as different representations of traditional Mexico: the original in Mexico City and its reconstructed counterpart at Disney World, Florida. Both used different boundary-markers depending on the context. For example, in Mexico City, catering to an elite market, the setting establishes Mexican heritage and its historic links with Europe. In contrast, within Disney World, with its audience of tourists, there is a strong emphasis on the Indian nature of Mexican culture alongside the Spanish colonial context. As Salamone explains, 'the elements which tourists expect, and upon which Disney insists, are elements transformed into romantic markers' (1997: 308). These not only link an image of Mexico with the wider world, but also highlight its uniqueness. In this context, both representations can be understood only within their socio-cultural settings.

The affirmation of authenticity within the tourist experience has been researched more directly by McIntosh and Prentice (1999) in their study of heritage theme parks in the UK. As part of the heritage industry, and a key element of post-Fordist forms of tourism consumption, the tourist experience

Table 6.3 **The reflections of tourists visiting heritage sites in the UK**

Visitors' reflections	Blists Hill	Black Country Museum	New Lanark
Experiences (thoughts)	Thought deeply about (%)	Thought deeply about (%)	Thought deeply about (%)
Past overall lifestyle			
What people's lives were like in the past	35.0	37.8	41.3
Hardships endured in past life	45.0	55.5	41.0
The standards of present day life	13.5	19.3	13.0
Comparisons between life then and now	30.8	37.5	32.8
The future	3.3	2.0	2.5
The inspiration of Robert Owen	n/a	n/a	49.5
Past industrial processes			
Conditions in which people had to work	25.3	28.5	21.5
How hard people had to work	23.0	26.5	19.0
How skilled people were	11.0	5.0	4.8
Health-related issues of the work	6.3	6.0	6.5
How technology has changed	17.8	11.5	12.3
Significance of the Industrial Revolution	5.8	3.3	5.8
Nostalgia/personal memories			
Memories relived	22.8	33.0	20.3
Could relate to a lot of things	13.3	21.5	16.0
Thoughts about ancestors	10.0	15.5	8.0
How everything seemed realistic or authentic	11.3	20.3	14.5

Source: modified from McIntosh and Prentice (1999).

at such sites embraces both 'fun seekers' and those interested in gaining cultural capital (see Chapter 5). This study of three major heritage parks attempted to 'test for authenticity through the definitions of experiences and benefits reported by tourists' (p. 595). Their research was based on a relatively large sample of 1200 adults across three sites: Blists Hill (West Midlands), the Black Country Museum (West Midlands) and New Lanark (Scotland). The survey focused on the thoughts and emotions experienced by tourists visiting these sites and, as Table 6.3 shows, these were based on 'past lifestyles', 'past industrial processes' and experiences of nostalgia or personal memories. These are not surprising, but what does emerge is that a good proportion of visitors were mindful of and sensitive to the heritage experience. Such findings suggest that tourists are affirming authenticity through both 'empathy and critical engagement in relation to the past' (McIntosh and Prentice 1999: 598).

In the context of this study, a number of key aspects of authenticity and the tourist experience are highlighted. First, many heritage tourism encounters represent a complex web of experiences and appear to be imbued with significant personal meanings. These relate to aspects of nostalgia and have been too quickly dismissed by some commentators as merely products of a contrived heritage industry. Clearly, as McIntosh and Prentice's study suggests, heritage tourists are far more 'sensorialy complex and emotion laden' (1999: 609) than past studies may have recognized. Second, the assimilation of information by tourists in such heritage settings leads to the 'production of their own experiences of authenticity' (p. 608). The tourists' appreciations, thoughts and insights about the past are therefore strongly characteristic of an authentic experience.

This latter aspect relates to the third type of authenticity within the tourist experience, namely, existential authenticity (Table 6.1). Unlike the other types of authenticity, this term refers to the state of 'being', where one is true to oneself (Wang 1999) and, as Berger argues (1973), it becomes an antidote to the loss of true self in many spheres of Western society. Turner and Manning (1988) support this notion of existential authenticity, as does Selwyn (1996) in his identification of so-called 'hot' as opposed to 'cool' authenticity. The former is, according to Wang (1999), a specific expression of existential authenticity, relating as it does to myths of the authentic self.

From the above discussion, it is clear that unlike the other types of authenticity, the existential version may have no association with the authentic nature of toured objects. Rather, the tourist experience is bound up with a search for an 'authentic self', which may be supported by certain holiday activities. Within this context, authenticity or discovery of the true self can be linked to many types of holiday experiences. Most obviously, the ideal of authenticity can be characterized by nostalgia, as McIntosh and Prentice (1999) and Sharpley (1994) argue it is in the form of romanticism. The latter is important because it emphasizes naturalness, and certainly the growth of certain forms of Nature tourism or ecotourism is important in this context. Similarly, the authentic discovery of self for other tourists is linked with the search for adventure as reflected in the rapid growth of adventure tourism (see Chapter 5).

In addition to these post-Fordist forms of tourism consumption, the notion of existential authenticity encompasses other tourist experiences. For example, Lefebvre (1991) sees the beach holiday as an example of the search for the authentic self. Here, the 'natural' liminal zone (Ryan 2002) of the beach provides an environment for playfulness or 'in short, authenticity in the existential sense' (Wang 1999: 361).

As Wang argues, existential authenticity can explain a much greater range of tourist experiences than other notions of object-related authenticity. Of course, this discussion also serves to highlight the complexity of the tourist experience and, more especially, the ways in which tourists regard authenticity.

MOTIVATION AND 'REAL' EXPERIENCES

As discussed in the previous section, there are established links between tourist motivation and the search for the real or authentic experience. MacCannell initially raised such links (1973), but it was Cohen (1979) who developed the ideas by proposing that there are five different reasons for travel that are embodied within the tourist experience. These are: recreational, diversionary, experiential, experimental and existential. For Cohen (1979) it was possible to recognize that different types of tourist have different motives for travelling. Others, notably Pearce and Moscardo (1986), have developed these ideas by suggesting that (a) tourists' perceptions of particular situations are important in determining their authenticity and (b) tourists' demands for the authentic also vary. As Waller and Lea (1998) explain, the enjoyment of a situation is mediated by both preferences and perceptions of authenticity. Early views of the search for the authentic or real experience saw it purely in terms of the experiential mode. However, as discussed earlier, the complexity of authentic experiences negates such a narrow perspective, since it is possible to find the 'authentic' in a range of tourism settings when viewed from an existential perspective.

Waller and Lea (1998) have pursued such ideas in an empirical context to research what tourists perceive as the authentic or real experience. As Box 6.1 shows, the research was based on the understanding of authenticity in the British tourist experience in Spain. The term 'the real Spain' was used to imply authenticity. The study also explored the relationship between motivation and the role it may play in mediating the association between 'authenticity' and enjoyment. However, before exploring this second theme in more detail, it is necessary to embark on a more detailed discussion of the concept of motivation.

The concept of motivation, as revealed in numerous papers (for general reviews see Ryan 2002; Shaw and Williams 2002), suggests that 'individuals constantly strive to achieve a state of stability, a homeostasis' (Goossens 1998: 302). Such homeostatis is disrupted when people become aware of a need deficiency. This has led to the creation of a needs-based taxonomy of tourist experiences, including the needs for: novelty, sensual enjoyment, stimulation, self-expression, relaxation and the sense of belonging.

The widespread literature on tourist motivation is characterized, according to Ryan, 'by the similarity in findings by many researchers' (1997: 28). While this is so, it is important to recognize that such similarities have emerged from different approaches. These various approaches can be categorized under three broad perspectives; reductionist, structuralist and functionalist. Such differences have led Pearce to argue that a good deal of the literature on tourist motivation is 'fragmented and lacking a firm sense of direction' (1993: 1).

Reductionist approaches have viewed tourist motivation as a tension between the search for the new or novel experience and the requirement for some degree of familiarity (Cohen 1972). More recent studies have been

---

Box 6.1 The 'authentic' Spain

The study by Waller and Lea (1998) is aimed at testing empirically the notion of authenticity as viewed by tourists holidaying in Spain. The case study was based on studies of undergraduate students aged between 20 and 23, and a sample survey of the general public. Four scenarios were used to examine the notion of authenticity:

- a beach holiday in a small unspoilt seaside village, taken by an independent traveller.

- a coach trip to a major cultural centre, staying in a modern hotel, based on a package organized by a tour operator.

- a stay with Spanish friends in a town with no significant tourism industry, participating in local events.

- a drive to Spain, with no pre-booked accommodation, and a stay at a campsite with other tourists in a small town.

For each scenario, respondents from the two main groups were asked to what extent each holiday would allow them to discover the 'real' Spain, as a measure of perceived authenticity. Some of the results from this research are detailed in the table below: with low scores representing less authentic ratings

Authenticity scores (mean ratings) as measures of the 'real' Spain

Scenario	Students (sample size: 90)	General public (sample size: 92)
Seaside Village	4.60	4.45
Coach Tour	3.40	3.31
Stay with Friends	5.95	5.52
Campsite	2.97	3.28

Source: based on Waller and Lea (1998)

---

preoccupied with attempting to measure the importance of the search for novelty. For example, Mo et al. (1993) developed an International Tourist Role Scale, using a 20-item scale to measure the novelty construct. At a broader level, Yiannais and Gibson (1992) classified holidaymakers according to their desire for novelty, social interaction or isolation.

By far the largest group of studies fall within the structuralist perspective, which has focused on identifying a series of underlying structures relating to both 'push and pull' factors. The emphasis has been on the former, which determine why people decide to take a holiday (Dann 1977). Similarly, Gnoth (1997) views motives as lasting dispositions, internal drives or push factors which cause the tourist to search for objects, events and situations. In these approaches, motives are linked to needs, with Maslow's (1970) influential work stressing a hierarchy of needs, from so-called 'deficit' needs through to 'being' needs. Ryan argues in the context of Maslow's work that 'holidays possess the potential for cathartic experience' (2002: 30). More specifically, Beard and Ragheb (1983) have identified four motivational components:

- An intellectual component assesses the extent to which individuals are motivated, by involvement in learning, etc., in terms of a holiday. This may be a specific educational/cultural trip or merely the visiting of cultural sites while on holiday.

- A social component concerns the extent to which individuals engage in activities for social reasons, i.e. friendship and esteem. The latter may be related to the notions of ego-enhancement through, for example, being seen as a seasoned, well-experienced traveller.

- A competence-mastery component concerns the extent to which individuals engage in an activity/holiday for achievement (see Ryan 2002).

- A stimulus-avoidance component concerns the drive to escape from over-stimulating situations or to seek rest and solitude.

As Ryan (2002) explains, these have formed part of empirical studies around attempts to develop a Leisure Motivation Scale by Beard and Ragheb (1983). Similarly, Weissinger and Bandalos (1995) have researched a scale to measure intrinsic leisure motivation, which they define as the tendency to seek intrinsic rewards within leisure behaviour. The scale comprises four components:

- self-determination – characterized by the awareness of internal needs along with a strong desire to have free choices based on such needs

- competence – characterized by attention to feedback from previous holiday experiences, which provides information about ability and skill

- commitment – characterized by the tendency toward a close involvement in leisure behaviours

- challenge – characterized by the tendency to seek travel and leisure experiences that stretch the individual's limits and provide novel situations

This approach takes the study of motivation and needs as outlined by Beard and Ragheb, and links it more firmly with the idea of tourists seeking intrinsic rewards within the holiday experience.

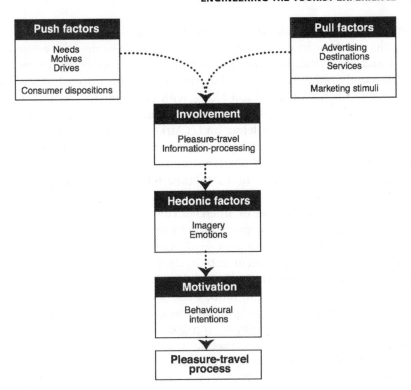

Figure 6.1 **The hedonic motivational model (source: modified from Goossens 1998)**

Models of motivation

Goossens (1998) has developed these ideas further in an attempt to explore the relationship between the push–pull factors of pleasure motivation. He argues that the concept of emotion is the psychological factor that connects the two sides of motivation. In this context, 'tourists are pushed by their (emotional) needs and pulled by the (emotional) benefits' of activities and destinations (Goossens 1998: 302). Therefore, emotional needs are important in leisure-seeking and choice behaviour. According to Hirschman and Holbrook (1982), such experiential processes as emotions, desires, imaginings and daydreams play a significant role in hedonic consumption. This term refers to consumers' multi-sensory images, fantasies, and emotional arousal in the use of products (Hirschman and Holbrook 1982). In the hedonic motivational model conceptualized by Goossens (1998), as shown in Figure 6.1, the role of marketing stimuli is fully recognized. As such, it takes a more realistic perspective on the interaction of push–pull factors, and on the importance of advertising and branding. This model recognizes the current state of consumption in postmodern societies, in which, as Schofield (1996) argues, consumers no longer consume products alone but also signs and images (Chapter 10). Similarly, Rojek (1990) highlights the fact that the superstructure of the media, in all their forms, is more important than economic structure in explaining motivation and behaviour. We would go further and argue that insufficient attention has been given to the media in not only conditioning motivation, but also in creating it (see Nielson 2001). The massive impact of television along with other media, including, more recently, the internet, has been a major influence on demand for travel and tourism. However, as Middleton and Clarke explain, 'the full effect on demand of growing media coverage is still not well understood, but there can be no doubt of its importance' (2001: 64).

In Goossens' motivational model (Figure 6.1) the motivation process is viewed in three main stages. The first is involvement, which is defined as an unobservable state of arousal or interest. This is evoked by certain stimuli from either, or both, of the push–pull factors, and leads to information processing (for a general discussion see Decrop 2000). In turn, this leads to hedonistic responses, which occur both in the phase of information-gathering and during consumption. Motivation is therefore strongly intertwined with hedonic consumption. Emotion, moods and experiential aspects of tourism consumption appear to play important roles in motivation and the tourist experience, which is increasingly recognized by destination marketing through the concept of 'mood marketing' (Morgan et al. 2002).

The third general approach to tourist motivation is that of functionalism, as proposed by Fodness (1994). This argues that the reason individuals hold certain attitudes is 'that these serve important psychological needs' (p. 558). In terms of the functional perspective, these inner needs may create tension of a psychological or physical nature. This may, in turn, relate to problems in the domestic–work environment of the type described in other motivational studies (see Table 6.4). Such tensions are released by actions that may involve holiday-taking. According to Fodness, the value of the functionalism approach to tourist motivation is based on its intuitive appeal as it attempts to address directly the questions relating to motivation, as well as the help it provides in understanding, and perhaps influencing, motivation and behaviour (Crompton 1979), providing links with market segmentation analysis.

As Fodness shows, it is possible to view a wide range of motivational studies in a functionalist mode (Table 6.4). Such a perspective may provide an organizational framework, but is far from helping to shape an overarching theory of tourist motivation.

Within the various approaches to tourist motivation, two key problem areas emerge. The first is what Pearce claims is the need for 'a blueprint for a sound tourist motivation theory' (1993: 115), while the second concerns difficulties in measuring tourist motives. Both are contested and have received considerable attention, as the work by Pearce (1993), Ryan and Glendon (1998), Todd (1999) and Ryan (2002), testify.

The failure to develop a recognized 'theory of motivation', or indeed an agreed system of measurement, does not negate the importance of motivation in the tourist experience. Increasingly, the urge to travel and take vacations is being shaped by a global media industry. The images created through television and other media not only act as powerful tools in selling destinations or types of holidays, but also mediate the motives for travel (see Chapter 7). For example, Waller and Lea (1998) have noted that the idea of the 'real' Spain was familiar to most people in their sample because of marketing. The tourist search for 'authenticity' in this case was part of a set of motives which, we would argue, are strongly conditioned by the way holidays are packaged and marketed.

Partly related to these ideas are the concepts of intrinsic and extrinsic motivation as discussed by Iso-Ahola (1982). The former refers to behaviour motivated by self-satisfying goals or for its own rewards, as opposed to extrinsic motivation, which is more socially controlled through external rewards. Iso-Ahola's model of intrinsic motivation is framed within situational influences and social environments, although he warns against 'culturally supplied explanations of motives' (Pearce 1993: 128) that may mask individual motives for behaviour. This may be contested, given the increasing power exercised by a global tourism industry that is inextricably linked to global media. In this context, television is one of the foremost means of promoting travel and tourism as a way of life. As Miles explains, 'above and beyond advertising, the television programmes themselves actively promote the sorts of benefits and fulfilment that can be enjoyed through conspicuous consumption' (1998: 79). We would go further and argue that all media, including the internet, help condition motives for holiday-making as part of global consumerism.

Of course, Iso-Ahola is right to highlight the role of the individual, at least in terms of how the images of holiday consumption are translated into motives by particular individuals. The travel career-ladder as developed by Pearce (1988; 1993) suggests that individuals change their motivations over time and across situations. This model views tourism motivation in a series of levels, which individuals can move to at different stages. It describes five levels of motivation, based on Maslow's (1959) hierarchy of needs (Figure 6.2). The travel career-ladder gives a dynamic element to notions of tourist motivation. Individuals can start at different levels, or move through in a progressive way, or, indeed, stay fixed in their motivations. The ladder was, in part, developed to demonstrate the motives of visitors to a historic theme park (Moscardo and Pearce 1986) but has been extended to cover wider settings. The model is also significant in highlighting that individuals may have a range of motives for travel and tourism, with several levels of the ladder interacting.

The travel career model is contested, with Ryan (1997; 2002) pointing to both conceptual and measurement difficulties. For example, Pearce's model postulates that the motivation for stimulation can be viewed along a risk–safety dimension, but Ryan (1997) argues that this is based on weak empirical evidence and may be more to do with personality types than an evolving travel career. Nevertheless, it is certainly the case that tourists learn from previous travel experiences, a fact acknowledged by Ryan's model of the tourist experience. In this there is scope for individuals to evolve a travel career based on experience, income, family responsibilities and overall

Table 6.4 **Tourist motivation literature as integrated into a functional framework**

Author	Ego-defensive function	Knowledge function	Reasons for travel		Value-expression function	Social-adjustive function
			Utilitarian function: reward maximization	Utilitarian function: punishment avoidance		
Gray (1970)			Wanderlust; sunlust			
Dann (1977)				Anomie	Ego-enhancement	
Schmoll (1977)		Educational and cultural	Relaxation, adventure, and pleasure; health and recreation (including sport)			
Crompton (1979)	Exploration and evaluation of self	Education; novelty	Regression (less constrained behaviour)	Escape from a perceived mundane environment; relaxation	Prestige	Enhancement of kinship relationships and social interaction
Hudman (1980)	Self-esteem	Curiosity; religion	Health; sports; pleasure			Visiting of friends and relatives; pursuit of 'roots'
Iso-Ahola (1982)			Desire to obtain psychological or intrinsic awards	Escaping of one's personal environment, personal troubles, problems, etc.		

Epperson (1983)	Self-discovery push factor	Historical areas and cultural events – pull factors	Challenge and adventure – push factors; sports – pull factor	Escape, rest and relation – push factors	Prestige – push factor	Ethnic and family motives: to visit places one's family came from; to visit friends and relatives; to spend time with the family
Moutinho (1987)		Education and culture; the gaining of a better understanding of current events	Recreation, sports: to have a good time, fun, or to have some sort of romantic sexual experience	Relaxation: to get away from everyday routines and obligations; to seek new experiences; health – to rest and recover from work	Social and competitive motives: to be able to talk about place visited; because it is fashionable; to show that one can afford it	The need for social contact
Coltman (1989)	Self-esteem	Curiosity about other cultures, places, people, religions, and political systems, as well as the desire to see attractions	The romance of travel; sports and entertainment	The use of leisure time to escape; the desire for change of routine, or merely the wish to have a new experience or to do nothing	To be able to talk to others about a trip for reasons of ego-enhancement; to follow a trend for a particular destination; to be one of the first to visit a new destination	
McIntosh and Goeldner (1990)	Self-esteem	Cultural knowledge – to gain knowledge about other countries	Physical – sports, recreation	Physical – rest, health; interpersonal motives – get away from routine	Status and privilege	Interpersonal motives – to meet new people, visit friends or relatives

Source: modified from Fodness (1994).

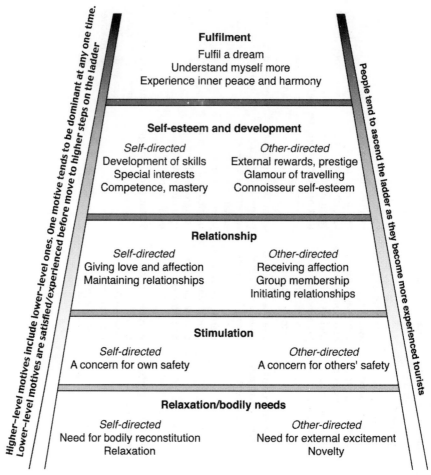

Figure 6.2   **The travel career-ladder (adapted from Cooper et al. 1998 and Ryan 2002)**

lifestyle. These expand the travel career model, with its original focus on personality and motivation, into a wider context. In addition, it is also widely recognized that, for many consumers, there is a discernible pattern of changing motivations, both within and between holidays (Ryan 1997; 2002). We commented in Chapter 5 on the trends toward more holidays being taken, especially those of a shorter duration. These reflect changes in consumption patterns and also permit the expression of different motives. For example, the same individual may take holidays that are family-based, perhaps in a theme park such as Euro-Disney, a cultural based short break, and an adventure-based trip. These various holidays are expressions of different motives, reflecting different needs for rest, relaxation or stimulation. In this sense, individuals may possibly assign different holidays to specific functions.

If this is so, then clearly there are implications for the travel career-ladder. It may not make it 'difficult to sustain', as suggested by Ryan (1997: 41), but rather it highlights the need to seek to modify the model and its application.

Such a modification has been based on the links between motivation, as expressed by recognized needs, and levels of satisfaction. This is, in part, based on Pearce's ideas that satisfaction with the holiday experience is determined by the relationship between need (motivation) and acquired experience (Ryan 1997). In this context there are three different possibilities:

- First, acquiring experience is important, as more-experienced tourists gain higher levels of satisfaction than inexperienced ones. This in turn relates to their position on the travel career-ladder.

- Second, satisfaction is gained from the perception by the tourist that a particular need has been met. Furthermore, it is unimportant whether such needs are self-actualization or social ones. More important is the fact that the experienced traveller is more able to satisfy their needs.

- Third, inexperienced and experienced tourists are motivated by the same needs, with the main difference being that the latter are better able to meet their needs and negotiate them in different ways. Nevertheless, it should be recognized that there are significant differences in experiences.

Earlier work by Gottlieb (1982) on the characteristics of the American vacation identified some of the differences discussed by Pearce and Ryan. She observed vacation types based on the 'American Dream', which utilized two opposing models – economic inequality (capitalism) and ideological equality (democracy). These were used to classify tourists on class grounds, and in this perspective experience was closely related to socio-economic class. While some of her ideas may have been too basic in nature, Gottlieb argued that 'what the vacationer experiences is real, valid and fulfilling' (p. 167). In other words, fulfilment or satisfaction of needs is clearly more than a question of acquired tourist experiences.

Significant alternatives have been put forward to the travel career-ladder, including Ryan's (1997; 2002) expectation–satisfaction linkage model, which identifies the linkages between the travel experience and the perceived gaps between tourist expectation and reality. In this context Ryan argues that 'all tourists want to have success on holidays, however they have defined success' (1997: 52). Success for some may be achieving certain basic needs as outlined by Pearce, while for others expectations may be very different. The importance of Ryan's model is that it attempts to link motivations with tourist experiences and behaviour.

TOURIST BEHAVIOUR: EXPERIENCING THE DREAM

Behavioural changes in tourism have formed a focal point of discussion in the literature and three key questions have been highlighted: Is tourist behaviour different from that experienced in the home environment, especially given the de-differentiation of activities associated with postmodern consumption (Ryan 2002)?; What characterizes tourist behaviour? (To this we may add: Do new forms of tourism, as described in Chapter 5, have different types of behaviour?); What factors influence tourist behaviour?

According to Krippendorf, 'there are several characteristics of tourist behaviour', an examination of which 'may help us understand the conflicting and strange nature of tourism' (1987: 31). He observed that tourists bring with them practices from their everyday lives, and argued that many tourists behave in much the same way as they would at home. Of course, while this may be true for some tourists – those for whom familiarity of routine is important – for others there are important variations on these behaviour patterns. Gottlieb (1982) has commented on these in terms of attempts to enact 'social reversals' as part of the vacations of middle-class Americans. While such work has been questioned by Currie (1997), the notion of reversal or inversion is seen as important in explaining tourist behaviour. Graburn highlighted the importance of behaviour reversals or inversions in the way certain 'meanings and morals of "ordinary behaviour" are changed, held in abeyance or even reversed' (1983: 24). This raises the question of how individuals select their changes in behaviour. In this context, Graburn viewed such inversions as simple binary opposites, along a continuum that reflected different aspects of behaviour (see Table 6.5). Graburn's work emphasizes two important themes.

The first is the fact that different types of tourism are 'characterized by the selection of only a few key reversals' (Graburn 1983: 21). It is also the case that different polarities of reversal are related. Taken at its simplest example, tourists may seek out isolated environments in search of tranquillity, reversing their normal patterns of home behaviour. It is clear that tourists are motivated by more than one kind of behaviour reversal (Graburn 1983).

Table 6.5 **Sociocultural inversions associated with tourist experiences**

Type	Contrasts
Environment	winter–summer isolation–crowds urban – rural
Lifestyle	simplicity–affluence thrift–self-indulgence slow–fast
Social norms	nudity–formal clothing sexual licence–sexual restriction
Health and security	tranquillity–stress slowness–exercise ageing–rejuvenation security–risk

Source: modified from Burns 1999.

The second key point is that there appear to be links between behaviour reversal and particular social groups, in the sense that particular groups tend to invert similar behaviours. Graburn identified three main determinants: discretionary income, cultural self-confidence and socio-economic–symbolic inversions. These are not only interconnected, but may also help predict patterns of tourist behaviour.

These latter ideas have also been pursued by Passariello (1983) in a study of three middle-class resorts in Mexico. This recognized the importance of discretionary income in limiting choices of style, distance and length of travel. In turn, this factor interacted with cultural self-confidence and educational experiences, which led to behavioural inversions when the meanings and roles of ordinary behaviour were suspended. Of course, as Burns (1999) observes, Passariello's work fails to address the role played by the travel industry in conditioning such behavioural changes. Certainly, the industry transmits powerful messages relating to tourist behaviour. These range from evoking relaxation to more hedonistic expressions of behaviour, or, more recently, to encouraging tourists to behave as more responsible visitors (see Chapter 7).

Notions of play in tourist behaviour

As part of the discussion on behavioural inversion, Lett (1983) and Currie (1997) have extended the debate by focusing on the notion of play (see also Chapter 5). Lett examined the behaviour of charter yacht tourists, which he regarded as a 'symbolic expression and an inversion of the central sexual and social ideologies' of the home culture of the visitors (p. 35). In Lett's terms, tourism could be viewed as an opportunity for expressive experiences, 'a stepping out of "real" life into a temporary sphere of activity' (p. 41), which has an associated sphere of behaviour. Lett's study focused on the sexual behaviour of tourists and it argued that, while on holiday, some individuals' customary courting behaviours were abandoned.

The notions of play have been conceptualized as 'liminoid' experiences, which involve an idiosyncratic symbolism expressed during holidays (Currie 1997). Such liminoid activities are those that appear to deny or ignore the legitimacy of institutionalized norms, values and the rules of ordinary life. Put simply, the norms of holiday behaviour become very different from other patterns of behaviour. Such ludic or playful behaviour is seen by some as restitutive or compensatory, making up for home and work routines. Ryan (1997; 2002) building on the work of Langer and Piper (1987) argues that in some tourism settings behaviour appears to be less goal-orientated and rational, and more characterized by mindlessness. According to this view, the 'act of mindlessness is part of a process of optimisation of experience' (Ryan 1997: 49).

Many of these ideas are encapsulated in the notion of tourism as a right of passage, as outlined originally by Turner (1974) and extended by Wagner (1977). This involves three stages:

- a separation phase, as represented by the social and spatial separation from the home/work environment and its routines

- a transition or liminoid phase, where conventional social ties are suspended. In this state there is the playful, non-serious behaviour that characterizes many tourists. Such ideas have been developed by Wagner (1977) through applying concepts of structure to tourism settings, and emphasizing that time is limited during holidays. As a consequence, time takes on a flowing quality and 'becomes free and unstructured to be disposed of at will' (Wagner 1977: 42). Tourists therefore exist in this liminoid or structureless state, at least in the sense that they tend to leave their primary mode of social interaction at home (Currie 1997)

- an incorporation phase, where the individual is reintegrated with his/her social group, but usually with greater social standing

The most important stages in the context of tourist behaviour are the first two, since the liminoid state occurs away from home, while on holiday. Of course, there is ample evidence to show that for many different kinds of tourists, holidays, and behaviour on holiday, are mere extensions of their home environments. Krippendorf observed that 'for many people the holiday experience exhausts itself in the feeling that they do not have to work and they are not at home' (1987: 32). The holiday destination is nothing more than an exotic backdrop, in front of which these tourists construct a familiar pattern of behaviour, different only in the use of time.

More recently, work by Carr (1997) has highlighted the way young people transfer patterns of leisure behaviour from their home environment to holiday situations. Thus, for both male and female tourists, on mass tourism holidays, holiday behaviour was seen to involve clubbing and drinking, just as did behaviour in their leisure time at home. In this context, evidence of behaviour reversal was somewhat minimal. Indeed, there is little evidence when one examines the extremes of youth behaviour on holiday and at home to suggest that there are very few clear, major differences. This is as much to do with shifts in the nature of contemporary consumerism as it is with tourism. In essence, this relates to Campbell's (1987) views of consumerism, as stemming from people's imaginative pleasure-seeking as part of the act of consumption (see Chapter 5). Campbell terms this 'imaginative hedonism', which he views as a somewhat autonomous feature of modern consumer societies. He argues that such a process is separate from certain types of social emulation and particular institutional factors such as advertising. Urry (1990) takes issue with the first of these assumptions, while we would certainly argue against the second.

It seems increasingly clear that what we have discussed as the experience economy in Chapter 5, with its attention to image, branding and lifestyle experiences, has, through the media – especially television – been an increasingly dominant force in shaping Campbell's imaginative hedonism. As Urry claims, 'it is hard to envisage the nature of contemporary tourism without seeing how such activities are literally constructed in our imagination through advertising and the media' (1990: 13). These processes are the ones underpinning the changing leisure patterns and the bringing together of liminoid behaviour in many types of environments, including, of course, the holiday scene.

Such a discourse hits at the limiting nature of past research on behaviour reversal and inversion. This is an issue raised, in part, by Currie (1997), who uses the spill-over/familiarity concept as developed by Burch (1969). This concept encompasses the notion that some individuals may want to participate in similar behaviours and activities in both their home and holiday environments. Currie's work contains two important ideas that may help in our understanding of tourist behaviour. The first is the notion of 'limen', based on the concept of the liminoid state. Currie explains that the 'limen is an imaginary transitional interweaving corridor that separates the home and tourism environments' (1997: 894). The second idea is that of recognizing two main types of behaviour: inversionary – which we have already discussed – and prosaic. The latter is behaviour that individuals feel is necessary to take with them from their home environment. This may consist of mere routines of the type Krippendorf commented on, or it may be particular forms of leisure behaviour. We can extend Currie's ideas to encompass not only different types of prosaic behaviour, but also at least two types of limen:

- a geographical limen, in which distance and the notion of travel are important for the individual to change patterns of behaviour

- a social limen, in which geographical distance is somewhat irrelevant – behaviour changes are invoked whenever free time is available. This coming together of behaviours is largely the result of tourism becoming indistinguishable from other social and cultural practices. Urry explains that 'pleasure can be enjoyed in very many places' as there has been 'a proliferation of objects on which to gaze' (1990: 102) (see Chapter 10).

FROM CARELESS TO CAREFUL TOURISTS: ENGINEERING BEHAVIOUR

In our discussion so far, we have recognized that tourism brings about behavioural changes that involve some setting aside of accepted norms and values. For many tourists, such a relaxation of social norms brings forth what could be considered extreme patterns of hedonistic behaviour in other settings. This may involve displays of nudity while sunbathing, sexual excesses and even drunken, loutish behaviour. Krippendorf regarded this 'have-a-good-time ideology' as often leading to an 'aggressive, reckless and colonialist phenomenon' (1987: 33).

As Ryan explains, the social contexts of tourism 'are pluralistic in nature and provide many opportunities for the expression of different behaviours' (1997: 25). Certainly the nature of postmodern tourism and the growth of the experience economy have extended these opportunities of expression. This expression relates to the ideas that 'the spatial and temporal barriers between tourism and non-tourism have become blurred' (Shaw and Williams 2002: 241), creating a range of pleasure opportunities. These are, in essence, expressions of the control exercised by the 'consequences of globalisation, and the management of the political economy' (Meethan 2001: 74). Hannigan describes such behaviour taking place 'within the context of programmed leisure experiences' (1998: 70).

Tourist and leisure experiences are controlled by a range of management systems and in a range of settings (see Chapters 9 and 10). Ritzer (1998) has drawn considerable attention to constraints and control exercised by organizations such as the Disney Corporation via their theme parks. In other settings, behaviour is, if not controlled, then certainly in part contrived, as in the case of packages related to the youth tourism market. Lewis has, for example, identified tour operators within this market, such as 2wentys, that sell packages to 'places that already have a reputation for partying' (1996: 33). Such tour operators, along with Club 18–30, both shape and reflect key elements of youth behaviour. Such organizations hold sway over a sizeable number of young people, accounting for around 60% of the British young people who use package companies for beach holidays (Carr 1997). The market is large: approximately 1 million young people per year were holidaying abroad during the mid-1990s (Wheatcroft and Seekings 1995).

Many of the tourists on package holidays have been identified, using Plog's (1977) typology, as psychocentrics, who are not particularly interested in active behaviour or adventure. They are rather, as Aramberri suggests, tourists 'whose idea of a satisfactory vacation is a break up of everyday life' (1991: 4). Even within such passive behaviour, hedonistic traits may play key parts, as these people want to 'experience' their holiday. As part of this holiday experience, they may visit theme parks, even heritage centres, provided, as Pearce (1993) observes, they are easy to get to and relatively close to their holiday residence.

The traditional package holiday was strongly engineered in the way experiences were offered to tourists. The British package-tourist holidaying in Spain during the 1970s could expect his/her behaviour to be limited to beach, hotel, and possibly some arranged excursions to contrived events. The post-tourist is more discerning and, as we have argued, searching for some degree of a 'real' or 'authentic' experience. For the non-institutionalized traveller, as described by Cohen (1972), such experiences can be gained by moving away from mass resorts and taking an independently organized holiday trip. However, even these types of holidays are being increasingly engineered by a range of specialized companies aimed at the post-tourist, whether he/she be interested in cultural, adventure or eco-tourism.

Commodifying the ecotourism experience

There is no doubt that many of those tourists taking cultural, adventure or ecotourism holidays are motivated by different needs and exhibit many

---

Box 6.2   Motivations of ecotourists: the Canadian case

Main characteristics

Canadian ecotourists are characterized by relatively high levels of education and income. They also have fairly high levels of environmental awareness informed by discussions with friends and educational material, including films and television. Their environmental interests encompass visiting national parks, wilderness and tropical rain forests, as well as experiencing new lifestyles. Activities include learning about the natural environment, bird-watching and photographing wildlife.

Main motivations

Major considerations are associated with the importance of a natural and undisturbed environment, especially wilderness. These were strongly held attraction factors within motivation. Similarly, there are significant social motivations associated with gaining experiences aiding personal development.

Sources: based on Eagles 1992; Page and Dowling 2002

---

contrasts in their behaviour compared with those on mass packaged beach holidays. Certainly, many would be regarded as 'allocentric' in nature, that is, active participants in the holiday experience. As Box 6.2 shows, Eagles (1992) was able to identify significantly different factors that motivated Canadian ecotourists compared with general Canadian tourists. Similarly, Palacio and McCool (1997) have found it possible to identify a range of ecotourism types holidaying in Belize, which exhibit different patterns of behaviour as measured by activity and participation rates. These ranged from 'nature escapist', who had the highest activity rates, through to so-called 'passive players', whose main motive was to 'learn about nature', but who had low activity rates. As McLaren declared, ecotravel covers a wide range of experiences from 'backpacking in special conservation zones to the purely hedonistic luxury vacations at typical resorts' (1998: 97).

The development of ecotourism is certainly becoming increasingly engineered, as more and more travel companies establish themselves in this area. Added to these controls are influences that condition behaviour in quite different directions. Of particular note are the various guide books that promote more careful behaviour by tourists. These range from early Green consumer guides through to more recent publications, such as *The Good Alternative Travel Guide* (2002), which claims to contain 'many examples of true ecotourism'. As publicity for this book explained, the guide 'proves that being ethical need not mean a dull holiday'. A number of NGOs have published guides to educate tourists in an attempt to modify behaviour and create a more careful, caring visitor.

Another form of tourism that is becoming increasingly institutionalized and engineered is that of backpacking. Cohen (1972) identified early forms of this in terms of so-called drifters, which he recognized as non-institutionalized tourists. However, as Spreitzhofer (1998) has demonstrated, the Australian-based Lonely Planet guide company's 1975 *South-East Asia on a Shoestring* has aided, along with many other factors, the rapid growth of the backpacker market. This submarket is characterized by 'budget consciousness and a flexible tourism style, with most participants travelling alone or in small groups' (Scheyvens 2002: 145). The backpackers have a tendency to eat and stay in low-cost establishments, and such economy-driven holidays are often couched within the terms of a search for 'real' experiences. Loker-Murphy and Pearce (1995), in a detailed study of backpackers in Australia, identified four main subgroups of backpackers with respect to motivation, namely: escapers/relaxers, social-excitement seekers, self-developers, and achievers. The heterogeneous nature of backpacker tourists has also been revealed by Uriely et al. (2002), who argue that it should be regarded as a form rather than a type of tourism. This form of tourism has become increasingly institutional-ized through the promotion of certain routes via guidebooks and, more significantly, through the influence of the general media. In the latter context, one of the most dramatic examples is Alex Garland's (1997) novel *The Beach*, written as a critique of backpacker culture in Thailand (Scheyvens 2002). However, the film of the book, starring Leonardo DiCaprio, brought global recognition to backpacking in Thailand, creating enormous ecological press-ures on the locations used in the film (Box 6.3). Other influential novels have been William Sutcliffe's (1999) *Are You Experienced?*, which recounts the travels around India of a British student backpacker.

The mass institutional form of backpacking has also been aided by the growth of specialized tour companies that market, mainly to students, adventure holidays or basic backpacking experiences (Box 6.4). Not surpris-ingly, the academic literature has commented on backpacking with increasing references to the characteristics of mass tourism. Some argue that these tourists are ego-tourists or self-centred tourists (Mowforth and Munt 2003). Aziz (1999), discussing backpackers in Egypt, claims that it is now just another form of mass, institutionalized tourism, while Noronha (1999) sees it as another facet of global tourism.

Increasingly, studies have also highlighted different patterns of behaviour associated with cultural tourism. As we saw in Chapter 5, this has formed an experience much sought after by many post-tourists. Here motives are often more related to the cultural capital gained from the experience rather than the experience itself. The engagement in this tourist experience appears to be strongly conditioned by levels of education. In this context, Stebbins (1996) invokes the notion of 'serious leisure' to explain the experiences of cultural tourists. As he explains, such cultural tourists are similar to hobbyists, in that they have a particular interest, and the holiday tends to be merely an extension of their specific leisure behaviours. The specialized cultural tourist usually focuses his/her behaviour on a small number of sites or cultural entities in their search for a deeper understanding (McKercher 2002). In contrast, there are many other visitors to cultural sites, perhaps the majority,

Box 6.3 In search of the ultimate beach: the consequences of Alex Garland's *The Beach* on Thailand

Background

The publication of Alex Garland's novel *The Beach*, and its subsequent release as a film starring Leonardo DiCaprio, popularized the beautiful Maya Bay on the island of Phi Phi Ley, in the Straits of Malacca, off south-east Thailand. The island, a marine reserve on which no building is allowed, was selected for the film location because of its picturesque character. Since the film, tourist numbers have increased dramatically in the cluster of small islands that make up the Phi Phi group. In 2001, in excess of 1 million tourists visited the islands, drawn by the lure of finding the 'ultimate beach'. As one student from Australia remarked, 'The film makes you want to come and see for yourself what it's like. But it also makes you want to go and find your own beach' (Chesshyre 2002: 2).

Consequences

- Most tourist stay on Phi Phi Don island in low-cost accommodation.

- The island has witnessed the growth of a large commercial infrastructure, including bars and clubs in Ao Tom Sai, the main settlement.

- Local communities are overwhelmed by large numbers of tourists.

- Development pressures are increasing, with a belief from local developers that more tourists equals more money.

- Tourist pressures are bringing environmental degredation, as pollution levels increase – including disruption to the coral reef's ecosystems.

Source: based on Chesshyre 2002

Box 6.4 Institutionalizing the drifter: an example of a backpacker tour company

The Moose Travel Network (www.moosenetwork.com)

The company was created to allow so-called independent travellers and backpackers an alternative way of exploring and experiencing Canada. As the internet site explains, it offers 'Canada's only national jump-on, jump-off adventure transportation network'. The company claims the majority of its customers are in the 19–34 age group, who are adventurous, independent and backpacker-orientated.

Tourists are offered a basic transport network around selected sites (see Figure 6.3) and 'pre-booked accommodation at hostels' is available. Flexibility comes from the fact that tourists can decide where, and how long, to stay on the prescribed circuit. A range of adventure tourism is offered, from whitewater rafting, through to wilderness camping and sky-diving.

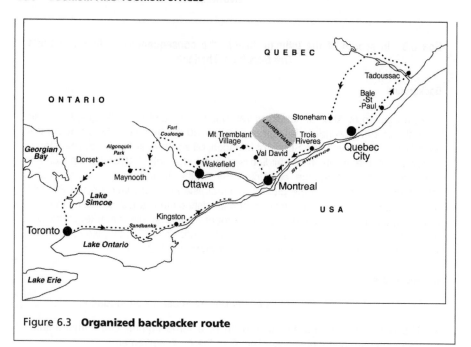

Figure 6.3 **Organized backpacker route**

whose motives are far more limited, and for whom the experience is merely one part of a wider set of holiday behaviours.

The post-tourist and holiday experiences

For the post-tourist, such behaviour sets are available either within one holiday, depending on the destination and length of stay, or by taking different types of holidays. The post-tourist experience has become highly eclectic, as different forms of behaviour can be sought through the 'sacred, informative, broadening, beautiful, uplifting or simply different sites' (Ritzer 1998: 141). In addition to the eclectic nature of tourism opportunities, post-tourists are also characterized by their playful behaviour. For such tourists, as we argued earlier in this chapter, there is no separately authentic experience, since all experiences may be viewed as authentic or real. To these characteristics of the post-tourist other commentators, most notably Rojek (1993), have added further ones (Table 6.6).

For some commentators, the Disney theme parks appear to have played an important role 'in stimulating the attitude of the post-tourist' (Bryman 1995: 177). In such strongly contrived settings as theme parks and themed shopping malls, the tourist experience can become heavily commodified. This is an extension of the way tourism was turned into a commodity by the package tour during the late 1960s and 1970s (Chapter 2). Both developments reflect what Ritzer viewed as the 'implosion of shopping and amusement' (1998: 142). Such arenas create environments for controlled social play. Tourists in such arenas are

Table 6.6   **Main characteristics of the post-tourist**

Feifer's perspective	Rojek's perspective
Increasingly home-focused, as media technology allows them to gaze on virtual tourist sites	Acceptance of commodification of tourism
	Regarding tourism as an end in itself
Possessing a greater range of choices for the tourist experience	Attributing importance to signs and signifiers
Focusing on play and touring cultures	

Source: based on Feifer (1985), Rojek (1993) and Ritzer (1998).

supposed to enjoy themselves – they are constantly bombarded with signs that are there to please them. These playgrounds of imaginary pleasures are central to the creation of self-illusory hedonism as described by Campbell (1987). In this form of tourist behaviour, shopping and recreation merge together, not by chance but through the engineered theme park environments (Chapter 10). As Lehtonen and Mäenpää explain, 'hedonistic fantasising presupposes an autonomy of play, which rests on being anonymous among other anonymous people' (1997: 164). Barry (1981–82) has drawn strong parallels between the flows of people in shopping malls and television and cinema, in that what people go to experience over and over again is their own desire.

Of course, there are very different motives and behaviours associated with the different forms of tourism. It is possible to recognize differences in behaviour between mass tourists and what some would term the post-tourist. More interestingly, there are recognizable differences within post-tourist behaviours, perhaps most clearly marked between the motives of 'hard' eco-tourists and visitors to holiday theme parks. This is hardly surprising, but what is becoming more apparent is the way that for many tourists, however we might define them, the holiday experience is becoming increasingly engineered.

GLOBALIZATION AND NATIONAL TOURIST CULTURES IN DEVELOPING COUNTRIES

It is easy when viewing the vast array of literature on tourist behaviour to conclude that holidaymakers are drawn almost exclusively from Western economies. As Ghimine notes, 'this "Northern bias" is not only reflected in government tourism policies, but also in the writing on tourism' (2001: 2). These distorted views are also increasingly outdated, in that national and regional tourism has grown rapidly in most developing countries in recent years. In this section of the chapter, attention is focused firstly on a brief review of growth trends, and secondly on a more detailed discussion of tourist behaviour among non-Western societies. This is partly to correct the inherent bias in many tourism texts and, more importantly, in the context of this book to explore the notion of a global tourist culture.

Trends in the growth of national and intra-regional tourism among developing countries are open to interpretation, as official data based on World Trade Organization definitions are problematic (Ghimine 1997; 2001). However, in spite of the limitations of official sources, it seems clear from a range of surveys that domestic and regional tourism has increased within many developing countries (Box 6.5). The growth of such tourism markets is

---

Box 6.5   Evolutionary patterns of domestic tourism in selected developing countries

Asia

The examples of India and Thailand highlight the importance of religious tourism. Cohen (1992) identified the importance of restored ancient temples in attracting large numbers of tourists, with the 1980s marking the take-off of domestic pleasure tourism. Within India, Rao and Suresh (2001) have identified four phases of tourism development:

- a traditional phase, dominated by pilgrimages and festivals

- a historical phase, with tourism restricted to the pleasure activities of the rich

- a colonial phase, where the British reconstructed familiar ways of holidaying

- a phase of rapid growth in mass domestic tourism, of a pleasure type, post-1980s. (In China, tourism growth only expanded during the 1980s as travel restrictions eased and more recent growth has been facilitated by economic liberalization policies.)

Latin America

Latin America witnessed diverse patterns, e.g. in Mexico, which traditionally have focused on visits to friends and relatives, and the reinforcement of family ties (Barkin 2001). In Argentina, Chile and Mexico, pleasure tourists emerged on a large scale during the 1960s and 1970s. In Brazil, such patterns evolved during the 1940s and 1950s, based on the urban working and middle classes of São Paulo and Rio de Janeiro (Dieques 2001).

Africa

Nigeria witnessed early developments in pleasure tourism during the 1960s and 1970s, but has subsequently seen decline because of economic crisis since the 1980s. In contrast, the domestic market in South Africa has expanded rapidly since the fall of apartheid in the 1990s. Similarly, economic investment has seen an expansion in domestic tourism to the resorts of the Mediterranean and the Red Sea coasts.

the product of a range of factors, including uneven improvements in socio-economic conditions and the growth of new industrial structures in selected developing economies (see also Chapter 10). These countries, in turn, are part of global economic systems and have counterparts in other countries, in a globalization process that involves the strong influence of Western mass media, especially through satellite television and channels such as CNN and MTV. Of course 'Western' international tourism and tourists have also helped spread new forms of leisure consumption and behaviour (Chapter 7). There are, therefore, a range of global and local factors that have combined to help promote leisure travel in a number of developing countries. For example, at the national level, the expansion of workers' rights, including, in some countries, paid holidays, has been influential.

Of course, development has not been uniform and it is possible to identify, as Box 6.5 does, different stages in the evolution of tourism. As can be seen, the starting points of growth are different, both in time and nature of the stimuli with, for example, the earlier tourism developments being recorded in Latin America compared with Africa and Asia. However, one common feature can be identified, and that is the more recent phenomenon of the growth of mass tourism. Ghimine claims that the development of mass tourism 'is already occurring on a significant scale in many developing countries' (2001: 15). In Thailand, for example, domestic tourism changed dramatically after 1987, following massive economic growth that accelerated in the following decade (Kaosa-ard et al. 2001). The diversification of domestic tourism away from mainly religious pilgrimages had started in the aftermath of the Vietnam War, following the influence of American servicemen. However, the drive to a fully fledged domestic tourist culture was largely a product of economic growth in the 1980s. Surveys by the Thailand Development Research Institute have highlighted such growth, and they claim that by 2003 an estimated 97 million trips will have been made by domestic tourists. Around 21% of these trips are focused on Bangkok, but tourist preferences include: sea and beaches, mountains and waterfalls – an estimated 12m visitors go to national parks; religious and historical attractions; cultural activities; shopping; and entertainment. Such activities are reflected in the pattern of tourist flows, which Figure 6.4 shows have distinct geographies for both domestic and international tourists.

Authors such as Ghimine (2001) argue that international, intra-regional and domestic tourism are merely different faces of the same phenomenon, as they all involve a high degree of leisure travel and are associated with socio-economic relations and exchanges. To this extent it is possible to talk of a global tourist culture, which embraces certain aspects of consumerism and commodification. There are variations, not least in the stage of evolution of this tourist culture in developing countries, and there are still some powerful national differences as exposed by the various authors in Ghimine's edited volume (and, as argued, by some theories of globalization – see Chapter 2). Nevertheless, there are also some important common features. As in Western economies, most domestic tourists in developing countries are essentially

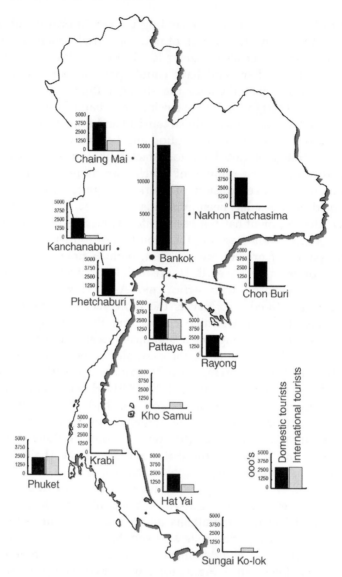

Figure 6.4 **Domestic and international tourist flows in Thailand**

urban dwellers and, as such, there are strong movements of tourists to rural environments. In many places, this phenomenon also embraces the beach holiday, as in Latin America, parts of southern Africa and some Asian countries such as Thailand. There is also a detectable growth of new forms of tourism consumption developing across a number of countries. In India for example, Rao and Suresh (2001) have drawn attention to schemes such as Off the Beaten Track, which promotes youth and alternative tourism activities, while Indian Holidays has attempted to develop new, untouched, destinations for explorer-type tourists. Similarly, in Thailand surveys have shown the increasing importance among younger people of natural attractions, notably the country's national parks. The attention of this group is focused on northern Thailand, which attracts some 95% of all domestic visitors to the protected areas there (Brockelman and Dearden 1990).

SUMMARY: TOURISM, GLOBALIZATION AND 'ENGINEERING'

This chapter has explored a number of key themes associated with tourist behaviour. We started and ended with a discussion of the nature of a global tourist culture, which can be recognized not only in Western economies, but also in an emergent form in a number of developing countries. In its full-blown form, such a tourist culture, we would argue, is characterized by two key sets of identifying factors.

The first is the search for the authentic experience. In this context, we argue that:

- the meanings of authenticity vary among tourists, just as the meanings of commodities are not fixed but open to different interpretation at both the individual and social level

- people may work at their identities through the constructions of personalized narratives of the self, mediated through commodity forms

- such perspectives (see also Chapter 5), however, may downplay the role of social and material contexts, especially the way in which experiences are constructed. This brings us to the notions of control and choice in consumption, especially within tourist behaviour, and in terms of behaviour, draws attention to the way needs and motivations are constructed.

The second, and perhaps defining characteristic, is that the tourist culture as represented by the tourist experience is itself strongly controlled and highly commodified. This engineering of tourist experiences is not new, but we would argue that the engineering processes are now more pervading and global in nature. All forms of tourist behaviour feel such control, including the so-called new forms of tourism consumption, as the tourism industry stretches to commodify more and more experiences.

7 Tourism and the Commodification of Local Communities: Impacts and Relationships

CHANGING PERSPECTIVES ON TOURISM IMPACTS

In the previous chapters we explored the tourism industry and the dimensions of tourist cultures. Both are characterized by increasingly engineered experiences and global developments, which have stretched the production and consumption circuits of Western society into developing economies. The cultures of tourism have become more economically oriented, while at the same time more goods and services are deliberately produced as 'cultural', that is they are designed with a 'specific set of meanings and associations' in mind (Meethan 2001: 118). We have discussed this broadly, and more specifically in terms of tourism, as the experience economy (Chapter 2). The cultural capital associated with these processes has a dual role: it is of symbolic importance, as we discussed in Chapter 5, and it plays a 'material role in moving financial capital through economic and cultural circuits' (Craik 1997: 125). Both dimensions of this cultural capital are charged with having significant and lasting impacts on host cultures.

Early commentators identified mass tourism as the key agency in the transformation of some traditional societies. During the 1970s and early 1980s, a series of studies sought to identify mass tourism as the main causative process in the transformation of certain 'traditional' societies (de Kadt 1979; Mathieson and Wall 1982). Many authors were critical of 'the tourist', as highlighted by MacCannell, who argued that it was 'intellectually chic nowadays to deride tourists' (1976: 9). For Turner and Ash (1975) and Crick (1989: 309) tourists were the 'barbarians, the suntan destroyers of culture'. In time, there were shifts in these perspectives, and by the 1980s so-called advocacy and cautionary perspectives on tourism impacts had been established, alongside a recognition of contingencies: that different forms of tourism and types of tourists had differing impacts (Shaw and Williams 2002).

More recently, the debate over tourism's role in cultural change has been re-examined and put into a broader context (Weaver 1998; Meethan 2001). Three main interrelated aspects are important in this re-evaluation. The first is that social and spatial relations have been redefined in the context of globalization. We have already discussed some of the main, and contested, dimensions of globalization (see Chapters 1 and 2), but these are explored further here in relation to sociocultural processes. Viewed in such a context, Waters argues that globalization is a process wherein 'the constraints of geography on social and cultural arrangements recede and in which people become increasingly aware that they are receding' (1995: 3). Arguably, one important consequence is that social relations become, in part, de-territorialized and less constrained by the various meanings of locality, although there is also a view that the global has given a new and often enhanced meaning to the local. Sharpley (1994) and, more especially, Ritzer (1998) have viewed global tourism systems as powerful agents in spreading cultural change. To some extent, Meethan takes issue with these ideas, arguing that the notion of 'the McWorld rests on a conceptualisation of culture as being composed of essential and unchanging attributes' (2001: 122). He argues that local cultures are not closed and fixed systems, but are fluid, with porous boundaries.

These issues relate to the second key aspect of the nature of cultural change: this is that, following Welsch (1999), the original notion of one culture being delimited relative to other cultures – homogenous and ethically bounded – no longer holds true. The prime reason, as identified, is that most societies have become multicultured; this is true of most tourism-generating societies and increasingly of tourism-receiving ones as well. Finally, and critically, Meethan raises the point that tourism should no longer be seen as a 'single and external causative factor' (2001: 145), as there are other powerful external influences, such as the media. This argument has been recognized in the work of several commentators, with Weaver (1995), Vellas and Bécherel (1995) and Rojek (1997) highlighting the cultural impact of global television. These broader agencies have not only contributed to the diffusion of Western consumerism, and the entrenchment of broader capitalist relationships (see Chapter 2), but have also helped to shape local consumer responses. These changes have contributed to the shifting of cultural boundaries and to increased cultural heterogeneity.

Of course, while these other social forces are significant, this is not to deny that tourism and tourists are also important in diffusing different cultural influences, especially in relation to Western consumerism. Mass movement of people from developed to developing countries, requiring some, if selective, direct contacts with host communities, has been particularly instrumental in this, although all forms of tourism have something of a role to play. We also note Crang's (2003) comment that tourism is not only a destroyer of places but is also a dynamic force in creating them.

In the main parts of this chapter, we explore the processes of commodification (see also Chapter 2) in terms of localities and cultures, before turning our attention to the assessment of the impacts on host communities. Finally, we examine the changing nature of the reactions of host societies to tourism and tourists, and the relationships between them. This is important as it demonstrates that host communities are not merely passive victims in the face of tourism developments; rather, they may be actively engaged in, and help to shape, them. Such discourses are significant in that the 'battlegrounds' of globalization are as likely to be cultural as they are to be political and economic.

THE LANGUAGE OF THE HOLIDAY BROCHURE AND THE CONSEQUENCES OF PLACE COMMODIFICATION

Chapter 2 set out the broad political-economy outlines of commodification, and here we explore further some of its more cultural dimensions. Meethan (2001) argues that commodification can be conceptualized as occurring on two interconnected levels. One concerns the images presented by the tourism industry in tour brochures and internet sites, which are representations of space. The second relates to the tourist experiences, which are acted out in destination areas, in local cultures. Of course, these two levels are strongly interrelated, both in the sense that they share the same spaces and that many tourists are attempting to live out the dreams created by image-makers. In addition, from the tourists' point of view, these 'symbolic representations of space are appropriated and incorporated as forms of personal knowledge' (Meethan 2001: 86).

The process of commodification starts, therefore, not with the arrival of tourists and their cultures, but rather with the way in which destinations are represented through the marketing system. As Crick (1989) explains, the imagery of international tourism is not concerned with reality, but with myths and fantasies. These fantasies, generated by marketeers and, for example, retailed in travel brochures, can have significant consequences for a destination by creating images that are alien to the identities and practices of the host communities. Such marketing processes aim to identify particular spaces and sites as being distinct arenas of consumption, and set apart from the 'ordinary' world.

Commodification implies uneven power relationships between international tourism on the one hand and tourism destinations on the other. Crick (1989) argues that the links between power and knowledge, the generation of images of the 'other', and the creation of 'natives' and 'authenticity' are significant in understanding the commodification of all tourism places. This is a proposition we would embrace, and our later review of tourism impacts will be broad-ranging. It is also worth reiterating that the processes inherent in the commodification of tourism places are not only culturally rooted, but also related to the growth of the advertising industry, with its interconnections with the media, especially television. These links are necessarily complex, in that the media both promote tourism and travel while also helping to define places as out of the ordinary. In this sense, signifiers of the exotic, as well as history, memory and tradition, provide important frames for the advertising industry (Goldman and Papson 1996). The processes of creating tourism spaces form the core theme of the third part of this book, where we examine in more detail how various places have been engaged in, and shaped by, tourism.

Image and the social construction of tourism places

Returning to the theme of image, Rojek argues 'that myth and fantasy play an unusually large role in the social construction of all travel and tourist sights' (1997: 53). There is an implication that tourists are semioticians, constructing their gaze around well-defined signs or markers (Dann 1996a). These are used to identify people, things and places, which enhance accessibility to tourist sites. Increasingly, places and cultures are being themed (see Chapter 10) and branded in an attempt to promote or establish tourism activities.

Rojek (1997) has summarized some of the main elements in the social construction of tourism places, and these include the observations that:

- All tourism sights depend strongly on the demarcation processes that help distinguish them from ordinary places. As we have seen, these are supported by signifiers in the landscape (Urry 1990; 1995), and the marketing industry.

- The nature of the demarcation between the ordinary and the extraordinary (tourism places) is cultural, as the relationship between the tourist and the sight is always culturally detailed and mediated (see Chapter 6).

- The distinction between the ordinary and the extraordinary has been undermined by what can be termed 'television culture'. Such representational codes of tourist sights have, in part, reduced the aura of tourism places, in that they can be viewed repeatedly in a range of contexts in the media. Television has also played another role in that, within strongly televisual cultures, the TV culture itself will increasingly be a tourist attraction. For example, many tourists are interested in visiting the locations of their favourite soap operas.

In the creation of a place image, the image-makers are presenting a vision that will appeal to certain types of visitor, and the connections with any objective reality may be tenuous at best. Within tourist brochures, tour operators are careful to sell to their target markets. Dann's (1996c) extensive study of the language and imagery of British holiday brochures allowed him to identify four different types of 'paradise' paraded before potential tourists (Table 7.1). On the whole, most mainstream tourist brochures examined by Dann did not encourage the idea that the holiday constituted a meeting ground between tourists and locals. Indeed, fewer than 10% of the images in the brochures showed locals and tourists together. In this context, the message seems to be 'enjoy certain elements of local culture', but in a controlled way. In Dann's

Table 7.1   **Dann's identification of types of paradise in holiday brochures**

Type	Brochure characteristics
Paradise contrived	No people shown Natives as scenery Natives as cultural markers
Paradise confined	Only tourists shown Presentation of tourist ghettos
Paradise controlled	Natives as servants Limited contact with locals Natives as entertainers Natives as vendors
Paradise confused	More contact with locals Views of locals-only zones Natives as seducers Natives as intermediaries Natives as familiar Natives as tourists, tourists as natives

Source: Dann (1996c).

study, the written text in the brochures that accompanied the photographs emphasized the qualities of the natural landscape, the opportunities for self-rediscovery and the exotic. The same images are often reinforced by journalistic travel-writers. For example, Hodson, under the title 'Thailand to die for', claims 'Asia's hottest new destination [Khao Lak] isn't paradise – but it's close' (2002: 57). Nowhere in this article – which discusses attractions, hotels, activities and services – is there any mention of local people and their customs. The only occasion on which locals are referred to is in the context of service quality in hotels.

The construction of touristic images by the tourism industry in the commodification of tourism sites is wide-ranging and obviously is not confined to fantasies of exotic places in developing economies. As Selwyn (1996) demonstrates, for example, myth-making also embraces the imagery of postcards, and can be found in all societies. In addition, image-creation is complex in that it involves a wide range of media and different agencies. Baloglu and Mangaloglu (2001), for example, have shown how US-based tour operators and travel agents play key roles in destination marketing. They go on to argue that the 'images perceived by travel intermediaries would reflect their clients' perceptions and be transferred by travel intermediaries over to their clients' (p. 7).

As Selwyn (1996) explains, the 'overcommunication' of the myths and images surrounding tourism destinations focuses on harmonious surroundings, both natural and cultural, while neglecting or concealing any fractures in local societies. Having constructed particular images of places and cultures, culturally informed market forces then create strong pressure for concrete changes in the destinations to match the images projected by the tourism industry (Williams 1998). The perceptions, motivations and expectations of tourists need to be confirmed by their experiences if interest in, and demand for, that destination is to be sustained. As Williams argues, 'in this way, tourist images tend to become self-perpetuating and self-reinforcing' (1998: 178).

There are inherent contradictions in these processes of place commodification, which operate at two main levels, the most salient features of which are that:

- There is spatial and social isolation of tourists from the local community, which can take various forms, including the extreme of resort enclaves (Freitag 1994) and the 'environmental bubble' of a package tour (see Chapter 9). In these circumstances, tourist places can, to varying degrees, become largely reflections of visitor expectations. The demand for more individualized holidays (see Chapter 5) has only partly mediated this physical separation of tourists, as images and expectations are still often retailed in the same fashion as they are for mass tourism (Chapter 6).

- The powerful expectations of tourists, as created by the tourism marketing machine, can only be fulfilled by concrete investments in particular developments. These may be large, Western-type hotels, theme parks and other elements of familiarity or – in a different form – locally-run backpacker hostels. But, in varying degrees, they all help to create new economic and cultural exchanges. Furthermore, in the face of mass tourism, and its post-Fordist forms (see Chapters 2 and 5), many destinations start to lose their original identities, becoming 'placeless and quite indistinct from other tourist places' (Williams 1998: 178) and unrepresentative of their original cultures.

Both of these contradictions have important consequences for local communities and local cultures, and these are effected through a series of sociocultural processes. To understand these we need to explore two important concepts: culture and commodification.

TOURISM, TOURISTS AND 'THE OTHER'

As argued earlier in this chapter, culture cannot be seen as static and bounded, but rather it changes in response to external forces, including tourism. Moreover, culture is a complex concept, which can be contextualized in terms of a number of interlocking components, as suggested in Figure 7.1. Viewed in this way, culture is a 'linked set of rules and standards shared by a society which produces behaviour judged acceptable by that group' (Burns 1999: 57). This general definition clearly embraces sociocultural components, which range from the level of ideal social norms through to culture at the level of observed transactions. However, as Meethan explains, culture, 'although

Figure 7.1 **The components of culture (source: modified from Burns 1999)**

being pervasive, is difficult to pin down' (2001: 115), as it is open to different interpretations. For example, culture can be viewed in terms of a 'high and low' dichotomy, the latter encapsulating mass, popular culture. Culture has also been seen in terms of material and symbolic production, and as part of a symbolic system (Martinez 1998). Our broad-ranging definition attempts to include these differing perspectives.

Conceptualization is made more complex by the nature of tourist cultures, which can be seen as both an act and as an impact (Burns 1999). Furthermore, the links between tourism and commodification 'are now considered more dynamic and complicated than was felt in the 1960s and 1970s' (Harrison 2001: 28). In response to such complexities, some authors have adopted a systems approach to studying tourism and cultural impacts (Wood 1993; Burns 1999). One dynamic element of the process is that symbolic meanings are attached to objects and actions, alongside the appropriation of culture by tourism. In addition, in the selling of culture, 'complexity is commodified and reduced to a recognisable formula' (Meethan 2001: 126).

Tourism impacts on local cultures operate through a series of processes, which are set within the broader framework of capitalist relationships (see Chapter 2). As argued at the beginning of this chapter, the first stage in these impacts concerns the marketing of local culture and its resources. However, the impact cycle is completed by the contacts between tourist cultures and the local community. The key influences shaping the impacts of tourism are:

- The relative cultural distances between tourist cultures and local ones. The non-ordinary world of the tourist is structured by both tourist culture and his/her residual culture. According to Jafari 'observable rituals, behaviours and pursuits' (1989: 37) bind tourists into a type of collective culture. However, the notion of residual culture highlights the potential differences among tourists, since it denotes the 'cultural baggage' that tourists bring from their home cultures. Residual culture shapes the behaviour of tourists, as discussed in Chapter 6. Central to this debate is the concept of cultural distance, which denotes the degree of cultural similarity between host culture and visitor culture (Shaw and Williams 2002).

- The type and number of visitors. In this context, 'type' most obviously refers to institutionalized and non-institutionalized tourists, as described by Cohen (1972), but we can also add to this the considerable variety of so-called independent travellers. The growing concern of many tourists has brought about the growth of ecotourism, as previously discussed, accompanied by new forms of behaviour. However, one of the critical influences is the 'numerical effect' of tourists. Harrison (1992) has drawn attention to such effects and sought to measure them in terms of tourist intensity rates. This measure relates annual tourist arrivals to local population size as a ratio. Such measures have been applied especially to small-island destinations, where extremely high tourist intensity rates have been recorded (Dann 1996b). More recently, McElroy and Albuquerque (1998) have expanded on those ideas through their development of a Tourism Penetration Index. This is based on three overlapping sub-indices, which measure sociocultural, economic and environmental penetration. High visitor numbers create pressure on local facilities and, in turn, on local communities, usually in the form of irritation to locals. Such impacts are widespread and have been well-documented (Doxey 1976; Jackson 1986; Harrison 1992; Mansperger 1995). In the wider context of resort decline, Russo (2002) has described the impact of high visitor numbers in historic centres such as Venice, as shown in Box 7.1.

As we have shown, tourists arrive with preconceived images and motivations, which shape their interactions with local communities. Moreover, the tourist–local relationship is also rather unusual, in that while the tourist is at play the local is usually at work (but see the discussion on occupational communities, and tourism work as performance, in Chapter 3). Other differences in this relationship are that many tourists have economic assets but only limited cultural knowledge, compared with locals with cultural capital but limited resources (Crick 1989). In most instances, the tourist does not become embedded in any long-term social relationships, as interactions are fleeting and usually superficial (UNESCO 1976). As Cohen (1982) explains, tourists are not guests, but rather outsiders to the local culture. They occupy the non-ordinary world of 'life in parenthesis' (UNESCO 1976: 85), where the media image-makers have distorted the experience of space and time (Crick 1989).

Box 7.1   The impact of visitor numbers on central Venice

Tourist numbers and impacts

- The overnight ratio of tourists to residents peaks at 50:1 in the historic core, rising to 175:1 if excursionists are included.

- Within the core area, there are strong imposition costs on local residents in terms of congestion, contested space (public spaces are dominated by tourists who account for 56.9% of users overall, peaking at 66.9% between July and October).

- Younger households are being forced out of the historic core because of congestion, house prices and a lack of specialized employment.

- The resident population is falling by 0.5% per annum and has reduced from 170,000 to 70,000 over the last 50 years.

- Carrying capacity has been calculated at 22,500 daily arrivals, but this is exceeded on at least 156 days per year.

- The area is being transformed into a tourism 'mono-culture', lacking other employment opportunities.

Source: based on Russo (2002)

Tourist–local encounters

The nature of tourist–local encounters encompasses various sociocultural processes, namely the demonstration effect, acculturation and cultural drift, internal and external change, and cultural assimilation. Some of these are important not only in terms of their cultural effects, but also as part of the process of commodification and the institutionalization of markets in capitalist economies (see Chapter 2). When moving from the abstract to the concrete level of analysis, however, it is sometimes difficult to distinguish the effects of tourism from other external forces of cultural change.

The demonstration effect is considered to be one of the major processes of change, and it describes the voluntary adoption by local residents – especially younger people – of some of the consumption patterns of tourists (Rivers 1973; Sharpley 1994). However, as Crick argues, 'tourists may have been chosen as conspicuous scapegoats' (1989: 335), as close attention suggests that very often other factors may be at work. Burns (1999) also suggests that empirical evidence for the demonstration effect is rather weak. Some studies, however, have found evidence of the demonstration effect, but it is conflicting. For example, Kousis (1989) examined the effect of mass tourism on traditional family life in Crete, and argued that most changes were because of economic processes rather than visitors' cultural intrusions. In contrast, Witt's (1991) study of Cyprus suggests that mass tourism has resulted in a modification of social attitudes, especially among young people toward sexual behaviour. Thus, the situation is more complex than first thought, and its complexity has increased because of the difficulty in differentiating between the sociocultural impacts of tourism and general processes of modernization (Burns 1999).

Of course, tourists' displays of wealth and hedonistic behaviour can arouse not only distress among locals, but also jealousy. Wealth disparities between tourists and locals raise unrealistic expectations among the latter. This can result in two responses, which are not mutually exclusive. First, local value-systems, attitudes, language, dress codes and demand for consumer goods change in response to tourist culture (Shaw and Williams 2002). Second, the tourist becomes regarded as a means of wealth provision via crime, with the tourist as victim. Cohen (1996) argued that in Thailand the structural features of the local society encourage certain types of crime against tourists. Similarly, de Albuquerque and McElroy have found that crimes against tourists have 'become a kind of routine activity of marginalised youth' (1999: 970). In a broader context, Ryan (1993) has constructed a typology of the conditions for crimes against tourists, as shown in Table 7.2, which highlights both intrinsic (i.e. tourism-related) and accidental scenarios. As can be seen, tourists may be victims or criminals, but either way local communities can face increased levels of crime and social breakdown in the face of large-scale tourism developments (Pizam et al. 1982).

The demonstration effect suggests that social impacts are permanent, but some anthropologists have identified more temporary transformations in local behaviour. This has been termed 'cultural drift', and is likely to be found in those local communities that have a strong tourism season, with tourist–local encounters being more intensive and disruptive during the main holiday season. Finally, attention should again turn to the idea of external or internal factors of change. Clearly, Western capitalist patterns of consumption tend to induce significant cultural impacts, with tourism being but one channel vector

Table 7.2   **Ryan's typology of conditions for crimes against tourists**

Type	Characteristics
Tourist as accidental victim	Tourists in the wrong place at the wrong time, but targeted because perceived as easy victim
Location as a criminogenic venue	Tourists in the context of nightlife culture, high on drink or drugs, making them easy victims
Tourism as provider of victims	Tourists as risk takers. As numbers increase, so does local hostility and willingness to victimize them
Tourists as imported demand for deviant activities	Tourists become both victims and also criminals in their loutish behaviour
Tourists as targets for politico-criminal action	Tourists singled out as hostages – symbols of capitalism by terrorist groups

Source: Ryan (1993).

Table 7.3 **Stages of cultural collision and intercultural interactions with local residents**

Stage	Characteristics
Toleration	Limited tourists, coexistence and toleration by locals
Segregation	Segregation – social and physical distance – affect interrelationships, as tourists and locals avoid each other
Opposition	Tourists rejected by locals or locals shunned by tourists: distrust and a degree of hostility
Diffusion	Tourist and local cultures converge, through acculturation and symbiosis.

Sources: Gee et al. (1989); Burns (1999).

of change. However, as Núñez (1989) points out, the tourist is the most ubiquitous representative of Western culture, since his/her behaviour can be observed directly by local people.

Acculturation and its derivatives – cultural symbiosis and assimilation – are the processes by which cultural borrowing occurs. Acculturation refers to the degree of cultural borrowing between two contact cultures (Núñez 1989); Meethan, following Hannerz (1996) and others, suggests that the way to view the convergence of cultures is to see them as new 'forms of creolisation or hybridisation' (Meethan 2001). This notion is contested by Friedman (1999) on the ideological grounds that it justifies the 'rootlessness' of metropolitan elites. It is usually most clearly evident – as with other concepts – in the relationships between Western societies and developing economies. The interaction between such differing cultures has been presented as a stage model by Gee et al. (1989), as shown in Table 7.3, although it has not been verified empirically (Burns 1999). There is also opposition to the unidirectional nature of the model and its terminology. However, in spite of these limitations, it does suggest a means by which tourist–host interactions change over time.

Most of the sociocultural processes outlined above are applicable mainly to tourism's impact on developing economies and traditional cultures. In contrast, the process of commodification is far more wide-ranging in its effects on local cultures. The general features of commodification have been identified by a number of authors, including Selwyn (1996), who highlighted the following:

- The process of commodification is part of consumer culture, with its emphasis on symbolic meaning. This is related to the fact that consumer culture constitutes the basis of Western society (see Chapter 5).

- Tourism-induced commodification and consumerism leads to states of dependency – including cultural dependency – in tourist destination areas.

- The commodification of social and ritual events leads to an erosion of meaning, accompanied by community fragmentation.

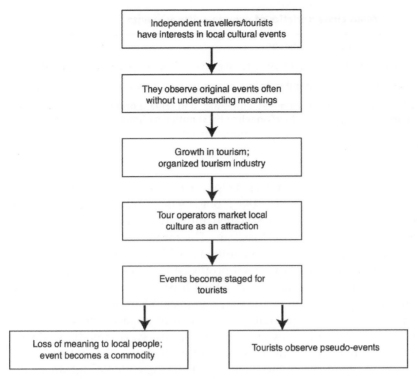

Figure 7.2 **Stages in cultural commodification**

The latter feature is at the heart of much of the debate on the sociocultural impacts of tourists and tourism. The search for the authentic tourist experience, discussed in Chapter 6, is a key element in the process of cultural commodification. For the cultural tourists, culture in all its forms is a significant element of travel. Meethan (2001) has drawn attention to the role of tour operators and policy-makers in stimulating this form of tourism. The production of heritage and cultural sites may be seen as another variant of cultural commodification. However, most attention has been focused on the impacts of mass tourism on local culture. A number of studies have encompassed the commodification of both traditional craft activities and local events – festivals, dances and ceremonies. The processes may be outlined as a series of stages, as shown in Figure 7.2, with the commodification process being initiated when tourism grows and becomes a significant activity. The pseudo-events created tend to share a number of related characteristics, as suggested in Table 7.4.

In reality, local situations and the process of cultural commodification are variable, and more complex than suggested in Figure 7.2. This is because of the range of influences at work, the nature of local cultures, and their reactions to tourism. In other words, tourists and the processes of cultural commodification are negotiated in various ways by different societies. In this way, host communities attempt to control access to local culture through the

Table 7.4 **Main characteristics of pseudo-cultural events**

Planned rather than spontaneous

Designed to be reproduced to order for the convenience of paying customers (tourists)

Have an ambiguous relationship to real events based within the local community

May become regarded over time as 'authentic', and replace original events that they represented. This relates to MacCannell's (1973) notion of 'staged authenticity'

Source: Williams (1998).

use of local knowledge, which helps to reinforce the distinctions between locals and strangers (Crick 1989; Boissevain 1992). These ideas relate to MacCannell's (1989) concept of staged authenticity and his division of cultural events into 'front' and 'back' regions. The former is where social interaction takes place between locals and tourists, while the 'back' region is largely hidden from tourists. MacCannell proposed that this 'front–back' dichotomy could be viewed as a continuum, with at least six identifiable stages marking the tourist's quest for an authentic experience (Sharpley 1994).

The research literature highlights a range of impacts and responses, from those in which cultural events have been largely commodified through to situations where locals have significantly mediated the potential impacts. Early studies tended to view such impacts in somewhat simplistic positive or negative terms, but the processes are far more complex than this. Such complexities appear to relate to types of activity, the cohesion of the local community, the stage of tourism development and the role of the state.

Unfortunately, there has been little comparative research on this theme, and we are largely reliant on a series of case-studies that utilize variable methodologies. Nevertheless, the case-study approach does help cast some light on the variability of the cultural impacts of tourism and tourists. Nimmonratana (2000) has, for example, described how the village of Ban Thawi, in the Chiang Mai area of northern Thailand, has been integrated into the global tourism system and been transformed. Ban Thawi is known as the 'carving village', because 150 of its 200 households are engaged in craft activities. This began in the late 1960s with the production of images of Buddha and other religious items. State intervention, banning the export of Buddha images during the early 1970s, combined with increasing demands from tourists, brought new opportunities for the craft village, which now produces a range of non-traditional artifacts, including images of Disney characters. Other studies of local crafts have identified the revival of craft activities through tourism, as pointed out by Graburn's (1976) study of the emergence of Eskimo soapstone carvings, which could be described as a form of emergent authenticity. In a similar way, Horner (1993) and Sindiga (1999) claim that there has been a revival of Kenyan art associated with the growth of tourism. There are, of course, counter-examples that highlight the more negative cultural impacts of commodification in the face of tourism. Sharpley argues that traditional designs and production techniques often disappear 'as simpler and less sophisticated replacements associated with the techniques of mass production are provided for tourist consumption' (1994: 150). Even if this does not occur, and the reproduction of cultural artifacts for tourists retains their quality and style, the meaning becomes lost in such stylized works (Sharpley 1994). (The economic relationships between commodification and craft work are also explored in Chapter 2, in relation to the production of souvenirs.)

Similar variations can be found in the case-studies of tourism's impact on local festivals, although there may be even more complexities here, reflecting how communities have negotiated change in different ways. As Sharpley (1994) and Williams (1998) have shown, numerous studies have exposed the creation of pseudo-cultural events in the face of tourist demands (Table 7.4). Again, there are contested views in that Daniel has argued – in the context of traditional dances – that cross-cultural studies of dance performance in tourism settings do 'not fully exhibit the usual effects of artistic commoditization' (1996: 781). More generally, Meethan (2001) has provided insights into the various forms of tourism impacts on local cultural festivals, which are summarized in Box 7.2. This suggests that there are a number of interconnected, but slightly varying, negotiated strategies.

Box 7.2   Negotiating the commodification of local culture: festivals and rituals

Controlling access

Laxson (1991) describes the ways in which pueblo ceremonials in New Mexico meet the expectations tourists have of their dance performance by wearing costumes and face paint from the culture of the Plains Indians. This meets some of the spectators' needs, but confuses and limits outside access to the pueblo culture. In this way, secrecy is maintained, which is an important element of rituals.

Staged authenticity

The pueblo case is a form of staged authenticity, which is also identified by Picard (1996) in Balinese dances, which are performed in a particular way for tourists who do not have access to the important elements of authentic ritual performances.

Changing and overlapping meanings

As festivals become commodified, the meanings of such events change, but they may remain authentic (Sharpley 1994). For example, in Bali ritual performances have taken on different meanings, especially among younger Balinese, who, according to McKean, 'find their identity as Balinese to be sharply framed by the mirror that tourism holds up to them' (1989: 132).

Source: based on Meethan (2001: 157–8) and Sharpley (1994)

WHY US, WHY HERE? LOCAL PERCEPTIONS OF TOURISM IMPACTS

As we have seen, local communities have different ways of coping with, and negotiating, tourism. Indeed, in recent decades, research on the impact of tourism has turned increasingly towards understanding how local residents perceive tourism and tourists. Early attention focused on a series of models or standardized views of resident reactions, as described by Doxey (1976) and Bjorkland and Philbrick (1975) (for a general review see Shaw and Williams 2002). As later researchers have observed, the models 'are best described as *post hoc* descriptive accounts' (Pearce et al. 1996: 18). Similarly, the initial studies of residents' perceptions of tourism tended to be limited in their explanatory power (Kayat 2002). The growing literature in this area suggests that local people are influenced by the perceived impacts of tourism in three basic categories of costs and benefits: the economic, the social and the environmental (Gursoy et al. 2002). Within these studies a range of influences have been explored, but usually in a rather descriptive way.

It is not surprising, given the differing methodologies used by many of the case-studies, that few consistent relationships can be detected (King et al. 1993). Indeed, Langford and Howard (1994) have demonstrated the considerable variety of measurement procedures and research paradigms that have been brought to these studies. In essence, however, there are two main categories of study, based on different scales of analysis (Williams and Lawson 2001). These are:

- Community-level studies, where influences on resident perceptions are examined in terms of local community attributes. Communities are defined in various ways, but usually in geographical terms. Of course, communities are complex entities and far from homogenous (Madrigal 1995), thus presenting significant measurement difficulties.

- Individual-level studies. These examine variations in the perceptions of individuals and the influences on them.

A few studies have attempted to take a middle way, combining both community- and individual-level attributes, as in Williams and Lawson's (2001) work on New Zealand. These studies allow the identification of some antecedents of residents' opinions of tourism, as shown in Table 7.5. From these and other findings, it is possible to recognize a number of key factors. These include the level of economic dependency or perceived community benefits derived from tourism, knowledge of tourism within the local area, and the considerations that local people have regarding their community.

A number of these influences have been linked together through social-exchange theory. The underlying assumption is that residents 'behave in a way that maximises the rewards and minimises the costs they experience' (Madrigal 1993: 338). Ap (1992) has extended this idea from the individual to the community level, that is to the benefits and costs that residents perceive the community will receive from tourism. There is certainly evidence of such

Table 7.5 **Suggested causal influences on residents' opinions of tourism**

Variables in residents' characteristics	Indicative studies
Distance of respondents' home from site of tourism activity	Belisle and Hoy 1980; Pearce 1980; Sheldon and Var 1984; Tyrell and Spaulding 1984.
Heavy concentration of tourism	Pizam 1978; Madrigal 1995
Length of residence	Pizam 1978; Brougham and Butler 1981; Sheldon and Var 1984; Liu and Var 1986; Um and Crompton 1987; Allen et al. 1988; Lankford 1994; Madrigal 1995
Locally born	Pizam 1978; Brougham and Butler 1981; Um and Crompton 1987; Davis et al 1988; Canan and Hennessy 1989
Personal economic dependency on tourism	Pizam 1978; Thomason et al. 1979; Murphy 1983; Tyrell and Spaulding 1984; Liu and Var 1986; Pizam and Milman 1986; Milman and Pizam 1988; Lankford 1994; Madrigal 1995
Ethnicity	Var et al. 1985
Stage in destination area life cycle	Johnson et al. 1994
Level of knowledge about tourism in local area	Pizam and Milman 1986; Davis et al. 1988; Lankford 1994
Level of contact with tourists	Brougham and Butler 1981; Lankford 1994; Akis et al. 1996
Perceived impact on local facilities	Perdue et al. 1987; Lankford 1994
Gender	Pizam and Milman 1986; Ritchie 1988
Perceived ability to influence planning decision	Lankford 1994

Source: Williams and Lawson (2001).

an exchange system, although it is complex and variable at the community level. For example, Ryan and Montgomery's (1994) study of local responses in the Peak District of England showed there was a degree of altruism among residents who had no direct economic benefit from tourism, with around 70% of interviewees agreeing with the statement 'tourism was good for the local economy'. These findings correspond with those of Prentice (1993), working in the north Pennine area of England. In both studies there was evidence of local support for tourism because of the perceived general economic gain for the community, despite some tangible disadvantages for the individuals themselves. These local findings also support the broader ideas of Lindberg and Johnson (1993), who argue that perceived economic and congestion impacts have a greater effect on residents' attitudes toward tourism than do perceived crime and aesthetic impacts, at least in developed countries. In this context, Getz's (1994) work on the Spey valley in Scotland confirmed elements of the social-exchange theory of resident attitude-formation, particularly the influence of economic benefits and problems.

Complexity and variability are evident, both between host communities and within them. This is related partly to the stage of resort development reached, but more especially to the types of tourism and tourists present (Pearce et al. 1996). In the case of different types of tourism, Ritchie (1988) discovered that levels of local support varied according to the types of proposed development, with casinos being the least-favoured option. Not unexpectedly, Simmons (1994) found most support for tourism developments that were small- to medium-scale and locally owned. With reference to tourists, Ross (1992) showed that the older residents of an Australian community were more accepting of American and Australian visitors, while Liu and Var (1986) also found evidence of the stereotyping of tourists.

The literature has focused mainly on intra-community differences and the social relationships that underpin these. More recently, interest has been directed at the role of 'power' in explaining variations in residents' attitudes. Ap (1992) highlighted this by proposing that disadvantaged locals' perceptions of tourism would be negative. However, detailed research by Kayat (2002), on Langkawi Island in north-west Malaysia, suggests that there is a more complex relationship between power and residents' dependency on tourism which in turn contributes to residents' values. The role of power, operationalized through the resources owned by local residents, was mediated by other factors, as highlighted in Figure 7.2. Kayat (2002) found, for example, that those dependent on tourism placed a high value on economic returns compared with religion or maintaining their culture. Respondents with fewer resources, and therefore little power, perceived that tourism had created economic opportunities for them. These residents were more willing to adapt to, and had favourable perceptions of, tourism.

Some studies have also drawn attention to more detailed differences in local attitudes to tourism development, distinguishing between the acceptance of tourism as an economic necessity and an open dislike of tourists (see Box 7.3). This is hardly surprising, given the social nature of tourism and the diversity of tourists. We have already examined some of the extreme outcomes of such responses in terms of crimes and anti-social behaviour against tourists. There is, however, another dimension to the dislike of tourists that concerns the notion of local–tourist adjustments, which are often asymmetrical in terms of economic power, certainly in most developing economies. Such adjustments concern not only residents' attitudes, but also changes in their behaviour, as they contest space and facilities with tourists, as illustrated in Box 7.3.

Another important way of understanding local, community responses toward tourism is through the framework of social-representation theory. This theory focuses on the content of social knowledge and, equally importantly, on the way in which such knowledge is created and transmitted throughout the community. According to Pearce et al. (1996), three main criteria help to establish social representation within a community:

Box 7.3   Residents' perceptions of and adjustments to tourists in South Devon resorts, England

Case-study characteristics

Surveys of residents (sample size 237) were undertaken in Dawlish and Teignmouth. In both resorts, the tourism product is based mainly on small hotels and self-catering holiday parks. The resident population includes retired incomers, and some professionals who commute to the nearby city of Exeter. There is low unemployment and a high proportion of seasonal employment related to tourism. The main survey findings are:

- Different attitudes exist towards tourism and tourists. In the case of the former, residents could see positive and negative impacts, but most accepted that the tourism industry was important for the area. They demonstrated a high degree of altruism, as reported in Ryan and Montgomery's (1994) work.

- By contrast, attitudes towards tourists were far less favourable. While they were seen as 'happy and lively' by 43% of the sample, 26% considered they were 'rude and pushy', while 73% stated they were the main cause of congestion.

- Residents responded by avoiding tourist areas (51% reported they did so), changing their use of local facilities (27%) and modifying their shopping habits (38%).

Source: Shaw and Curtin (2001)

- There is a shared commonality or consensus, suggesting that a search for similarities of responses can provide a basis for community problem-solving in tourism-related conflicts.

- A network of shared experiences, beliefs, values and explanations of tourism impacts binds together a community. Social representations are conceptualized as systems of related attitudes and values. In its simplest form, this may be how locals define tourism and tourists (Lea et al. 1994).

- The existence of a need to understand how different beliefs and attitudes are interrelated, which re-emphasizes the importance of complex networks of beliefs about tourism, as illustrated by Murphy (1988).

Of equal importance to the above criteria are the influences that help to shape the beliefs and attitudes of residents towards tourism: the media, discourses and social interactions at different levels within the community, and the direct experiences of tourism and tourists. The media – for example, the local press – can establish agendas, and provide individuals with content for their social representations. For example, newspapers may run stories about young tourists behaving badly because of excessive drinking, which in turn establish a major talking point in the community. In addition, the media can set up situations of conflict between the visitors and locals, or alternatively may present tourism in a favourable light. For example, Timmerman (1992) has explored the role of local newspapers in Queensland, Australia, where the majority of tourism-related articles were positive in respect of economic, social and environmental impacts.

It is significant that the so-far limited applications of social-representation theory in tourism have been used largely to understand 'how hosts understand, define and evaluate the future of tourism' (Pearce et al. 1994: 178). Such ideas build on the early, innovative work of Murphy (1985), which drew attention to the importance of community participation in the tourism planning process. Subsequently, such ideas have been fully embraced by the rhetoric, if not always the practice, of sustainable tourism.

MEDIATING CHANGE: TOURISM–COMMUNITY RELATIONSHIPS

Earlier in this chapter we illustrated how some local cultures negotiated the sociocultural impacts of tourism. Here we want to expand on this theme and look more closely at the ways in which communities mediate change and, more especially, the process of community participation in tourism developments. This also needs to be seen in context of the mode of regulation in particular national states (see Chapter 2) and the discussion of place (Chapter 8).

Krippendorf (1987) emphasized the need to harmonize tourism developments with their local surroundings, and that for this it is necessary to 'unravel the tangle of often conflicting interests and set up clear priorities' (p. 117). This certainly applies to tourism–community relationships, as these hold the key to local people exerting some control over tourism developments within their communities. Furthermore, integrating local community needs and ways of life with tourism developments is essential to avoid the problems and conflicts associated with the erosion of local cultures.

Pearce et al. (1996), following Painter (1992), identified three main types of community participation: protest, information exchange and negotiation. Faced with the threat of impending tourism developments, local communities may engage in all three forms of action. The broader ideals of community involvement in tourism planning are also increasingly part of state and NGO agendas, billed under 'community tourism' or 'sustainable tourism'. For example, the UK-based NGO Tourism Concern (see www.tourismconcern.org.uk) argues that community tourism should:

- be run with the involvement and consent of local communities, which of course links directly with the ideas of community participation

- be in a position to share profits 'fairly' with the local community

- involve communities rather than individuals

Similarly, the former English Tourism Council, which helped market and facilitate tourism developments in England, has embraced sustainable ideals, as is evident on their internet site (www.wisegrowth.org.uk). At the international level, the UN Environment Programme gave early recognition that alongside the need for profitable tourism investments there 'must be added consideration of social and cultural effects' (1983: 53).

In practice, local community involvement is still somewhat limited and variable, as highlighted by Pearce et al. (1996). As a consequence, we have a series of case-studies that demonstrate varying degrees of success, rather than an established blueprint of best practice. It is also relatively easy to point to increased state concern over issues relating to tourism and local communities than to ascertain their practical importance with any certainty. Part of the problem in the relationships between tourism development, the state and local communities lies in the uneven and fragmented nature of control. Hall (1996) goes further and argues that the main difficulties are the top-down approach and the fact that many decisions have often been prescribed by central government (see also Chapter 8 on the nature of the local state).

Some of these issues can best be illustrated through specific examples. Nimmonratana (2000) has shown in north-west Thailand that, in spite of the government and the Tourism Authority of Thailand's increased emphasis on cultural and environmental conservation, local people feel excluded from decision-making. This exclusion contributes to widespread concerns over what is considered to be the decline of local culture. Nimmonratana suggests that the engagement of the local community can be achieved through both better cooperation with regional tourism authorities and increasing local self-awareness with increased education. The complexities of community participation have also been highlighted by Brennan and Allen (2001) in a series of case-studies in Kwa-Zulu-Natal, South Africa. These show clearly how political and organizational fragmentation have serious consequences for community involvement. In some instances, as at Kosi Bay in the far north of Kwa-Zulu-Natal, the degree of participation was relatively high for the stakeholder communities. Unfortunately, internal rivalries among these groups impeded progress, contributing to poor-quality tourism products. Brennan and Allen indicate the need 'for developmental models based on notions of self-interest and diversity within communities rather than obvious notions of community spirit' (2001: 215). Their work points to a somewhat different and more varied approach than the 'one size fits all' solution.

As we have argued, state authorities (local and national) and private enterprise continue to set the agendas, not only for tourism development but also for the involvement of local communities. The array of studies on residents' perceptions of tourism, along with the application of various conceptual frameworks such as social-exchange theory and social-representation theory, all point to the importance of increased community participation. The host community support models developed by Gursoy et al. (2002) may provide a further organizational framework for tourism planners. However, as with other perspectives, this will be of use only if communities are actually asked about their concerns, and are engaged in the processes of tourism development. The conceptual frameworks for such participation have been partially constructed in abstract terms, while more specific frameworks tend to have been disproportionately researched in the context of ecotourism developments in fragile environments. The shift from theory to practice remains a major issue in community tourism planning in all types of environments.

SUMMARY: COMMODIFICATION, IMPACTS AND COMMUNITIES

This chapter has focused on the impacts of tourism on local communities. Within this context notions of culture are explored, especially in relation to commodification processes and tourism. The processes of commodification relate to localities, cultures and host communities.

- Significantly, commodification starts long before the tourist arrives in a destination and is established at two interconnected levels: the images created and presented by the tourism industry, as reflected in brochures and internet sites, and the way tourist images and experiences are acted out within particular destinations. At both levels, images are of key importance and determine the social construction of tourism places.

- The commodification process is also constructed around different power relationships between tourists and locals. Further there is a range of factors identified within the literature, which influence the extent of tourism impacts, including the degrees of cultural distance between tourist and locals, and the types and numbers of tourists as expressed through tourist intensity rates.

- A number of processes of sociocultural exchanges have also been identified, although many of these are strongly contested. However, what seems most likely is that local situations and the processes of change are variable – with changes and potential impacts being constantly negotiated in different ways by local communities.

- These processes of negotiation are stimulated by the reactions of local communities toward tourism and tourists. Research from a range of case-studies suggests that the attitudes of local residents are conditioned by costs and benefits across economic, social and environmental implications. Increasing attention has been directed at social-exchange theory and social-network theory to help understand the reactions of local communities.

PART 3 CONSTRUCTING AND RECONSTRUCTING TOURISM PLACES AND SPACES

8 Tourism Places, Spaces and Change

PLACE AND SPACE

Chapter 1 set out the importance of analysing tourism in the context of globalization. One of the features of globalization, as expressed by Harvey (1989a), is that time–space compression has enhanced the significance of space and place rather than diminished it. In this chapter we explore how the relationships between tourism and space/place are shaped and reshaped. First we consider some of more pertinent aspects of production and consumption, especially concentration processes, resort and product cycles, local labour markets, and capital embeddedness. Then we shift our attention to the mode of regulation, and to how this relates to space and place. We examine, in turn, tourism interest groups, the shift from government to governance (including partnerships), and the role of the local state. The following two chapters will explore these themes through more detailed case-studies.

The starting point is the need to distinguish between place and space. There are many different understandings of the nature of space, including the idea that it is the setting for human activities, or that it consists of spatial relations. However, definitions have increasingly tended to be relational, that is they have focused on the ways in which space is understood, used and produced. This takes us towards the concept of place. Hudson (2001: 5) states that 'recognizing a distinction between space and place introduces a greater complexity . . . focused on the ways in which (re)producing places created by socialized human beings with a wider agenda than simply profitable production relate to industrial (dis)investment'. In other words, echoing the arguments in Chapter 1, although the starting point of this volume was a political-economy perspective, this does not mean that everything can be reduced to material relationships. Instead, production and consumption are infused with culturally symbolic processes, which are territorially embedded (Thrift and Olds 1996; see also Chapter 7). Places have to be viewed in a context of material relationships, but are not reducible to these. Moreover, places are complex mixes of material objects (the outcome of previous rounds of investment – see Massey 1995), companies, workers, local civil societies, the local state with the co-presence of other forms of the state, and all kinds of practices, values, and multiple identities.

Turning to tourism, Gordon and Goodall, deliberately echoing Massey's views, comment that 'tourism places are shaped by the sequence of roles which each has played in the spatial division both of tourism and of other economic activities' (2000: 292). For them, tourism interacts with place characteristics, and is both place-shaped and shaping. Moreover, tourism places are complex mixes of: the material objects produced by past investments in facilities, such as piers, marinas, promenades and parks; various forms of tourism and non-tourism companies; host communities (encompassing those who do and do not work in tourism); the local state (where tourism departments interrelate with non-tourism departments, such as education and housing); and various tourism and non-tourism practices. Over time, even green-field sites (completely new centres) of tourism production develop place identities, which mean that they are far more than centres of production. Local populations and tourists inscribe them with values, while places contribute to identities. In other words, the people who live there have relationships and identities other than those stemming directly from tourism production. And the tourists who visit such places are informed not only by immediate holiday motivations, but also by wider sets of values and identities. Some of these complexities have emerged in the discussions of the socio-psychological and sociological literatures on tourist motivation and behaviour (Chapter 6), and on host–guest relationships (Chapter 7), but the focus on place allows us to locate these in the context of wider material and non-material relationships.

Places are open, not closed, and the degree of openness has changed with globalization. Two important points follow from this. First, places do not exist as such but are actively constructed by social processes, including tourism. While the underlying dynamic is capitalist, many social processes are at work (Harvey 1996). These do not coincide conveniently so that neat lines cannot be drawn around places; rather they are discontinuous and overlap (Allen et al. 1998). Hence, when moving from abstract to concrete levels of analysis, we should not expect to be able to define places as discrete entities. Indeed, that would kill the very notion of place. This links to the debate about the definition of tourism destinations. For Davidson and Maitland a tourism destination is 'a single district, town or city, or a clearly defined and contained rural, coastal or mountain area' (1997: 4) with shared characteristics. This is a highly pragmatic definition, responding to the needs of concrete analysis. But the notions of 'clearly defined' and 'contained' are problematic, if tourist destinations are also viewed as places, which are open and at the centre of diverse social processes.

The second point is that places are constantly changing over time, through both their internal dynamics and the manner in which these interact with external and increasingly globalized processes of change. Places, or more precisely local communities/societies and their political forms, are not passive in these processes of change. They negotiate their engagement with tourism, albeit in the context of unequal power relationships, and they contest their roles in the world (see Chapter 7). As Amin and Thrift argue, 'local initiatives structure responses to processes of globalization and themselves become part of the process . . . of globalization' (1994: 257). Local actors shape not only the economic trajectories of places, but also their social and cultural trajectories. Of course, places are not homogeneous and are not necessarily harmonious. The vision of any place's role in the world is contested, as is illustrated by the shift from growth-machine to growth-management politics (discussed later in this chapter). People or social groups can be organized and energized to campaign around particular place identities and to contest their futures. This may take different forms – exerting pressure for more, or different, state intervention (at different levels), through private investment, or voluntary or community initiatives. But the capacity to organize such resistance partly depends on place identity, as well as on resources (including human and material ones, and social capital).

In most places, the local state is at the heart of the response to change. The challenge for the local state is threefold. First, it has to respond to competing demands. This is problematic because the demands of tourism capital may not be the same as those of non-tourism capital, and both may differ from the demands of different groups of residents. Tourism capital may demand further investment in tourism infrastructure, but this may be resisted by non-tourism capital concerned about the increased competition for labour. The views of local groups are likely to depend on whether they are directly employed in tourism, whether they are retired, where they live, and their alternative visions of the future of that place (see Chapter 7). Second, as discussed in Chapter 2, globalization – especially its implications for competition and the mobility of capital – has challenged the role of the state as a site of regulation (of places). Third, the power of the local state will also be constrained by the historic power compromise between the local and national state. There are vast differences between the powers residing with the communes in Switzerland compared with those held by the more emasculated entities of the local state in the UK.

The trajectories of change, encompassing the shifting relationships between tourism and place, are represented in idealized form in Figure 8.1. It posits two starting points: the tourism-dependent place (resort) and the non-tourism place. Over time, places can respond to the challenges to restructure by diversifying away from or increasing their engagement with tourism activities. Type A is the classic tourist resort, established on a green-field site. When faced with a crisis in the tourism industry, it can seek to reinvent itself as a tourist resort (A1), or develop other functions (A2), such as becoming a dormitory for commuters to nearby cities or a centre for retirement migration.

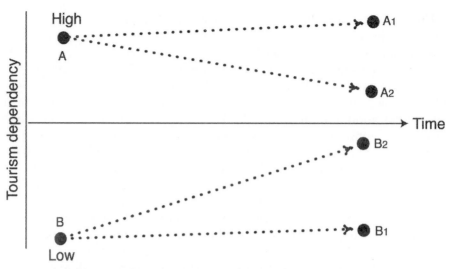

Figure 8.1 **Tourism places: trajectories of change**

Type B is typified by the single industry, manufacturing town which, in the face of an industrial crisis, can restructure its economic base through developing either new non-tourism industries (B1) or seek to develop tourism (B2).

Types A and B both represent mono-industrial structures, and in moving towards the horizontal axis they become more-diversified local economies, or multi-industrial structures. In reality, most places occupy positions somewhere between these polar types. However, there are some places – high-altitude ski resorts, tropical island resorts, etc. – that do approximate to type A. These have been specifically designed as centres of tourism production/experience (see Chapter 9). Environmental and economic conditions make it difficult for other activities to take root in these places, other than to serve those directly working in tourism. But, even in these instances, it would be wrong to see these places as totally tourism dominated. Tourism is only rarely grafted onto a blank social page. Usually, there are inherited legacies – in land ownership, in the built environment, and in social relationships. For example, even the high-altitude ski resort may be built on land where there are historic summer grazing rights, or forestry, or rights of way for hiking in the summer. And the migrant workers who come to live and work in such places bring with them particular cultural baggage, constituted of values and identities formed elsewhere. Also, over time, place identities are created even in new green-field resorts. The people who live there have relationships and identities other than those stemming directly from working in tourism (see also the notion of occupational communities, Chapter 3).

The succeeding chapters will explore, through concrete examples, the many ways in which tourism and places and spaces are interrelated, and how these relationships change over time. The remainder of this chapter explores the general framework of these changes.

PRODUCTION, CONSUMPTION AND CHANGING PLACES

Regimes of accumulation, understood as 'a systematic organisation of production, income distribution, exchange of the social product, and consumption' (Dunford 1990: 305) are essentially national systems (Chapter 2). However, the local manifestation of the regime of accumulation will be shaped by place characteristics, such as the history of entrepreneurship, class conflict and unionization. Tourism contributes to these local characteristics, depending on the extent to which the local economy is diverse or tourism-dependent.

The tourism contribution is likely to be most marked, of course, where it dominates local production. This raises the issue of clustering or concentration as a feature of many tourism localities. There are three general causes of concentration/specialization, in the words of Gordon and Goodall:

- the comparative advantage arising from inherited local/accessible resources

- scale economies in the provision and use of key items of infrastructure, notably transport links or terminals, but also major attractions

- economies of scale and scope in the operation of tourist services – and key supports such as mobilization of a suitably skilled labour force, and place marketing – which may be achieved either internally within large monopolistic enterprises, or externally through agglomeration of related businesses (2000: 296)

Some locations (e.g. at the seaside) or landscapes provide comparative advantages, which attract clusters of tourism investments. These are reinforced by scale economies in infrastructure: for example, railways and air terminals are critical elements in tourism scapes, that contribute to the channelling of tourism flows to particular destinations. The impact of the car is perhaps 'concentration-neutral', facilitating both dispersal and concentration. There are also economies of scope, evident in the way that large resorts are able to invest in massive marketing campaigns to reinvent their place images. The economies of scope may also extend to the creation of a pool of experienced labour and networks of producer services, which support tourism production, as well as 'unmarketed interdependencies in relation to information dissemination, product reputation, labour availability and the ambience of the place' (Gordon and Goodall 2000: 297). Part of that ambience is created by the spatial concentration of tourism consumption of course. The presence of other tourists is a marker of the significance of a tourist site, and their practices also contribute more generally to the constitution of places.

Tourism concentrations are not static. Their comparative advantages may change over time, perhaps because of shifts in consumption (changes in tourist preferences) or because of the inherent contradictions of the process of accumulation: 'over-development' may make the destination less attractive because of noise, pollution and the sheer pressures of the increasing numbers of tourists. The nature of host–guest relationships, and the complex mix of tourist motivations, mean that in tourism – more than in most other sectors – there is a need to consider both the material and the non-material relationships that underlie these changes. Product and resort cycles provide a starting point for disentangling how material (capital, labour, technology, etc.) and non-material (cultural, motivational, socially interactive, etc.) factors influence changes in spaces and places.

Product and resort cycles

The product cycle rests on the proposition that any product has a limited life. This is related to technology changes, fashion changes, and the processes of (non-paradigmatic) competition, which rest on, among other things, product differentiation and innovation (Chapter 4). The growth and decline of sales follow a characteristic bell shape, and the marketing literature tends to divide this into four phases: development, growth, maturity and decline (Kerin and Peterson 1980).

Inevitably, a number of problems are encountered when trying to apply the product cycle to concrete tourism examples, and here we highlight two of these. First, the classic bell-shaped distribution of sales over time (Figure 8.2) may not be encountered in reality. The speed of technological change means that product cycles are becoming shorter and more peaked. Massive marketing may also mean that sales peak almost instantly, followed by a gradual decline in sales, as exemplified by blockbuster movies. Second, there are difficulties in defining discrete products; or more bluntly, when does product modification become a new product? For example, a number of versions of a particular car model may be produced, and a number of editions may be produced of a book. In some cases, the modifications are so minor that they constitute merely extensions to the existing product. But they can be substantial enough to signal the introduction of a new product.

Both problems are encountered when applying the product cycle to tourism. First, the distribution of sales can follow a number of patterns (Figure 8.2). Rather than the classic bell curve (A), there may be a short and highly peaked curve (B); classically this represents event tourism, such as the Olympic Games or the Millennium Dome in London. Where a tourism attraction has received massive media attention or promotion before opening, the curve may have an asymmetrical form (C); arguably this applies to the Eden Project in the UK, although it is too early to know this for sure. And finally, the product may be subject to constant renewal, so that its life is extended repeatedly (D). Agarwall (1994) argues that this latter trend applies to resorts, because the coalitions of local interests are unable to accept the consequences of decline. These are only a sample of the possible product cycles. Further complexities enter the analysis of product cycles, and their

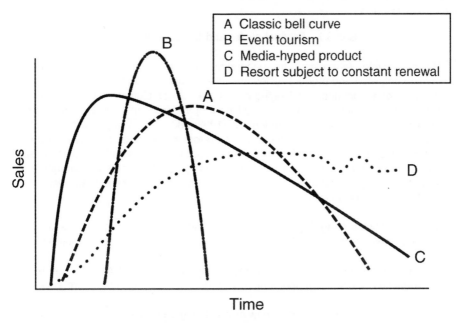

Figure 8.2 **Tourism product cycles: contrasting examples**

shapes, when interactions with the business cycle are taken into account (Box 8.1). The shape of the curve is of more than academic interest, however, because of the consequences for how, where and when the negative and positive impacts of tourism are distributed (Chapter 7).

Second, defining the tourism product is problematic because of constant adjustments, e.g. does the addition of a new ride in a theme park mark the start of a new product cycle? It does in terms of the individual ride, but not if the product is considered to be the theme park as a whole. This highlights the way in which many individual products combine to produce the tourism experience. The same problem is found when considering resorts. Are tourists attracted to a particular resort because of its place characteristics, or because of individual attractions within the resort? This issue leads us to the concept of the resort life-cycle.

Although there have been a number of attempts to analyse the resort life-cycle, this literature is still dominated by Butler's (1980) model (Figure 8.3). It proposes that the development of a resort passes through a number of stages of growth from exploration to consolidation, followed by stagnation and decline. The decline stage, however, is not deterministic and places can adopt rejuvenation strategies with varying degrees of success. The model rests on a number of assumptions:

- *Tourist demand*: this will grow steadily from a low base, until the power of the attraction declines.

- *Capital*: there is a shift over time from local to external capital, with an associated redistribution of power.

Box 8.1 Product cycles and the business cycle

Business cycles (changes in the general business environment – in confidence, investment, etc.) and product cycles are linked. However, as Haywood argues, in the case of accommodation lodging cycles are not completely in synchronization with economic cycles. . . . They are driven by periods of imbalance between the growth in room supply and room demand (1998: 280). The swings in the business cycle contribute to overall demand and therefore to the sales for any individual product. But the two sets of cycles are not coordinated. Essentially, there is so much fixed material capital (and probably human capital, in terms of management and skilled employees) locked into a particular hotel that there are lags between the two. This has several implications:

- The removals of ageing stocks of rooms and hotels is very slow, despite the clear messages sent at the low point of the business cycle about their future.

- The effects of changes in average room prices to eliminate short-term surpluses or deficits of space are constrained – occupancy rates are sticky.

- Developers persist in starting new projects long after average room rates have started to fall during the business cycle.

Source: after Haywood (1998)

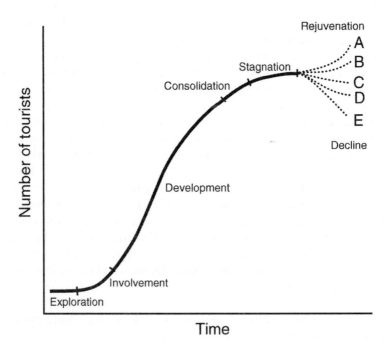

Figure 8.3 **The tourism resort cycle (source: Butler 1980)**

- *Host–guest relationships*: these tend to deteriorate in response to the growth in tourist numbers.

- *Reflexivity*: places, or at least the communities that inhabit them and their political organizations, can contest the future trajectories of the resorts.

The model has been subject to exhaustive debates and critiques, many of which are summarized in Haywood (1998). In part, these echo the earlier critique of product cycles. First, there is the problem of the *definition of the tourism area*: decentralization of facilities to the urban fringe may simply mean that growth has been displaced beyond the formal borders of the resort, rather than the resort being in decline. Second, a resort *may miss one or more stages* in its development, with consequences for the distribution of sales over time, and the shape of the curve. Third, *how is tourism to be defined*? If retirement migration linked to earlier tourist visits is included, then the total tourism complex may continue to grow long after the narrowly defined tourism product has stagnated. Fourth, *tourism markets are segmented*, and while one segment may be in decline, others may be in the growth or consolidation phases. Fifth, *the critical turning-points* are ill-defined, which makes it difficult to apply to concrete analyses.

Despite these limitations, there have been a number of attempts to apply the model to the evolution of individual resorts. Table 8.1 summarizes Gonçalves and Aguas' (1997) analysis of the evolution of the Algarve in Portugal. Putting aside reservations as to the extent to which the Algarve can be considered a single resort, let alone a product, it does illustrate the general descriptive value of the model, while exposing the limitations of using it mechanically and of data constraints. The model is perhaps most useful in indicating the different policy challenges faced by resorts, investors and local communities, at different stages of development. Some of these policy issues relate to the way in which the mode of regulation is manifested in particular localities, and we return to this theme later in the chapter. Before that, we turn to the distinctiveness of the labour process in particular localities.

Local labour markets

The labour process was discussed in Chapter 3, focusing on recruitment and the organization of work within the firm. Here we focus more on the external labour market, although the two aspects are related. For example, the labour process will be informed by the availability of particular types of workers in the local labour market (either immediate, or taking into account migration into that labour market). Hence the availability of skills, willingness to take casual or part-time jobs, traditions of collective labour organization, and prevailing wages are all important. Similarly, gender, and migration-based and other social divisions will inform labour-market segmentation, both externally and within the firm. The alternative sources of employment in local labour markets are also

Table 8.1 **Application of the resort lifecycle to the Algarve**

Product Life Cycle stages and indicators	Facts about the Algarve
Exploration Small numbers of allocentrics Little or no tourist infrastructure Natural and cultural resources	*1920 to 1960* Nationals, Spaniards (Andalucia) and British Absence of infrastructures Good weather and beaches
Involvement Local investment	*1960 to 1970* Construction of the first hotels – 1960–65 Vilamoura project begins – 1966
Public investment in infrastructure Emerging tourist areas	Opening of Faro Airport – 1965 Example: Albufeira changes from a fishing village to a tourist resort
Advent of a tourist season	High season – summer
Development Rapid growth in the number of tourists (number of room nights) Visitors outnumber residents	*1970 to 1985/90* Room nights grow from 1,040,000 in 1969 to 7,399,000 in 1985 366,911 guests in 1972 compared with a host population of 258,040 in 1970
Well-defined tourist areas	Albufeira, Lagos, Monte Gordo, Praia da Rocha, Quinta do Lago, Vale do Lobo and Vilamoura
External investment	Example: Quinto do Lago, Pine Cliffs, Four Seasons, Trafalgar House, Bovis Abroad, and Forte hotels
Manmade [sic] attractions development	Examples: Aldeia das Açoteias (athletics), Vilamoura Marina, water parks and golf courses
Mid-centrics replace allocentrics	Increase in the number of packages
Consolidation Slowing growth rates	*Post 1985/90* Growth in annual number of overnights in hotels: 1969–84 = 12.7%; 1985–90 = 5.5%
Developing of new markets; attempts to overcome seasonality Residents approve of the importance of the activity	Defined as priorities Tourism is considered vital to the region (since the 1970s)
Stagnation/Stabilization Peak visitor numbers reached Capacity limit reached Resort image divorces from the environment Areas no longer fashionable Heavy reliance on repeat trade Low occupancy rates	 Probably not reached yet Not yet Very limited spots For some Not yet Decrease in rates because of excessive supply compared to the evolution of the demand
Rejuvenation Complete renewal of the attraction	 No signs
Decline Decrease in markets A move out of tourism activities Tourist infrastructures are replaced	 No signs No signs No signs

Source: Gonçalves and Aguas (1997).

important, as are the possibilities of combining part-time or casual tourism work with other paid and unpaid work, for example seasonal farm work.

Simms et al. (1988) have analysed the relationships between internal (to the firm) and external labour markets. They focus on the notions of strong versus weak internal labour markets. Strong internal labour markets are characterized by a high-skill specificity, clear criteria for recruitment and promotion, and continuous training. These tend to be relatively closed to external labour markets, because of the hierarchical nature of the labour process, the clear progressive pathways to promotion, and the development of human capital. In contrast, weak internal labour markets are characterized by a low-skill specification, limited on-the-job training, and informal recruitment and promotion procedures. They are relatively open to external labour markets, because of the low barriers to entry, lack of internal promotion opportunities, and high rates of labour turnover. Tourism firms tend to be characterized by weak labour markets, and are therefore relatively open to external labour markets, which consequently are of particular significance. However, there are some exceptions, such as airline pilots, who work in relatively strong internal labour markets.

Turning to the characteristics of external labour markets, a number of elements are important. First, one of the distinguishing features of local labour markets is their degree of openness to migrants, whether temporary or permanent, and of national or international origin (Williams and Hall 2000; 2002). This has particular importance for tourism, because of the rapid growth of, and often isolated (enclave) nature of, many tourism destinations. The extent to which these generate migration rather than reliance on local labour depends on four main considerations: the scale of demand; the nature of demand in terms of skill and educational requirements; the speed of development (and whether labour can be attracted from other sectors of the economy, or from non-waged sectors such as household work); and the degree of enclavism and hence the availability of local labour reserves.

Migration serves to fill absolute shortages in local labour markets, as well as shortages of particular skills. For example, by the late 1980s, migrant workers accounted for some 30–40% of chefs, 65–75% of skilled waiters, and 20–25% of hotel/club managers in Australia (Cooper 2002). This has a number of implications. Potentially, it creates greater fluidity in labour markets and may depress local wage rates, at least in the short term, although it stimulates economic growth and therefore wages in the longer term. It may also facilitate weak labour markets, and it is no coincidence that many major tourism destinations have high rates of in-migration.

Care must be taken, however, not to exaggerate the extent of labour mobility. While high rates of in-migration are characteristic of labour markets in large resorts and major cities, this does not necessarily apply to smaller resorts, or to diffuse rural tourism. Moreover, the uneven power of capital and labour rests partly on the supposedly greater mobility of the former. In contrast, labour tends to be more tied to particular places by localized social networks, knowledge, and simply inertia. Storper and Walker memorably have written that the local embeddedness of labour means that there is a 'fabric of "communities" and "cultures" woven into the landscape of labour' (1983: 7). It can be added that they are also woven into the landscapes of large swathes of capital as well. Many owners of firms are tied to particular localities not only by the social ties of community and culture, but by the economic logic of their knowledge of local economies and their networks of 'untraded interdependencies'. Therefore, many – perhaps most – firms have to adapt to local labour markets, irrespective of whether or not these are characterized by high levels of in-migration.

The effects of migration, and indeed the general characteristics of external labour market, are also influenced by the mode of regulation. This operates in different ways. In most of the more developed countries, formal regulation of labour markets (minimum wages and hours, etc.) are determined at the national level, although actual wage levels may be determined more by prevailing local wage levels (Riley et al. 2002; Doherty and Manfredi 2001).

Trade unions play an important role in the regulation of external labour markets. In general, their role depends on the economic context, tradition and culture, institutions and political parties (Martin et al. 1994). While concrete evidence about the role of trade unions in tourism is thin, they are generally assumed to be relatively weak, because of a number of reasons. The fragmented nature of production means that firms tend to be relatively small (with personalized and often paternalistic, rather than formalized, management–labour relationships), the temporality of demand leads to intense pressures to use informal and casual labour, and migrant workers tend to be weakly unionized. Royle's work on McDonald's restaurants provides one of the most detailed case studies of the reasons for low unionization in the tourism industry (Box 8.2).

While there has been little concrete research on trade unions in tourism, the scant evidence available confirms the picture of weak collective organization. For example, in the UK Doherty and Manfredi (2001) found that trade unions were rarely present in hotels in either Blackpool (a major resort) or Manchester (a diverse urban economy, with strong traditions of collective organization). Go and Pine (1995) report that in the USA, 12% of hotel and restaurant workers were unionized compared with 20% in the labour force as a whole in early 1980s. And Milne and Pohlmann (1998) found that in Montreal 5% of small hotels, 32% of medium hotels and 100% of large hotels were unionized. A note of caution is required, however, in interpreting these data. Baldacchino reminds us that to focus on trade unions is to neglect the significance of 'all other forms of labour organisation and action, whether individual or collective' (1997: 49). For example, in 2001 (largely) self-employed bus drivers, in the Balearic Islands, won significant improvements in their contracts after a short, well-organized strike, mostly outside the framework of formal unionization. Collective action by non-unionized workers can also win significant wage and working-condition improvements from the owners of individual firms.

Box 8.2 Working for McDonald's: reasons for low unionization

One of the most detailed studies of unionization relating to the hospitality industry is Royle's study of McDonald's. Of course, many – if not most – McDonald's outlets serve mainly local markets, but others are located in centres of tourism. Royle considers that low levels of unionization are a function of general conditions in the hospitality industry as well as the particularities of working for McDonald's.

The general features of the hospitality industry are summarized as follows:

> The often geographically dispersed, small unit, temporary, part-time and low skills base of the jobs in the wider hospitality industry have typically fostered high levels of labour turnover. These factors, together with the employment of ethnic minorities, young and female employees, make union organisation very difficult. (p. 109)

Moreover, unions have not always been willing to focus resources on building up union organizations within the hospitality industry because of these perceived difficulties.

In addition, factors particular to McDonald's include:

- The McDonald's culture, which encourages managers to regard unions as a threat to good management practices.

- Franchisees' unwillingness to attract criticism from the corporation, which is strongly antagonistic to unions.

- A lack of interest among many workers, who are young (lacking in industrial experience) and who see their employment as only temporary. But many other workers either feel intimidated by the corporation or simply are ignorant of their rights.

Source: based on Royle (2000, Chapter 5)

Capital embeddedness

One of the distinguishing features of tourism production is the extent to which capital is locally or externally owned. This links to three major strands of theoretical debate concerning tourism. The first concerns globalization of production and the continual global scan by capital for profitable locations (Chapter 1); this in turn links to questions about the shifting locus of power between the national and the global (Chapter 2). The second relates to the power of transnational corporations, especially in terms of mediating economic relationships between cores and peripheries (Bianchi 2002). And the third strand concerns local economic linkages. There has been considerable research on tourism multipliers, which effectively measures the extent to which tourism expenditures create jobs and income within a particular locality, rather than being lost through 'leakages' to the external economy (Archer 1982). This has now been widened into concern with the broader concept of embeddedness, defined by Turok as the 'creation of a network of sophisticated interdependent linkages which facilitate growth in endogenous firms' (1993: 402).

Embeddedness is a much broader concept than the multiplier, for it also emphasizes the quality of local linkages, their endurance, and whether they involve mutual cooperation and networking. The debate is particularly polarized where the ownership of capital is not only external but also foreign. One of the key issues is additionality. Does external capital represent additional investment that would not otherwise have been forthcoming in a particular locality? Or is it a form of substitute investment, displacing local capital either directly, through mergers and acquisitions, or indirectly, by truncating opportunities for local investors? If the external investment is in lieu of local investment, then the issue becomes whether alternative use is made of this local capital. Is it invested in other local tourism activities, in some other sector of the local economy, or is it diverted externally? The answers to such questions necessarily depend on concrete analyses of particular places, but below we set out a summary of the major advantages and disadvantages.

The economic advantages of foreign direct investment (FDI) for a locality may be as follows:

- Foreign intermediaries, such as tour operators, who have specialized marketing knowledge, provide market access. This opens up not only larger markets, but potentially higher income markets.

- Inward capital flows can fill gaps in local or national capital markets. Moreover, FDI usually consists of more than a financial transfer, being accompanied by embedded knowledge, including technology. Capital for tourism development may be in particularly short supply in less developed countries (LDCs) or less developed regions, or where there has been historical antipathy to tourism as a 'non-productive' or 'candyfloss' industry (Williams and Shaw 1988).

- Local employment is generated, although this depends on the degree of additionality and the extent of intra-company transfers of managerial and skilled workers.

- Income is created, although this depends on additionality. Company policies are also important. Does the external investor simply wish to exploit low-cost local labour, or does corporate policy require that it offers minimum wages and conditions higher than those prevalent in locally owned companies?

- FDI, almost by definition, has a direct impact on the current account of the recipient country, and may also contribute to attracting foreign tourists, whose expenditures further contribute to this. But this is balanced by 'leakages' in the form of profit remittances, and possibly higher levels of external sourcing of inputs because they are centrally controlled (Chapter 4).

- Technology and knowledge transfers. FDI in tourism can be a vehicle for the transfer of capital, such as advanced IT systems, innovations in hotel construction or new forms of sporting/entertainment equipment. Much of the knowledge transfer is tied up in the company transfers of human capital. Baldacchino (1997) considers that one of the beneficial effects of FDI, therefore, is the creation of greater diversity in tourism management practices, which may have a demonstration effect for locally owned firms.

- External investment makes a fiscal contribution to the state, although its real impact is subject to the additionality issue, the capacity of the state to collect taxes efficiently, and the distribution of such revenue between the local and the national states. Some local states are empowered to levy local taxes, including tourism taxes.

- FDI can contribute to regional economic convergence if it favours less developed regions. This rests on the argument that remoter and more inaccessible areas are the preferred destinations of tourists driven by a desire to seek peace and solitude, and to escape the pressures of urban-industrial life (Williams and Shaw 1998b; see also Chapter 6).

The disadvantages of FDI are in many ways inverse interpretations of relationships and condition that are considered advantages by other commentators:

- FDI potentially increases corporate power in relation to the state or local communities, and is therefore a constraint on the capacity of the latter to mediate their engagement with tourism (see Chapter 7). This is particularly pronounced where the foreign company has a degree of control over access to markets, or to local tourist attractions. Britton (1989) has conceptualized this in terms of a three-tier hierarchy, with capital accumulation flowing up and control flowing down the hierarchy. The transnational company sits at the apex of the hierarchy, and its relationships with local capital are seen as exploitative. Some national states have responded by trying to develop national champions to challenge the power of transnational companies, for example national airlines. This needs to be seen in relation to the discussion of globalization and the limits of national regulation (Chapter 2).

- Capital has become increasingly mobile, perhaps hypermobile. It is involved in a constant global scan for profits, and may rapidly relocate as the landscape of profit changes. Places become highly vulnerable to external decision-making.

- Foreign capital may be less embedded than local capital, although as noted in Chapter 4 (Box 4.6) this is questionable. There is strong evidence that many economies have become dependent on external sourcing. For example, in Fiji 53% of the food consumed by tourists, 68% of hotel construction, and 95% of tourist-shop goods are imported (Varley 1978). It is not clear from this study whether transnational companies are more or less likely to import goods. However, Britton (1989) shows that where package holidays use only foreign airlines, the destination countries receive only 45% of the retail price of the inclusive tour; where the hotel is also foreign-owned, only 25% is received in the destination country.

- External investment can result in regional economic divergence. This is based on the argument that transnational companies in LDCs favour core regions because of the availability of infrastructure, skilled labour, and access to domestic markets. In practice, it depends on the tourism product of course, but the fact that capital cities such as Bangkok are global tourism destinations reinforces this argument.

In practice, there are complex patterns of embeddedness. Any assessment depends on corporate strategies, the level of sophistication in the local economy, and the ability of local capital and labour to respond to opportunities. Thus, Dwyer and Forsyth (1994) argue that even if all tourism assets in Australia were in foreign ownership, this would result in an increase of just 10% in the income lost through international leakages. The leakages in less developed, and smaller, economies are likely to be far greater.

Much of the debate about the embeddedness of external investment in LDCs has centred on the relationship between tourism and food production. Telfer and Wall (1996) argue that these relationships can be put on a continuum from conflict through to coexistence. The two key issues are the extent to which there is conflict between tourism and agriculture over the use of land and labour, and the degree of sourcing of food from local agriculture as opposed to external suppliers. Bryden (1973) takes a largely negative view of the effects of tourism, but others, for example Hermans (1981), argue that labour has been drawn from agriculture for a number of reasons, including a general decline in the prestige of such work. Cox et al. (1995) argue there is a trade-off: tourism increases the costs of agricultural production (via land and labour markets) but creates markets for non-traditional higher-value products, and leads to infrastructure improvements. In reality, of course, there are likely to be changing relationships, and Lundgren's classic model (1973) proposes that the degree of local linkages will increase over time. Simpson and Wall (1999) demonstrate how complex these linkages can be (Box 8.3).

The extent to which external capital is embedded also depends on how it is inserted into a local or national economy. In part, this is an issue about the organization of capital. For example, Sadi and Henderson (2001) report that there have been four main types of (tourism) FDI in Vietnam (Table 8.2). These are characterized by different degrees of property rights, creation of a separate legal entity, and control. These clearly have very differing implications for embeddedness, both in the short and the long term, as well as for the degree of external versus local control. The choice of organizational form

Box 8.3 Local economic linkages: resort hotels in Indonesia

Simpson and Wall (1999) demonstrate the complexities of the local embeddedness of external investment. They compared the development of two resort hotels, the Paradise and Santika, in Indonesia. They have had very different economic, environmental and cultural impacts:

Paradise hotel:
- Distant from existing settlements

- Local employment benefits mostly limited to the construction phase

- Lack of training opportunities, so that few local people secured jobs in the hotel

- Local people lost land they had traditionally rented, with little compensation in return: this reduced their capacity to grow food for themselves, or to to sell to the hotel

- Little consultation with local people, who were not engaged with the aims or the opportunities provided by the project

Santika:
- Located in the village and therefore easily accessible for potential workers and for tourists

- Local people had mostly owned the land that the hotel was developed on, and received compensation for its loss

- The hotel provided local job opportunities

- Training opportunities were provided to help recruit young local people

- Local people were able to sell goods and services to the hotel

There were four main reasons for the differences in how external capital was embedded in these places:

- Location: enclavism was negatively associated with embeddedness.

- Property rights: residents in Santika owned the land they farmed in the area, and were able to extract some of the benefits of its commodification for tourism purposes.

- Human capital: there were higher levels of education in Santika.

- Human agency: the attitudes of the hotel owners to local economic development differed.

Source: based on Simpson and Wall (1999)

Table 8.2 **Different forms of foreign direct investment in Vietnam**

Type	Property rights	Legal entity created	Control
Contractual business cooperation	Blurred	No	Blurred
Joint ventures	Yes	Yes	Joint
Sole owned ventures	Yes	Yes	Complete
Build and operate ventures: transfer of contracts	Leased	Yes	Complete (over fixed period)

Source: based on Sadi and Henderson (2001).

is itself partly shaped by the mode of regulation, at the national level and in its manifestation at the local level.

REGULATION AND CHANGING PLACES

The mode of regulation was discussed in Chapter 2, and was understood in terms of 'a specific local and historical collection of structural forms or institutional arrangements within which individual and collective behaviour unfolds' and a system of coordinating individual decisions (Dunford 1990: 306). These 'arrangements' provide stability and ensure the reproduction of economic systems, and this book contends that the national remains the principal site of regulation. However, in the face of the neo-liberal agenda, there has been some withdrawal of the national state from economic intervention in recent years in developed, newly-industrializing and less developed economies. Local and regional states have partly expanded their roles to fill this gap, especially in the more developed economies, although there are a number of reasons for this, including: legitimation, making more effective use of economic resources, and responding to local and regional political pressures and social needs. The local manifestation of the mode of regulation does, of course, extend beyond the role of the local state, but this provides our starting point.

Local communities/places are not passive in the face of globalization and other challenges. They contest their role in the world, and the local state is instrumental in this. In pluralist democracies, the local state does not represent a single set of interests, but is the site of struggles between competing views and groups. The extent to which individuals or social groups can be organized and energized around particular issues depends on the nature of the challenge, the resources of local communities (their human, social and material capital), and the strength of place identities. It also depends on their interaction with the local state itself. There is a considerable debate about the nature of the state in capitalist societies, which we will not pursue here, except to assert the following: first, that the state is an arena where conflicts between capital (in fact different fragments of capital), labour, and other interests (e.g. environmentalists) are played out and attempts are made to reconcile these; second, that the (local) state has its own powers and agendas. This does not mean that it is autonomous or independent, for interest groups do have access to the levers of powers within government. But equally, it does not mean that the local state is simply reducible to the outcome of the competition between different interest groups.

Tourism interest groups

Greenwood confirms the role of interest groups in the mode of regulation, arguing that they can be 'stable mechanisms for negotiation, bargaining and the resolution of conflict' (1992: 237). Furthermore, he argues that in advanced capitalist economies there are three types of settings for interest group activity. In the pluralist setting, a weak state is unable to arbitrate between competing interests. In the neo-pluralist, clientelistic policy communities are formed between government and tourism interest groups. And in a neo-corporatist context, interest groups play a strong role in policy formulation and implementation, constituting key components in 'private interest government'. This is increasingly important as the state withdraws from tourism policies and looks to the private sector and non-governmental organizations (NGOs) to fill the gap.

These settings vary among countries, within them (place differences), and over time. Given the growing recognition of the role of tourism in the economy, as well as its sociocultural and environmental impacts, the setting for tourism tend to be either neo-pluralist or neo-corporatist. This leads to the question of the differential power of interest groups within the arena of the (local) state. The strength of interest groups depends on a number of considerations:

- Style of government/governance. The traditional view holds that there is 'sub-government' constituted by a triangular structure of legislative committees, executive agencies, and interest groups (Hall and Jenkins 1995: 57–8). The key issue is the extent to which these triangles are open to all interest groups, or have been 'captured' by particular groups. Are they relatively open to the issues raised by other interest groups, or are they locked into highly selective, even clientelistic, relationships? And when does consultation with some groups become a closing-up of the policy process to other groups (Hall and Jenkins 1995)?

- Resources. The size of the group, its relevance to its membership interests, its past effectiveness, its finance and other resources, and the sanctions at its disposal all influence the extent to which it can mobilize its members, and the pressure it can exert on the state.

- Type of interest group and competing interests. Lindblom (1980) argues the primacy of producer interest groups over consumer interest groups because of two factors: first, the likely coincidence of their interests with those of the government on issues such as growth, employment creation, income, and foreign exchange earnings; and second, producer interest groups can draw on the often vast resources of large corporations. While this view has some weight, it is clear that well-organized and determined consumer, environmentalist and other groups can also be influential, either through direct action or via their sheer numbers and potential electoral power. The anti-globalization protests of the late 1990s and early 2000s bear powerful testimony to this.

In general, tourism interest groups have been relatively weak compared with other sectors for a number of reasons. First, tourism is a composite industry, and tourism activities may be a low priority for some of the companies involved. For example, tourism may be of minor interest to local bus companies, although transport services may be critical to local tourism. The host community itself may be deeply divided in its attitudes to tourism, depending on the degree of dependence on the industry for employment. Second, there are varied interests within the tourism industry, with differences between, say, airlines and tour companies, large and small hotels, and those services providing services to domestic, inbound and outbound tourists. For example, the latter will have very different views on exchange-rate movements, which have contrasting impacts for firms operating in domestic and international markets. This may mean that tourism speaks to governments with plural rather than a single, effective voice. Third, tourism is a fragmented industry (Jeffries 2001), and organizing large numbers of small tourism firms poses logistical problems. Fourth, there tends to be weak collective organization of labour or trade unionism (discussed earlier in this chapter). And fifth, in some tourism regions the owners of capital are motivated as much by lifestyle motivations as by business goals, and therefore will not be motivated to contribute to interest groups, as long as the aims for their personal satisfaction are being fulfilled.

The power of interest groups will vary in different settings, and they need to be investigated through concrete examples. Tyler and Dinan (2001) provide an example of interest groups in England (Box 8.4).

From government to governance

By the 1990s, traditional notions of government were being undermined by three main challenges. First, there had been pressures to reduce the role of the state in capitalist economies. Second, many non-state bodies were involved in the 'management' of particular areas: arms-length government agencies, NGOs, and major companies, among others. Third, increasing numbers of hybrid forms of public–private agencies and partnerships were involved in 'managing' the economy and society. Jessop (1994) explains the emergence of governance in similar but more theoretical terms. He argues that the nature

Box 8.4   Tourism interest groups in England

Context

The state is the fulcrum of the policy network. It has a core executive (ministers, departmental representatives, etc.), and is to some extent autonomous (having its own agendas and goals). The influence of tourism within this fulcrum is partially diffused, because responsibility for tourism is distributed among at least nine different ministries. Yet only one of these has a direct link with the English Tourism Council, which is the principal national agency for tourism policy. Overall policy tends to be fragmented, given the lack of an established mechanism for cross government coordination of tourism policies. Over time, the focus of policy has shifted from employment generation to sustainability, then to competitiveness and quality issues. The policy style that has emerged is one of consultation and reliance on regulation, rather than legislation. This means that the state has been relatively weak in policy-making and implementation, and has instead been particularly reliant on policy networks.

Interest group activity

The author's detailed survey identified the following features of tourism interest groups in the UK:

- The larger, better resourced groups had more collaborative arrangements and a wider range of lobbying activities.

- There is broad overlap in the issues that concern interest groups, but these tend to be approached differently. The more focused groups tend to have greater influence.

- Collaboration tends to be limited within tourism subsectors.

- The groups use both formal and informal means of communication and coordination.

- Key individuals often play a pivotal role within the interest groups, parliament and the Civil Service.

Conclusions

Networks have proven to be the most effective way to coordinate policy given the complexity and diversity of the tourism industry. The most effective strategy for tourism interest groups is to be 'useful' to government in pursuit of its agenda. Effectiveness within networks depends on building trust, having access to information, budgets, and frequent interactions with other key players.

Source: based on Tyler and Dinan (2001)

of the state in capitalist societies has changed from Keynesian to Schumpeterian (see Chapter 2). This has three main characteristics: denationalization (the 'hollowing out of the state' as power moves outwards, upwards and downwards from the national state); destatism (public policy increasingly relies on non-state agents); and internationalization (the influence of extra-territorial actors).

In response to this critique, there was growing recognition of the need to think not of local government but of local governance. The latter involves not only the institutions of local government but also 'institutional and individual actors from outside the formal political arena, such as voluntary organizations, private businesses and corporations' (Goodwin and Painter 1996: 636). Governing also has to be seen as embedded in wider practices (Painter and Goodwin 1995), and therefore cannot be represented by simple models: instead, it changes over time and over space. As Massey (1995) emphasizes, governance stems from the social relations that are constituted in, and constitute, particular places (Massey 1995).

Where does tourism fit into this picture of governance? Gordon and Goodall write that:

> Much of the local governance literature has a distinctly functionalist character to it ... implying both that there are clearly identifiable local economic interests and that the institutional capacity will be forthcoming (from authorities, agencies or partnerships). (2000: 304)

Empirically identifying local economic interests in tourism is problematic, and their influence is constrained, as the earlier discussion of interest groups indicates. Tourism interests tend to be weakly represented in local governance, and tourism partnerships are undermined by lack of trust and shared goals, even among businesses let alone between businesses and other potential partners. And yet the growing importance of tourism to national economies has meant that national and local governments have increasingly sought to build partnerships for local development that are focused on, or incorporate, tourism. Partnerships are often presented as a counterpoint to traditional, hierarchical bottom-up policy-making and implementation. In reality, the weakness of tourism interest groups means that partnerships are often imposed on them through initiatives originating from the national or local state.

Whatever the origins of such initiatives, the outcome is a blurring of the boundaries between the public and the private (Bramwell and Lane 2000). The local state has become more 'enabling' and less directive. This is linked to the idea of the new entrepreneurialism, based on public–private partnerships, 'in which a traditional local boosterism is integrated with the use of local governmental power to try and attract external sources of funding, new direct investments, or new employment sources' (Harvey 1989b: 7). Such partnerships have a number of advantages, including greater capacity to tackle complex problems that cannot be resolved by individual bodies or agencies. This is especially important, because problems and underlying social processes do not neatly follow the boundaries of local government. Moreover, building partnerships that involve the principal interested parties helps to reduce conflicts and speed up implementation. It is also a vehicle for securing political legitimacy for particular projects.

The effectiveness of an individual partnership is, of course, dependent on several considerations (Box 8.5). Perhaps of even greater importance are the limitations of partnerships:

- They reflect inequalities in society, rather than transcend them. Power mediates attempts to influence tourism policy (Hall 1998), and this is as true of governance as of government. Moreover, partnerships may not involve all interested parties, or at least not on equal footings. Jamal and Getz (1995) emphasize that there is a minimum level of skills and resources required to participate in partnerships, so that the weakest and most marginal social groups are often excluded.

- Partnerships may not involve all the important bodies or agencies, e.g. if the key player is a transnational company, the vital decision-makers may be located elsewhere.

- There are problems in sustaining partnerships over the medium and long term because of 'partnership fatigue'.

- Partnerships have only a limited autonomy, and therefore are not immune from shifts in the agendas of individual partners.

- Effective partnerships may need to set real priorities in the use of resources, which may be difficult given the emphasis on consensus-building. Above all, consensus requires an ability for collective learning, and for building trust and confidence (Hall 2000).

One important strand in the research on collaboration is the 'new institutionalism'. Although explicitly concerned with understanding differences in the economic trajectories of regions and cities, this literature also has relevance for partnerships. In particular, there is a well-rehearsed argument that 'institutional thickness' is a key to economic development. MacLeod (2001) summarizes the ideas of Amin and Thrift (1994) and others on this subject. 'Thickening' is enabled through five critical conditions: first, the strong presence of organizations and institutions, including firms, banks, development agencies, trade unions and voluntary agencies; second, the embedding of weak and strong ties between organizations, so that permanent innovation is linked to an atmosphere of trust; third, the existence of a mixture of coalitions and structures of domination, so as to corporatize economic life; fourth, a local society characterized by cosmopolitan openness; and, fifth, a progressive sense of place. This argument resituates partnerships in relation to the debates on place and the nature of local societies. It can also be stated in simpler terms: there is no clear

Box 8.5 Critical factors in the effectiveness of collaboration

Scope of collaboration

- extent to which it is representative of all stakeholders
- extent to which relevant stakeholders can identify positive benefits in participation
- whether the collaboration includes a facilitator
- whether the collaboration includes the stakeholders responsible for implementation
- extent to which individuals representing the stakeholders are fully representative of that group
- the number of stakeholders involved (representation versus group efficiency)
- the extent of initial agreement about the general scope of the collaboration

Intensity of collaboration

- degree to which participants accept that collaboration is likely to produce qualitatively different outcomes
- degree of commitment to consensus-building
- when and how stakeholders are involved – as formulators, managers or receivers of the work of the partnership
- effectiveness of information dissemination and consultation techniques.
- whether participation involves information dissemination or interaction among stakeholders
- nature of dialogue among partners: degree of openness, honesty, trust and respect
- capacity of participants to learn about one anothers' values, etc.
- degree of control exerted by facilitator

Degree to which consensus emerges

- realization that actions will not satisfy all participants equally
- extent of consensus among stakeholders about issues, policies and their assessment
- extent to which consensus emerges across inequalities among stakeholders or reflects them
- extent to which stakeholders accept systemic constraints on what is feasible
- whether stakeholders are willing to implement policies

Source: based on Bramell and Shurma (1999)

model of how partnership should work; they will be shaped by both the nature of particular projects and the contingencies of policy environments. Furthermore, we would concur that the emphasis on institutions, including partnerships, tends to downplay the regulatory role of the state (MacLeod 2001).

The local state and tourism policies

As the previous sections demonstrated, tourism interest groups have been relatively weak. Tourism was sidelined until comparatively recently in overall policy priorities in most developed countries (Jeffries 2001). Instead, the major clashes of interest were mainly among the competing claims of financial services, agriculture and manufacturing. The fragmented tourism industry, often lacking a single overall ministry within government in many countries, had a relatively weak voice. This varied in extent among countries, however; for example, tourism interest groups have been significantly weaker in the UK than in Spain.

Tourism policy is highly variable in focus across both time and space. It is even difficult to pin down exactly what constitutes tourism policy, and Hall, for example, settles for 'Tourism public policy is whatever governments choose to do or not to do with respect to tourism' (2000: 9). Despite these reservations, Hall (2000) and Getz (1987) do identify some of the main approaches to tourism planning. These are summarized in Figure 8.4. Boosterism, or growth maximization, is rarely encountered in advanced capitalist economies, and the other forms are probably found in combination, rather than constituting alternatives. In reality, tourism policy and planning are likely to have evolved piecemeal over time. Moreover, shifts in approaches are unlikely to be linear; instead, they are likely to be contested by different interest groups, and policy directions may shift backwards and forwards. There is also a difficulty in that this schema does not reflect the overall strength of tourism policy, taking into account the impact of the neo-liberal agenda for reduced government.

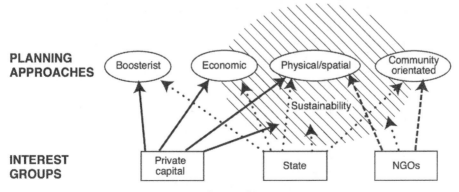

Figure 8.4 **Tourism planning approaches and interest groups**

Table 8.3 **Approaches to tourism planning**

Boosterism	Economic approach	Physical/spatial approach	Community orientation	Sustainability
• Led by short- or medium-term business interests • Uncritical of tourism growth • Aims to reduce obstacles to growth • Invests in promotion and infrastructure • Typified by some mega-events, but rarely found in its 'pure' form	• Motivated by planners using tourism as an economic instrument • Gives more attention to economic than to social and environmental impacts • Aims to maximize economic benefits • Invests in promotion and marketing, and development incentives • Typified by national and urban policies using tourism to generate external income, growth or urban regeneration	• Motivated by the wider land-use planning goals of tourism, and by conservationist pressure groups • Emphasizes land use and, increasingly, environmental planning • Aims to protect or enhance aspects of the physical environment • Uses visitor management techniques • Has policies for pressurised urban and rural environments	• Motivated by aim of recognizing the interests of the local community • Emphasizes community participation and, more generally, bottom-up planning • Aims to satisfy the needs of local communities, with the added value that this will improve the tourism experience • Uses various mechanisms to engage the local community in planning tourism	• Motivated by need to take long-term holistic view • Emphasizes building partnerships of stakeholders, and pursuing economic, social and environmental aims • Aims to balance the needs of present and future generations and of different social groups within generations • Adopts varied measures, but advocacy to the forefront, with emphasis on education of tourists, tourism firms, etc. Is appealing to both idealism and self-interest

Source: after Hall (2000).

It is not possible to review all the myriad types of tourism policy that exist at the local and national levels. Instead, we end this chapter with a consideration of three key dimensions of recent developments: images and place wars; growth machines versus growth management; and sustainability.

Kotler et al. (1983) coined the phrase *place wars* to signify the competition among places for economic development. They argued that localities compete against one another economically, at a global scale:

> All places are in trouble now, or will be in the near future. The globalisation of the world's economy and the accelerating pace of technological changes are two forces that require all places to learn how to compete. Places must learn how to think more like businesses, developing products, markets, and customers. (1983: 346)

There has been the emergence of the 'entrepreneurial state', at least in advanced capitalist states, and – by extension – of 'entrepreneurial local states'. Harvey writes of the emergence of a 'new entrepreneurialism' (1989b), in which traditional 'local boosterism' is combined with the use of local-government powers. He considers that public–private partnerships are often vehicles of this new entrepreneurialism. However, he does not view such partnerships in an idealistic way, but emphasizes that they are less accountable than local government and that the powers of the partners reflect the inequalities of capitalist societies.

One of the most powerful instruments available to compete in these 'place wars' is selective place imaging. Holcomb writes that 'the primary goal of the place marketer is to construct a new image of the place to replace either vague or negative images previously held by current or potential residents, investors, and visitors' (1993: 133). Hall (1998) links this to flexible production – the need to adapt the product to the changing market – but place-imaging is not necessarily tied to any particular regime of accumulation. Place-imaging involves more than just employing public relations and advertising consultants. To be effective, it needs to be linked to real changes in the tourism product – here understood to be the place as viewed, and experienced by tourists. One of the foremost instruments at the disposal of the entrepreneurial local state is the overlap between cultural policies and tourism policies. Hall writes:

> Cultural policies and tourism policies are becoming almost inextricably entwined, while funding for the arts, culture, sports and recreation, and amenity improvements are usually justified primarily in terms of the contribution they will make in economic attractiveness via tourism rather than their social contribution to all the inhabitants of a region. In this light, the institutional arrangements for tourism are perhaps best understood as instrumental arrangements of the local state which serve a narrow range of global and local interests. (1998: 216)

Critics stress the danger inherent in this approach, of reducing places, or at least their presentation, to 'bundles of social and economic opportunity' in competition for investment (Philo and Kearns 1993: 18). There is also the danger of 'serial reproduction', i.e., the creation of too many places and place-images, with similar tourism products chasing limited markets. These policies are often backed by 'growth coalitions' (Harvey 1988), constituted of those who stand to gain directly from developments (see Chapter 10).

The *growth machine* argument emphasizes that the potential beneficiaries from land development – those with property rights over it, investors, etc. – constitute 'local growth elites', who support local-government politicians and bureaucrats, who also see growth as their primary goal (Logan and Molotoch 1987). They push through policies that can be categorized as 'boosterist' or 'economic', following Getz (1987) and Hall (2000). Because of changes in the nature of governance, and greater emphasis on public participation in planning, there has been a shift in recent years to 'growth management' strategies, which facilitate growth while mitigating its negative consequences. More fully:

> Growth management is inherently a governmental process which involves may interrelated aspects of land use. The process is essentially coordinative in character since it deals with reconciling competing demands on land and attempting to maximise locational advantages for the public benefit (Cullingworth 1997: 149–50).

Bosselman et al. (1999) argue that there are three main objectives of growth-management strategies. First, some strategies focus on the quality of development, usually with the objective of encouraging only development that meets certain standards. Second, other strategies manage the quantity of development by regulating the rate of growth or ultimate capacity for development. Third, many strategies emphasize the location of development by expanding or contracting the existing areas that attract growth, and diverting the growth to new areas. Some of these are illustrated in Gill's study of the policy shift in Whistler (Canada) from a growth-machine approach to a growth-management one (Box 8.6).

Whether 'growth'- or 'management'-led, the effectiveness of local economic interventions has been criticized by number of commentators. Gordon and Goodall (2000), for example, consider that the significance of local economic strategies is more symbolic than substantive. Not least, they argue that strong promotional coalitions are likely only where there are a few major beneficiaries, where the local economic base is relatively homogenous, and where there are strong local traditions of cooperation or the local state provides strong leadership. Consequently, they concur with Hudson and Townsend (1992) that the social distribution of the benefits of tourism is questionable. It was partly in response to such social critiques, combined with environmentalist critiques, that the ideology of sustainability has taken such a strong hold on tourism policy in recent years.

Sustainability has generated such a vast literature in recent years that it is no longer necessary to provide a detailed discussion here. It stems from the Brundtland Report proposition that 'sustainable development is development

Box 8.6   From growth machine to growth management: Whistler

Whistler has developed rapidly since the 1990s into a major international ski resort, which, in the late 1990s, was host to 1.5 million visitors annually. A major investment in a sealed road opened up the area, made the construction of a ski lift viable, and attracted investors from nearby Vancouver. As a result, Whistler evolved from a small settlement of fewer than 600 people in 1976 to a town of 8,700 in 1998. From 1975 it acquired its own mayor and council, a development that was to become the focus of struggle between competing interests in the following years. Alison Gill argues that these struggles can be classified in terms of three time-periods, although she warns against assuming there was simple unilinear progression from growth machine to growth management.

The growth-machine years

The first mayor and council members, elected in 1976, strongly supported development. Only one council member could be considered 'neutral' on this issue, while three were actually developers. The new council set up the Whistler Village Land Co. to acquire and service land, and then sell it on to private-sector developers. Despite a hiatus in the early 1980s, when recession undermined growth, there followed two decades of strong growth, with investments in ski facilities, golf courses, hotels and second homes, and houses for permanent residents. At the peak of the boom, the development limit was raised for the town from 40,000 to 52,500 beds. All the councils and mayors in this period, with one brief interruption, were strongly pro-growth, and these years fit the description of a 'growth machine'.

Local contestation

The surge in growth in the late 1980s coincided with the emergence of interest groups among the expanding resident community. As a result, a period of local contestation was entered, and this was signalled by the election of a number of anti-growth or growth-neutral councillors in 1988. Responding to the expressed demands of local residents, the council used its burgeoning property and development taxes to finance the construction of a new school and well-equipped leisure centre. Growth continued, but 'it became clear that growth coalitions could no longer engage in "value-free development" with respect to the social and environmental concerns of residents'. There had been a significant shift in the relationship between local elites and the growth machine.

Growth management

The shift to a growth-management orientation was marked by a number of council decisions in 1994, with the purpose of relating new developments to housing, environmental and transport needs, and engaging in wider community consultation. This was symbolized two years later by the election of a pro-community mayor and – for the first time – a council where pro-growth interests did not hold the balance of power. Among the measures implemented was the Whistler Environmental Strategy, which set a new paradigm for future development. As a result, in a period of growth management, developers have had to seek out new forms of coalitions with local interests and elites.

Box 8.6 *Continued*

This case study provides insights into the nature of political struggle over the development process, and also illustrates a useful methodology for researchers. However, the author is careful to warn about the time- and place-contingent nature of her research and against the desire to draw neat boundaries around the time-periods. Instead, they should be seen as a level of abstraction designed to deepen understanding. It is also important not to allow the attractive periodization to blind us to the continuing and fluctuating struggles among competing interests within each of these time-periods.

Source: Gill (2000)

---

that meets the needs of the present without compromising the ability of future generations to meet their own needs' (WCED 1987: 49). Four basic principles are embedded in sustainability: development should be holistic; it should preserve essential ecological processes, protect cultural heritage and promote economic development; it should promote intergenerational equity; and it should also promote intragenerational equity.

Sustainability has been subject to an extensive critique. Mowforth and Munt for example, consider that it is 'a concept charged with power' (1998: 25), and that in context of Third World tourism, 'sustainability is ideological in the sense that it is largely from the First World that the consciousness and mobilisation around global environmental issues have been generated and in the sense that sustainability serves the interests of the First World' (p. 39). There are also criticisms that it fails to balance local and global needs – tourism practices at a destination may approximate to sustainability, but travel to the destination may not be so. Much of the sustainability literature can also be criticized as being disembedded from broader analyses of power, globalization, social exclusion and other key social-science processes.

Nevertheless, despite such criticisms, the notion of sustainability still pervades tourism planning and policies (see also Hall 2000), although more at the level of rhetoric than of practice. But that, in a way, creates new problems. It becomes an unchallengeable mantra, and at worst a substitute for analysis and discourse.

SUMMARY: PLACES, SPACES AND REGULATION

Globalization has enhanced rather than diminished the significance of space and place. The difference between these is understood in terms of place being the product of socialized human beings, i.e. as more than the outcome of profit-seeking activity. Places are open, not closed, and they are actively created and re-created by social processes. They should be seen as the outcome of processes whereby individuals, social groups, and the local state seek to contest the trajectory of change in particular places. The major points to emerge in the discussions in this chapter are that:

- There are distinctive economic reasons for the clustering of tourism activities, which are significantly different from those driving clustering in some other industries.

- Product cycles provide a useful perspective on long-term changes in tourism production. This leads to the concept of the resort life-cycle. Its operationalization is problematic, but it still provides an useful organizing concept.

- Tourism has relatively weak local labour markets, which means that entry barriers are low and firms' access to external labour markets is strong. Migration and mobility help shape these external labour markets. Trade unions are relatively weak in most tourism sectors.

- The extent to which capital is embedded is understood in terms of the type, duration, and quality of local linkages. Much of the literature on embeddedness in tourism has focused on foreign direct investment (FDI), especially the relationships with the agricultural sector.

- The local state is a focal point for the way in which places contest their role in the global economy. Control of the power of the local state is contested.

- Interest-group activity is relatively weak in many branches of tourism, because of the industrial structure and the nature of tourism capital and labour.

- Tourism has been affected by the generalized shift from government to governance. This is evident in the emphasis on partnerships, although there have been difficulties in making partnerships truly representative, effective and sustainable.

- The local state has become increasingly entrepreneurial, and this extends to tourism. We focus on issues relating to 'place wars' or inter-place competition, the growth-machine versus growth-management debate, and sustainability.

9 Established Tourism Spaces in Transition: Changes in Coastal Resorts

THE SEASIDE AND THE BEACH

The seaside, and more especially the beach, has been signposted as a liminal and carnivalesque tourism space by a variety of commentators (Ryan 1997; Shields 1991; Walton 2000). According to Ryan 'beaches are margins of experience' (p. 167). Such marginality is not only geographical, but also defined by social and psychological experiences. Within this context the beach has often been seen as a space for display, the 'other world which lifts the curtain upon suppressed profanities' (Ryan 1997: 161). In this marginal space, tourists can express themselves and indulge in the types of fun and play recognized by Podlichak (1991). This use of the beach is not new, as 'common people had been congregating on the beach for centuries' (Lencek and Bosker 1998: 93). Corbin (1992), for example, describes a prehistory of popular sea-bathing in locations such as Santander and Oporto that was usually linked to religious festivals (Walton 1997). Similarly, Travis (1997) draws attention to the regular custom of working people in the north of England visiting the coast to bathe in the sea, in the years before the eighteenth century.

The representation of the beach as a longstanding leisure space and its symbolism as a liminal zone may also be illustrated in the numerous attempts to create artificial beaches. Perhaps the most extreme case is that of the socialist mayor of Paris, who spent £1 million creating a temporary beach on the Right Bank expressway of the Seine, to enable local people to laze, play, dance and generally have fun (Bremner 2002). This action emphasizes the symbolic nature of the beach as a significant pleasure space. It is important, however, to make a distinction between two interlinked tourism spaces: the beach and its associated seaside resort. While in most cases they are physically linked, there are many instances of separation. The beach is an enduring zone of pleasure and fun, while the resort is a more contrived space, which has varied in its popularity. Such links and differences have been highlighted by Jeans (1990), who distinguishes between the resort, which is trolled by function and class, and the beach/shoreline, which promises ntial danger and pleasure (Figure 9.1). The mixing of culture and nature therefore symbolic of this transition area. In this chapter, attention is sed on the development of seaside resorts, while Lencek and Bosker's

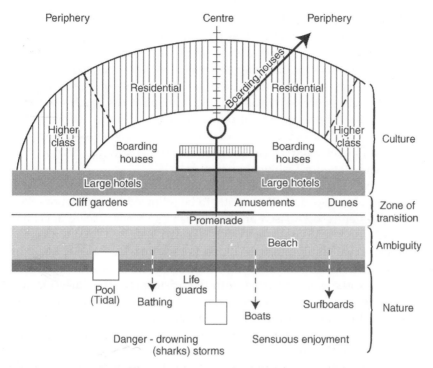

Figure 9.1 **Zones of control and pleasure in the seaside resort (source: modified from Pearce 1995)**

(1998) wide-ranging history of the beach should be consulted by those interested in its cultural history.

This chapter focuses on a number of themes. First, we consider the various pathways and forms of resort development. Second, the nature of resort development is explored in a range of environments, with particular attention to the creation of large-scale international resorts. A third, and perhaps more important, theme in the context of this book examines the characteristics of resorts in transition, more especially the nature of resort restructuring in the face of changing consumption, production and regulation (see Chapter 8).

CREATING PARADISE: THE RISE AND DIFFUSION OF THE SEASIDE RESORT

In England, during the second part of the eighteenth century, the spontaneous and haphazard use of the beach was increasingly supplanted by more organized and fashionable sea-bathing. This marked the start of the transformation of public behaviour along the seaside (Walton 1997a). More importantly, it was also the beginning of an institutionalization of the seaside, with the creation of distinct coastal resorts. 'Brighton was the jewel among British seaside resorts, polished to dazzling perfection through its association with the Prince of Wales' (Lencek and Bosker 1998: 89). The Prince's visit in 1783 cemented the town's growth as a fashionable resort, and the completion of the Pavilion in 1820 confirmed its position as market leader (Gilbert 1954). It also signalled the start of a long and much-copied process of what Lencek and Bosker term the 'architecturalisation' (1998: 90) of the seaside. This involved the construction of fashionable housing areas, such as the Royal Crescent, and the development of public recreation leisure spaces, including sea terraces. Visitors could view the sea, meet together and stroll in an organized and controlled space. The seaside promenade was born to provide the space and stage for fashionable display.

These ideas were soon copied throughout England in resorts such as Scarborough, Margate and Weymouth, to be followed by many more (Walton 1983). Moreover, resorts based on the English experiences were developed in France, Belgium, Spain and the North German states. In Spain, members of the Spanish royal family patronized and helped develop San Sebastian in the early nineteenth century (Walton 1997). Similarly, in France, the early growth of seaside resorts was associated with the aristocracy. The early nineteenth-century American seaside resorts tended to be smaller and less sophisticated versions of Brighton. Nevertheless, the early resorts of Newport (Rhode Island), Long Beach (California) and Cape May (New Jersey) catered for the wealthier classes (Lewis 1980).

The transformation of seaside resorts from the preserves of the wealthy to mass holiday centres occurred during the late nineteenth and early twentieth centuries. Across Europe and in North America there was a combination of factors that transformed these early tourism spaces into centres of mass pleasure. These factors embraced technological, social and structural changes in a series of interrelated movements (Walton 1983; Williams 1998). According to Lencek and Bosker, seaside resorts were 'the children of cities and the great town-planning movements of the nineteenth century' (1998: 115). We can expand on this by highlighting the key influences in resort expansion, which in different ways all contributed to the evolving scapes of tourism (see Chapter 1):

- Improvements in access, through the development of railways. These shortened journey times and made resorts more accessible to large urban centres. In England, the industrial cities of Lancashire and Yorkshire were the sources of much of the demand for new leisure spaces, as was metropolitan London. Similarly, in Spain, Madrid was the major source for the growth of coastal resorts in places such as Alicante.

- The impact of the railways was also accompanied by gradual improvements in social access to leisure time and travel (Walton 1981). And travel was facilitated by the establishment of organized excursions, the first being initiated by Thomas Cook in 1841. Visits to the seaside became a focal point of Cook's early business. These organizational changes also became embedded into the scapes of tourism.

• Significantly, major financial institutions and private individuals invested in the railways to develop new coastal resorts. Such investments brought two other processes into being. One was the creation of an increasingly sophisticated destination-marketing industry, controlled initially by railway companies, but later also by hotels and local authorities (Morgan 1997). The second was the transformation of the structure of resorts in physical and social terms to cater for mass tourism.

These changes, especially the last one, transformed seaside resorts as well as creating new types of seaside destinations. They were all 'dedicated to the proposition that life is to be enjoyed in a place and time entirely removed from the messy business of survival' (Lencek and Bosker 1998: 115).

The transformation of some seaside resorts into spaces for mass pleasure involved the construction of new forms of entertainment. For example, between 1897 and 1904 three large-scale amusement parks were constructed at Coney Island, near New York, at a cost of $5 million. As Lewis explains, 'this was the "New Coney Island" with lavish display and family entertainment' (1980: 48). As one developer explained, 'the aim was to manufacture a carnival spirit and offer fast-moving elaborated childrens play' (the words of Frederick Thompson in 1908, the developer of Luna Park; quoted in Lewis 1980). In this instance, the notion of play and carnival originally associated with the beach was commodified and located within a more regulated environment. Within this resort, located close to New York, attractions had been created with 'an architecture of fantasy and escape, which gave even familiar pleasures an exotic allure' (Lewis 1980: 48).

Similar features were found in English resorts, but usually on a more limited scale and mainly located on the piers, at least until the early twentieth century. These investments singled out many seaside resorts as liminal environments, 'where the usual constraints on respectability and decorum in public behaviour might be pushed aside in the interests of holiday hedonism' (Walton 2000: 96).

Seaside resorts took different pathways to growth, and to some extent sought different segments of the holiday market – or what some commentators have called social tone (Walton 2000). Lancek and Bosker (1998) express such variations in the American context more forcefully, claiming a 'world of difference separated the beach experience of urban workers from that of the wealthy entrepreneur' (p. 149). The very wealthy visited resorts such as Newport, Rhode Island. Other resorts, like Atlantic City with its seven-mile long Boardwalk, which had some 4000 hotels in 1900, claimed the middleground, attracting the new middle class American worker. Within Europe, the casino and the grand hotel marked out the leisure spaces for the more wealthy classes, especially along the French Riviera.

The most complex resort system was found in Britain, which, in the twentieth century, 'satisfied a wide range of aesthetic preferences ... and catered for almost a complete cross-section of society' (Walton 2000: 27). This complex set of tourist spaces is considered in more detail later in this chapter.

Of course, it is possible to categorize resorts in terms of their evolution, morphology, and tourist elements as suggested by Pearce (1995) in his wide-ranging review. However, for our purposes, it is not necessary to explore this theme in detail, but rather to note that resorts not only evolved, usually from small fishing villages or ports, but also that many were planned in their entirety, representing highly specialized forms of production and consumption. Examples of the latter include the aforementioned Atlantic City in New Jersey, Eastbourne in England, and Deauville in France.

These variations are significant in that they correspond to different scales of development and also mark different patterns of investment (see Chapter 8). Consequently, they are critical to understanding the capacity of resorts to contest their roles, and to re-invent themselves, as will be explored later in this chapter.

THE RESORT TRANSFORMED AND TRANSPLANTED: THE GROWTH OF INTERNATIONAL COMPETITION

The growth of resorts in the United States, Britain and mainland Europe during the nineteenth and early twentieth century was very much based on the exploitation of domestic demand. Of course, there was a degree of international travel for, as Lencek and Bosker explain, the wealthy British 'came to the Mediterranean to cast their net of affluence around the pleasure ports of the French Riviera' (1998: 131). But, on the whole, international travel was limited, reflecting a tight regulatory framework, as was competition. The resort systems that had evolved in Britain and North America competed largely at the national level and for domestic tourists. Within Britain, seaside resorts enjoyed a period of growth and investment during the first half of the twentieth century. Walton comments that the 'surge of growth which ushered in the new century was unprecedented' (1997b: 22) in resort development. Resorts such as Blackpool and Bournemouth sustained their early growth to become even larger tourist centres. A number of key trends emerge in Walton's analysis including:

- increased differentiation of resorts in terms of private investment as disproportionately more external capital was being drawn to the larger centres (see Chapter 8)

- growing interventionism by the local state, both in terms of resort marketing and the development of recreation facilities

As a consequence, the period from 1950 to the mid-1970s has been termed the 'Golden Years' of English resorts by Demetriadi (1997), as the seaside resort remained the most popular form of domestic holiday. In 1968, for example, seaside resorts accounted for some 75% of all main holidays in Britain (a main holiday is defined in British Tourism statistics as four or more nights away

Table 9.1 **Changing levels of international tourism undertaken by British residents, 1955–75* (millions)**

Holiday breakdown	1955 (%)	1965 (%)	1970 (%)	1975 (%)
Holidays in Britain	25 (92.6)	30 (85.7)	34.5 (85.7)	40 (83.3)
International holidays	2 (7.4)	5 (14.3)	5.75 (14.3)	8 (16.7)
Total holidays (millions)	27	35	40.25	48

*Refers to holidays of 4+ nights.
Source: British National Travel Survey 1976.

from home; British Travel Association 1968). However, as Walton (2000) has shown, tourists were already starting to be attracted to some resorts at the expense of others. For example, resorts in Devon increased their share of the domestic market between 1950 and the 1970s, peaking at 3.5 million visitors in 1978 (Walton 2000). The period witnessed increased domestic competition, as resorts started to develop in different ways and market themselves to different groups of tourists; in other words, there was competition within changing paradigms (see Chapter 4). In extreme cases, as at New Brighton near Liverpool, resorts failed disastrously, when they could no longer compete within changing taste (Walton 2000).

Probably the most significant paradigm shift, however, was the strong and steady growth of competition from international resorts, already evident in the late 1960s, but becoming more pronounced in the mid-1970s (Table 9.1). The reaction to such competition has spawned extensive academic perspectives, including attempts to model the resort life-cycle (see the reviews by Shaw and Williams 2002; and Prideaux 2001) and a series of belated official and quasi-official reports. Cooper (1997) has attempted to summarize the main factors influencing change in his anatomy of resort decline. We want to examine these factors of decline, starting in this section with the growth of international competition. This phenomenon represents the transplanting of the seaside resort and, as we shall see, its reformation into a global tourist space.

The earliest evidence of the impacts of competition and resort decline were to be found in Britain and, especially, North America. Some of the older American seaside resorts, on the north-eastern seaboard, had started to lose their appeal by the 1950s, and their market share declined. In Coney Island, a series of major fires hastened the switch in demand and the resort's demise (Lewis 1980). Similarly, Atlantic City started to lose much of its main tourist market to the rapidly growing Miami Beach – the name given to this manufactured resort by its three main developers. Of course, Miami Beach had a climatic advantage, being in southern Florida, but it was also socially constructed as a very different development; it was 'to the seashore what the constructed swimsuit of the 1950s was to bathing attire: a highly engineered setting' (Lencek and Basker 1998: 234). In spite of, or perhaps because of, its artificial environment, the resort set new trends in holiday accommodation and destination-marketing. The strong internal competition in the United States among seaside resorts, and the rapid decline of older resorts such as Atlantic City, prompted academics such as Stansfield (1978) to consider an early version of the 'resort cycle', later to be developed by Butler (1980) (see Chapter 8).

Given the longevity and complexity of British seaside resorts, it is hardly surprising that considerable attention has focused on their fortunes. Perspectives on their decline fall into economic (Cooper 1997) and cultural (Urry 1997) discourses, although these are strongly interlinked. For example, the changing values and tastes in the tourist gaze identified by Urry (1990) show clear links with the general socio-economic shifts in demand observed by Cooper (1997) and in more detail by Seaton (1992). Much of this relates to a shift in domestic competition since the creation of new tourism spaces (see Chapter 10) and, more especially, the growth of international competition. The broad trends in the number of holidays taken overseas by British residents are shown in Table 9.1. This, however, masks the complexities in the composition of travel and the ever widening nature of international competition, as seaside resorts were transplanted into new, exotic locations. The rise of international competition and the transplanting of the seaside resort are linked through the growth of mass tourism. Significantly, this arose at a domestic level in the early twentieth-century United States (Lavery and van Doren 1990). However, it was the creation and development of the overseas package holiday that brought the largest impacts. The details of mass tourism have been discussed elsewhere (see Shaw and Williams 2002; and the discussion of the Fordist regime of accumulation in Chapter 2), and here we focus more on the consequences of resort development. Of course, the mass package holiday was more than an innovation in the organization of production, for it was also associated with technical improvements in air travel and built on the established tradition of seaside holiday-taking. In this context, the opening-up of the Mediterranean to a new form of holiday resort was part of a 'social phenomenon which is deeply embedded in European society as in the built landscapes of the Mediterranean coast' (Williams 1996: 133).

Resort development: the Spanish case

Spain witnessed some of the more spectacular early developments of the new purpose-built, mass seaside resorts, especially along the Costa del Sol. These place (re)constructions offered the tourist affordable holidays, new hotels with a range of amenities and, above all, familiar surroundings and a warm, sunny climate. Some of the earliest developments occurred in Torremolinos, where the 150-room Hotel Pez Espada was opened in 1959 by developers from Madrid (Barke and France 1996). The number of bed spaces for tourists increased by 287% between 1964 and 1976, mainly in hotels based around package holidays, and usually three-star ones (Barke and France 1996). As tourism continued to develop, so the range of holiday accommodation expanded, especially with the growth of self-catering apartments. This created a new form of tourist accommodation not previously available to the package market. According to Prunster and Socher (1983), this form of mass self-catering development became increasingly popular with holidaymakers seeking cheaper alternatives to the hotel, especially as European economies went into recession. In conceptual terms, we have argued earlier that this represents a shift to more flexible large-volume production of tourism services, rather than post-Fordist production (see Chapter 2).

Developments along the Costa del Sol have been shaped by its proximity to the airport at Malaga, with the earliest resorts being those that were most accessible (Barke and France 1996). Investments in hotel development were largely unregulated and, even when they were locally based, the influence of international tour operators on hotel developments was still strong. As Barke and France explain, a common strategy employed by the tour companies 'was to advance low-interest loans to the Spanish developers, but on the condition that rooms were guaranteed to the tour operators at fixed prices for five, or even ten years ahead' (1996: 278). These public- and private-sector invest-ments played a key role in the definition of scapes linking the markets of northern Europe with the arenas of tourism consumption in the Mediterra-nean regions.

In the initial phases of growth, resorts were strongly beach-orientated, in that proximity to the beach and, at the very least, views of the sea were of prime importance. This represented something of a rediscovery of the significance of the beach as a key tourist space, compared with those subsequent developments in British resorts. There were two important factors at work: the forcing-up of land values and the influx of foreign capital. During the 1980s, foreign investment was increasing and much of it was focused on resorts such as Marbella, although the role of national capital should also not be discounted (Williams 1995). Similarly, in the Balearic Islands foreign-owned tour operators helped shape development (Buswell 1996; Sàlva-Tomàs 2002). Both factors have strongly conditioned the built form of Spanish resorts, most noticeably in the creation of high-rise hotels, high-plot densities and a spread of tourist *urbanizaciones* (planned residential areas) into agricultural areas. In the most rapidly expanding resorts based on package tourism, such as Torremolinos, the flood of investment capital led to overdevelopment, creating 'formless and untidy built-up area[s] usually polluted by characterless buildings' (Williams 1998: 60). The rapid expansion of the resorts also meant that they had to rely on in-migrant labour (Sàlva-Tomàs 2002).

These resorts were not only created physically by the economics of mass tourism and the demands of tour operators, but the same organizations also shaped tourists' images of these places. The Costa del Sol, along with other Spanish coasts, was retailed heavily in the holiday brochures of the major tour operators, who were effectively the 'gatekeepers' to northern European markets. During the early 1970s, the Spanish coastal resorts dominated holiday brochures – accounting for 50% of coverage in Thomson's 1973 summer holiday brochure (Barke and France 1996). These brochures were selling an image based on a beach holiday, including sunshine and cheap alcohol (see Chapter 7). The theme was somewhat different from, and more exotic than, what had become seen as the rather predictable British resorts. The availability of cheap alcohol aided the development of the night-time economy, bars and nightclubs. Increasingly, the resorts acquired a more racy image, which emphasized this night-life. During the late 1980s and 1990s, the importance of the night-club scene increased as more tour operators exploited the demands of the youth tourism market segment. In this context, a number of Spanish resorts, along with other Mediterranean ones, are strongly marketed as centres of the club scene. British tour operators such as '2wentys' promise 'Great value for money, the liveliest resorts across the Mediterranean ... [and] A guaranteed seething, roaring, shouting mass of energy and excitement in exotic locales' (2wentys Holidays 2002: 1). As Box 9.1 shows, holidays in the Balearic Island of Ibiza are firmly rooted in the club and music scene. Similarly, in Thomson's *Club Free Style* brochure, San Antonio in Ibiza is described as 'the club capital of the universe in big clubs, big name djs and the best music' (Thomson 2002: 10). For many of the tourists attracted to such holidays, the holiday experience is reduced to a relatively limited set of practices on the beach and at the poolside in the day-time, complemented by extensive night-life practices – often dominated by drink, music and the

---

Box 9.1   Ibiza and the marketing of the club scene

Many large-scale tour operators have identified the youth market as lucrative and so promote aspects of the club-music scene. This is especially important on Ibiza, where the emphasis is on excitement, continual partying and ludic forms of behaviour. On Ibiza, San Antonio is branded as 'the most famous clubbing destination in the world' (2wentys Brochure 2002: 11). This brochure includes a short Ibiza club guide, which highlights the most famous clubs and their key features. These are sold in terms of the brand names, disc-jockeys and associations with the British club scene. The descriptions are evocative of the holiday youth scenes as illustrated by the following:

- *Eden*, located in San Antonio where 'nights here are always very messy!'

- *Ministry of Sound* at Pacha, 'the most cosmopolitan of Ibiza's clubs, attracting a glammed up, European crowd'

- *Cream* at Amensia, in central Ibiza, 'the Liverpool superclub'

- *Manumission* at Privilege, in central Ibiza, marketed as one of the largest clubs in Ibiza, with a capacity for 8000 people, which 'has hosted some of the biggest parties on the island'

Source: based on 2wentys Holidays (2002: 12)

promise of sexual encounters. This is, in part, a reinvention of the seaside resort, in that it marks out a new form of tourist space, where extreme forms of consumption and behaviour can be experienced. Andrews (2002) has described such tourist experiences in the case of Magaluf, Majorca. Here, the 'symbols of identity speak of "Britishness", characterised by unfettered consumption and freedom from responsibility' (Andrews 2002: 22).

In effect, what Andrews is describing is a form of resort enclave development, where tourist activity is internalized and limited to a prescribed, commodified space – the hotel, pool, beach, bar and club. Of course, to parody this as the only type of seaside resort within Spain or, indeed, the Mediterranean, is clearly simplistic. Throughout the Mediterranean resort system, there has been an increasing market orientation of resorts as different market segments have been exploited, in pursuit of competition within existing and changing paradigms. Once again, the marketing of resorts has largely been set by major tour operators selling some destinations to the youth market, while others are sold to families. In addition, the spread of the resort culture towards the eastern Mediterranean has brought increased international competition (Apostolopoulos et al. 2001). The growth of beach holidays in Turkey, for example, reflects considerable levels of foreign control and all the characteristics associated with this, so that, to a large extent, these new resorts have tended to imitate the typical features of previous developments. In the case of Turkey, economic liberalization between 1980 and 1990 provided key incentives for prospective investors in tourism schemes and resort growth (Var 2002). The effects of this were to increase the number of foreign tourists from just under 1.4 million in 1982 to almost 8 million by 1995.

RESORTS AND THE PLEASURE PERIPHERY

As competition for international tourists has intensified, in the context of globalization (see Chapter 1), tour operators and investment companies have fed the demand for new destinations by combining, sometimes with local/national states, to create new resorts. This globalization of the resort into almost every type of beach environment is a recognized phenomenon of tourism development, although the detailed geography of this growth has received only scant attention (Meyer-Arendt 1990; Pearce 1995; King 1997; King 2001). Boers and Bosch (1994) draw some limited attention to this process by describing the 'resortisation' of the world. In a more detailed approach, King (2001), following the early work of Turner and Ash (1975), views these new resort zones as constituting a pleasure periphery. As King explains, other commentators – such as Cazes (1992) – have taken more extreme views and see tourism in developing countries 'as a *de facto* transformation of sovereign states into holiday resorts' (King 2001: 178).

This so-called 'pleasure periphery' of developing countries heavily involved in resort tourism comprises parts of South-East Asia, East Africa, Latin America and the Pacific. The spread of resort tourism into these areas has been conditioned by a number of factors, constituting the refashioning of scapes, including:

- improvements in air travel, in which larger and faster commercial aircraft have led to long-haul tourist destinations, such as Thailand, becoming economically viable

- the creation and subsequent modification of package tours, easing the purchase of international holidays

- the development of global hotel chains offering recognized and standardized forms of accommodation

- the increased importance of image in tourism and the creation of a large-scale tourism-marketing industry linked to the major tour operators; this is supported by the travel media's output of books, articles and television programmes

- the active development of new resort destinations by national governments and international property companies

Purpose-built resorts, large hotel complexes and holiday villages typify developments in many developing countries. Ayala (1991) has described the process as an international megatrend, which has produced a massive transformation of the environment through the creation of a hotel landscape. This, in turn, is associated with what Ayala calls the focus on the creation of images, which stresses 'placenessness', especially the lure of the beach. This latter notion links with the extent to which resort developers and tour operators use particular settings or themes in order to create product differentiation. King (2001) has sketched out a basic typology of resorts, based on their characteristics and degree of specialization. King's review identifies a diverse range of resorts, from sporting and beach-orientated ones, to theme-park resorts and golf resorts, to what Thomas and Fernandez (1994) have termed 'mangrove-resorts'. The growth of so-called golf resorts has been especially marked in parts of South-East Asia, such as Thailand, where they have successfully attracted large numbers of Japanese tourists (Pleumaron 1992). Other recent developments in this region include the growth of large-scale water-park resorts, which, as Turner (1996) shows, also incorporate hotels and shopping malls. These developments have been especially focused on South Korea, Thailand and, more recently, China (King 2001). Such diversity of developments leads King to argue that 'the type-casting of resort tourists as interested in the three Ss [sun, sand, sea] seems increasingly outdated' (2001: 186). We would agree partly with this view, in that resorts are becoming more diversified, in terms of both market segments and appearance. There has, for example, been a growth in the use of new types of design, which involve 'a postmodern melange of local tradition and contemporary styles' (King 2001: 186). However, set against this is the dominance of the mega-resorts with their massive hotel complexes, strongly orientated towards beach holidays.

Resorts in the pleasure periphery may be diversifying, but they also share certain important features because of the nature of their development. These concern the role of individual capital and the state in the overall development process. In most cases, this process has involved large-scale, capital-intensive projects, which have depended on state initiatives and foreign investment.

The state's intervention in tourism development within developing countries has often been viewed in a rather simplistic fashion. For example, in such economies the state has been seen as merely aspiring to rapid economic development through the use of tourism. Lumsdon and Swift's (2001) review of tourism development in Latin America adopts such a stance and thereby ignores the complexities of the resort-development process. As Clancy explains, it is useful 'to consider the relationship between state actors and societal groups to be contingent' (2001: 23). Furthermore, this relationship may vary over time and be dependent on the policy area, which resonates with the stress placed on the national state and national economic space within regulation theory (Chapter 2). For example, Grindle (1986) has argued that it is often much easier for the state to engender change in industries where class or group interests are not well organized. Clancy (2001) follows this line of reasoning in his detailed analysis of tourism development in Mexico, where the state played a leading role in the construction of tourism around beach resorts.

Planning for mega-resorts in Mexico

Resorts have been developed in Mexico in distinct phases relating to state policy and intervention. The major watershed is 1968, when the Mexican central bank, the Banco de México, published the findings of its two-year study, which identified tourism as a potential major export activity. The plan called for development of a number of new resorts, along with an aggressive marketing strategy (Clancy 2001). Of course, there already existed popular beach resorts such as Acapulco, Matzatlán and Puerto Vallarta, but these were also upgraded after 1968, witnessing the development of new hotels, apartments, marinas, shopping malls and golf courses (Lumsdon and Swift 2001). The central bank's study called for the building of five large, new resorts:

- Cancún, on the eastern coast of the Yucatán peninsula (see Box 9.2), developed as a mega-resort

- Ixtapa, in reasonable proximity to Acapulco in the state of Guerroro, developed as a mega-resort

- Los Cabos, on the western Baja California peninsula, developed into a large scale, golf-orientated resort at Cabo del Sol (Jesitus 1993)

- Loreto, in a similar location to Los Cabos

- Las Bahías de Muatulco, in the southern state of Oaxaca, which is predicted to develop into a resort larger than Cancún (Ayala 1993)

These planned resorts were developed by INFRATUR, the National Trust Fund for Tourist Infrastructure, established in 1969. This was based within the central bank and was also charged with encouraging private investment. The latter operated through FOGATUR – a trust fund created to subsidize loans to tourism-related projects within the private sector (Clancy 1998). During the 1970s, the administrative structure changed, as INFRATUR became an agency of the Ministry of Tourism.

The state, via INFRATUR, focused resources on developing the two mega-resorts of Cancún and Ixtapen. In the case of Cancún (Box 9.2), state agencies expropriated land (and associated property rights), cleared and drained areas for development and more or less built a completely new city (Bosselman 1978). Clancy argues that government agencies 'played central roles in virtually every aspect of building Cancún' (2001: 55). Similar forces were also at work in Ixtapa, which, unlike Cancún, was built adjacent to the fishing village of Zihuatanejo, which already had a small tourist industry. In both Cancún and Ixtapa, the state developed workers' settlements close to the resorts, in recognition of the need for labour in-migration to support the

---

Box 9.2   The structure of a mega-resort: Cancún, Mexico

Tourism spaces

Cancún comprises two main settlements, Cancún City and Cancún Island, with the latter being the main purpose-built tourist area (Figure 9.2). The resort proper is based around a narrow island, 14 miles in length, known as the 'Zona Hotelera'. It is dominated by beachfront hotels and condominiums, which act as an almost uninterrupted backdrop to the resort. The lagoons, which enclose the island inland, also provide further potential for tourist facilities, with the Laguna Nichupté being lined with waterfront restaurants, shopping malls and golf courses. A further tourism space is known as the 'Party Zone' in many guide books, as it contains most of the nightclubs and bars, along with major shopping malls.

Spaces for 'the other'

Cancún city represents a very different functional space, largely devoid of the tourism glitz and hype of the tourist zone. This is an area for local people who work within the tourist zone. This has none of the major shopping malls, high-class stores and fashionable restaurants. It is a much more of a support area for the main resort.

Sources: Lonely Planet 2002; www.mexonlin.com/qr/ver5.html; www.allaboutcancun.com/allaboutcancum.html

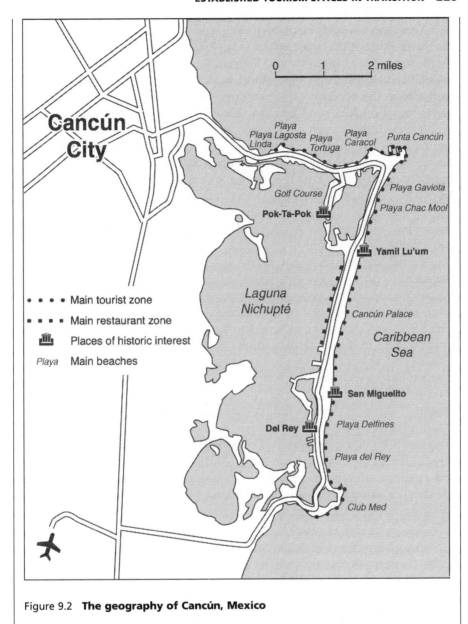

Figure 9.2 **The geography of Cancún, Mexico**

emerging low-cost labour process. This created what would become, in the words of Hiernaux (1999), the classic resort model in the pleasure periphery. The model is characterized by the physical separation of work and residential space for employees, with the latter segregated from tourist spaces. As Clancy explains, 'the bulk of the Megaprojects . . . . are directed at those tourists who are especially affluent and want a degree of social and physical distance from poorer, local inhabitants' (2001: 67). Indeed, many of the hotel complexes are gated enclaves, which have few physical links with the local area.

The resultant development of these so-called integrated or mega-resorts was the product of two inter-linked factors. The first consisted of the drive for a high-export based industry and the state's involvement in the development process, initially through the central bank. These requirements placed the emphasis on the necessity to attract large numbers of overseas tourists, with resorts based on sun, sand and sea. Such developments required large-scale facilities, including high-rise hotels catering to an international market dominated by US tourists. Even in the late 1990s, over 61% of all visitors to Cancún were from North America, compared with just over 8% from Europe (Lumsden and Swift 2000). The second, but closely related factor, concerned the need to focus on up-market international hotels, with identifiable brands, which would appeal to North American tourists. This led to a high degree of dependency on hotels run by transnational corporations, although in Mexico their precise development is contested (Clancy 2001). Indeed, the picture is complicated by the fact that many hotels are owned by Mexicans, but transnational groups have established control through contractual means. In this context, there has been a growing tendency for Mexican hotels, especially at the more luxurious end of the market, to become affiliated with foreign-owned chains. For a more general discussion of the relative advantages and disadvantages of transnational capital for destinations, see Chapter 8.

These patterns of hotel ownership are significant, as they play a major role in controlling the nature and scale of resort development through the levels of investment. As Box 9.3 shows, Mexico's beach resorts are underpinned by a complex series of changes in hotel development.

The need to understand the nature of resort development in terms of the role played by entrepreneurs and property developers also applies to other parts of the pleasure periphery. King (2001) has provided a review of some of the main studies in this context and, more specifically, presented one of the few detailed comparative studies of island resort development (King 1997). This is based on island resorts in two different settings: the eight resorts of the

---

Box 9.3 Main characteristics of the operation and control of large hotels in Mexican resorts

- The hotel sector has become increasingly segmented and centralized in that luxury hotels have become more closely linked to international capital.

- The relationship between foreign hotel operators and domestic ones has become increasingly blurred. For example, some local organizations have become more involved in hotel management, franchising and support services.

- Recent trends in the market highlight the strategic advantages of transnational groups – especially in terms of brand recognition.

Source: based on Clancy (2001: 91)

Whitsunday area of the Queensland coast, Australia, and the twelve resorts in the Mamanucas islands of Fiji. In both case-studies, key individuals played important entrepreneurial roles in the resort-development process, and various institutional factors affected infrastructure and destination-marketing. Clearly, both structural factors and human agency are significant in this context.

TOURISM SPACES AND RESORT LANDSCAPES

The vast majority of the resorts developed within the pleasure periphery are created, engineered landscapes, which reflect a range of economic and cultural influences. Just as the early English resorts were manifestations of production and consumption systems operating in the nineteenth century (i.e. the prevailing regime of accumulation at that time), so too are the so-called mega-resorts, such as Cancún. In both circumstances, resorts attempted to set themselves apart from other places (of work and residence) where everyday routines dominate. This perspective fits with King's (1997) view of Pacific resorts as elements of postmodernism. Here we can recognize a number of important aspects of the postmodernist perspective. The first is the fact that the built environment is a key component of postmodernism (Urry 1990). King includes elements of this in his study of the Whitsundays and Mamanucas resorts, where 'the pastiche of styles so typical of postmodern-ism' is evident (1997: 210). In the Whitsunday resort of Hamilton Island (Queensland), high-rise apartments, Polynesian-style villas and even a nine-teenth-century chapel are all juxtaposed. This diverse built landscape is, as Parry explains, the 'architecture of pleasure' (1983: 152) and, in this sense, is nothing new but rather an extension of the pleasure landscapes of earlier resorts. The second key aspect is the commodification of the resort landscape, which links with Sack's (1992) ideas of consumption places. According to Sack the 'resort is not only a place in which things are consumed, but whose landscape is arranged to encourage consumption' (p. 2). These consumption spaces are usually carefully contrived, certainly in the mega-resorts. This is certainly the case in Cancún, where shopping is sold as a major holiday activity, and where shopping malls dominate the resort's landscape (Box 9.2). Both Sack (1992) and King (1997) have drawn attention to the ways in which landscapes in resorts become a function of commodification, which orientates spaces towards the selling of goods and experiences. In such landscapes, these consumption spaces take on clear symbolic meanings, they become represen-tational spaces (Meethan 2001). This process is not, however, simple and unidirectional, in that 'anticipated tourist consumption plays a determining role in the layout of facilities' (King 1997: 215). The acts of tourist consumption become, in a sense, place-creating, both shaping the resort landscape and the way in which it is perceived. Within the resort systems of the pleasure periphery the impact of globalization has been to transform such destinations and, according to many commentators, 'engender placelessness' (King 1997:

215; see also Ayala 1991; Sack 1992). To counter this, Ayala argues that resorts should reorientate their market positions by developing a sense of 'placeness' in order to differentiate themselves from competitors (see Chapter 8).

King (1997), within the context of the resorts of Whitsundays and Mamanucas, argues that there are variations among consumption places. The larger resorts may be clearly designated as such places of consumption, dominated as they are by numerous retail facilities. However, many of the smaller resorts are more like sanctuaries, away from the full array of shopping spaces. This is not to say that such resorts have not developed commercial facilities, but rather they are less brash and blatant in their presentation.

A third key aspect of resorts is their attempts to present themselves as exclusive spaces, where tourists can relax in safety away from social elements unlike themselves. This is at its most extreme in resort enclaves, although many resorts in the pleasure periphery have various formal and informal means of erecting boundaries between themselves and the 'other'. These may be physical: for example, many hotel complexes are gated and access is controlled. Such boundaries divide up places and highlight different forms of space. Examples of the first include the physical division between Cancún city, initially constructed for service workers, and Cancún resort. The division of space by such boundaries is also significant for the tourist, since it demarcates differences between private and public space. Within the resort, behaviour is controlled by individuals or private organizations (the owners/ managers of the resort). In contrast, space outside the resort boundaries is usually controlled by the norms of the local community – it is public space. As discussed in Chapter 7, the interaction between tourists and local residents is played out in such spaces, and increasingly these spaces are becoming contested.

The dimensions of exclusivity also encompass differences among tourists. The tourist media sell the notions of exclusive resorts, which, in the case of island resorts, are isolated from mass tourism by market mechanisms (cost grounds). This represents another form of product differentiation and competition, largely within existing paradigms. However, as King (1997) points out, such exclusivity has its limits, as most large resorts need to attract a range of tourists. Within these larger resorts, there are obvious signs of social segregation between types of tourist, as greater numbers of hotel complexes have been developed to cater for more up-market visitors.

RESORTS IN TRANSITION: THE ANATOMY OF RESORT DECLINE

Earlier in this chapter we started to examine the characteristics of international competition, both globally and in terms of its impacts on traditional resort areas. We now want to return the discussion to the latter theme and focus on two key debates. The first concerns the nature of resort decline in traditional resort systems, in particular in Britain. The second, and major, debate concerns the nature of resort restructuring, again largely focusing on British examples.

The growth of international competition and the rise of the foreign package holiday initially had significant implications for British seaside resorts. As stated earlier, this competition has represented a series of shifts that, since the mid-1980s, have been bound up with the changes associated with post-Fordism. Agarwal (2002) has categorized these as:

- the search for capital accumulation opportunities, characterized by the globalization of tourism (see Chapters 2 and 3), which, as we have shown, has created a new range of resorts in the pleasure periphery

- changes in consumption, which include demands for new types of holidays (see Chapter 5)

- shifts in production mode as part of a search for capital accumulation potential, creating new consumption spaces (see Chapter 8)

- flexible production, which has allowed the customization of tourism products

In addition, there have been a number of supply-side factors, which, according to Cooper, 'reinforce the resorts' difficulties' (1997: 87). Associated with these, Cooper has identified the major threats to British seaside resorts, as shown in Table 9.2.

The decline of British resorts started in the late 1970s, and between 1978 and 1988 39 million visitor-nights were lost (Wales Tourist Board 1992). However,

Table 9.2   **Major threats to British seaside resorts**

Nature of threat	Characteristics
Changes in demand	Growth in low-status, low-spend visitors Highly seasonal destinations High dependence on long-stay market or day-visitor market Demand for overseas holidays
Changes in supply	Outdated, poorly maintained facilities Lack of high-class attractions High dependency on small-scale entrepreneurs with low capital and skills base
Environmental factors	Poor access and traffic problems Lack of investment in local environment Poor interpretation and information
Performance of the local state	Low priority given to strategic thinking Short-term planning horizons, because of local-government planning and budgeting deadlines Lack of confidence in tourism business community Political interference in decisions

Source: modified from Cooper 1997.

Table 9.3 **Changing market share of English seaside holidays, 1973–98**

Year	Seaside holidays as a share of all tourism* (%)
1973	21
1980	21
1986	16
1988	17
1992	17
1995	16
1998	14

*All tourism including overseas travel.
Source: English Tourism Council (2001).

the pattern and pace of decline has been uneven. For example, within England, eight main resorts – Blackpool, Brighton, Bournemouth, Great Yarmouth, Newquay, the Isle of Wight, Scarborough and Torbay – account for around 75% of the volume and value of seaside tourism (Ventures Consultancy 1989). In contrast, the loss of staying visitors has affected the smaller resorts, which are estimated to have lost at least 50% of such visitors during the last 20 years (Shaw and Williams 1997), a trend that has largely continued through the 1990s (Table 9.3). Furthermore, there has been a change in the types of visits to English resorts. Thus, while the long stay market has continued to decline from 3 million trips in 1993 to 2.5 million in 1999, over the same period short-break holidays (1–3 nights) increased from 4.8 million to 8.8 million (English Tourism Council 2001). Similarly, the day-visitor market has become increasingly important to many resorts, but especially to the medium and small-scale ones. In total, some 179 million leisure day-trips were made to the seaside in 1998, representing an expenditure of £1.9 billion. These shifts in demand have had important consequences for the structure of British resorts. The removal of large numbers of staying visitors has impacted on the revenue base of resorts and the supply of accommodation. For example, Scarborough, on the Yorkshire Coast, saw its bed-spaces decline from 78,000 in 1978 to 51,000 by 1992 (Scarborough Borough Council 1997–8). The combination of low occupancy patterns, seasonality and a dominance of small firms have conspired to drive down profitability and subsequently investment (Cooper 1997). Investment levels fell off sharply during the early 1990s, although insufficient data make it impossible to track current trends (Shaw and Williams 1997). This decline has led to falling standards of accommodation, especially in the serviced sector. During the period from the late 1970s until 1990, it was estimated that only three four-star hotels were built in English and Welsh resorts (Association of District Councils 1993).

Various studies by Stallinbrass (1980) and Shaw and Williams (1990; 1997) have outlined the problems of the small-firm sector in the resort economy. Their work highlights a lack of capital for investment and barriers to firms' growth because of limited entrepreneurial skills (see Chapter 5). Many of the small firms also find it difficult to respond to changing market conditions.

The declining economic base has also reduced revenue to the local state and, as a consequence, there has been less money available to invest in the local infrastructure. This has led to a decline in resort environments. As the English Tourism Council report points out, 'it only requires one element of the product to be below standard to reduce the overall appeal of the resort' (2001: 18). In the case of many British resorts, several elements of the product have failed to meet tourist expectations. This process has led to image problems and a general feeling that 'resorts no longer appeal to the modern consumer as a chic destination' (English Tourism Council 2001: 19). Recent research by the English Tourism Council has attempted to identify the main market segments and their propensity for visiting seaside resorts (Box 9.4). Such survey work clearly highlights the trends of certain market segments

---

Box 9.4   Market segments and English seaside resorts

The English Tourism Council undertook research in 2000 based on a survey of 3500 adults. This was used to identify patterns of behaviour, related market segments and their propensity to visit coastal resorts. The main segments are:

Conformists – seek sameness and familiarity. Prefer activities centred around the beach, castles and historic sites. Both short and long break holidays are important.

Sentimentals – seek constancy and sameness and not response to changing trends. Prefer expensive hotels and value good service. Take longest holidays.

Seekers – image driven, following trends and fashion. More likely to stay in self-catering accommodation, and take short breaks.

Radicals – Like to travel to exciting places, sense of discovery. Have high expectations and demand distinctive holidays. Short breaks are important as is off-season travel, with a preference for budget hotels and campsites.

Independents – seek different cultural experiences. Prepared to pay for good service. Availability of upmarket restaurants, heritage sites, and beaches is important.

Pragmatists – highly cultured but settled in their tastes, enjoying the outdoors and related activities. Seek out value for money. However, 20% of this group do not take holidays.

Figure 9.3 shows the importance of these segments to seaside resorts and highlights the dominance of the so-called 'conformist' group, estimated to be around 3.4 million people, of whom between 25% and 30% are likely to go on a seaside holiday.

Source: English Tourism Council (2001)

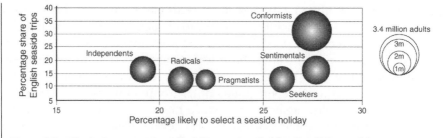

Figure 9.3 **Market segments and visitor potential for English seaside resorts.**

away from the traditional resort and particular types of accommodation. Nevertheless, the research estimates that some 8.5 million adults make trips to the English seaside (Figure 9.3).

This image has also been damaged by the downward drift of surplus tourist accommodation, much of which has been converted into low-quality hostels and flats for socially and economically vulnerable groups. There are also much higher levels of unemployment in seaside resorts, well above the national average in the early 1990s (Agarwal 1999). The combined effect of the traditional patterns of seasonal employment, along with an increase in general unemployment and an influx of economically vulnerable people, has created problematic conditions in many resorts. This is reflected in the number of seaside resorts targeted for economic assistance in the mid-1980s through their designation as Assisted Areas (Pattinson 1993). These included Great Yarmouth on the east coast of England, while Clacton, Ilfracombe, Bideford, Dover and Deal, Hastings and Skegness, and the Isle of Wight were designated Intermediate Areas (Agarwal 1999). This represents a significant revamping of direct national state interventionism in tourism resorts in the UK.

Of course, British resorts have not been alone in experiencing a decline in fortunes, as the shifts in tourist demand, because of stronger international competition, have also been felt in some of the early mass tourism resorts of Spain. In many cases, other factors appear also to have been at work within the Spanish *costas*. For example, Pollard and Rodriguez (1993) believe that the poor environment created by an over-production of mass tourism facilities, together with what they term a falling 'quality' of tourists, has contributed to the decline of Torremolinos during the 1990s. Of course, such problems have affected other resorts along the Spanish coast and 'negative images of the crumbling costas' have been well documented (Barke and France 1996: 301). Much of this decline is linked to behavioural problems associated with mass tourism (or at least with a minority of mass tourists), which include loutish behaviour, aggression, drunkenness, and an increasing use of drugs (Economist Intelligence Unit 1990). Similarly, the fall-off in visitors to Majorca has focused attention on the problems facing this mass tourism resort, underlining the inability of individual tourism capitals to guarantee their own survival, and the need for state intervention to mediate such crises of accumulation (see Chapters 2 and 8). According to Morgan (1991) these problems include:

- Over-commercialization and a poor image. This was often associated with the Anglicization of resort facilities, linked to its over-dependence initially on the UK market, but increasingly on the German market.

- The deterioration of the physical environment because of the impact of mass tourism. This includes the loss of natural habitats, and increased pollution (waste and noise) and damage to historic places (Buswell 1996). Sustainable tourism strategies have sought – with limited success – to address these ideas.

- A lack of investment, because of an over-reliance on the package holiday market characterized by high volume and low yield. In this context, tourists are drawn from a narrow range of socio-economic groups; for example, 57% of German tourists and 43% of British visitors were drawn from skilled and unskilled manual and clerical groups (Buswell 1996).

In the case of the Balearic Islands, and especially Ibiza and Majorca, various environmental-impact studies have taken a pessimistic view of tourism, while Buswell claims that the 'beach-line resources are almost consumed in Majorca and Ibiza' (1996: 333). As we have seen (Box 9.1) in the case of Ibiza, recent marketing has concentrated heavily on aspects of mass tourism associated with the youth market. This clearly does not represent a solution to the island's problems, although state intervention may impose spatial limits on these patterns of holiday-making.

RESORT REDEVELOPMENT: THE LIMITS TO INTERVENTION

As we have seen throughout this chapter, resorts are extremely dynamic tourist spaces, which have often changed form a number of times. Moreover, the changing fortunes of resorts depend on a range of interlinked factors, including:

- changes in consumer tastes, reflecting shifts in consumption of the types discussed in Chapter 5

- changes in the quality of the tourism product, associated with shifts in capital and changes in modes of production (see Chapter 8)

- changes in the level of competition, and the globalization of the resort as discussed in this chapter (see also Chapter 2).

The circulation of capital within the global tourism system is the means by which resort development or decline takes place, as tourists and investments shift geographically over time. Such shifts, for whatever reasons, will inevitably marginalize some resorts and change the characteristics of these tourism places.

The classic resort life-cycle model (see Chapter 8) attempts to provide a descriptive framework of these changes at the resort level. More recent perspectives have argued for a resort development spectrum based on the operation of the market (Prideaux 2000b), while others have sought to explore resort change in terms of theories of restructuring (Agarwal 1994). It is not our intention to discuss these theoretical perspectives, but rather to focus on the endgame of these resort models, which suggest that resorts face choices of redevelopment, rejuvenation, or stagnation and decline (Agarwal 1994). Especially, we are interested in the processes of redevelopment, particularly in terms of a case-study of British resorts.

The responses to resort decline are varied and tend to be located within specific local/regional settings. Throughout many of the Mediterranean mass tourism resorts, the reactions have been toward curbing the main destructive effects of large-scale tourism. In particular, emphasis has been given increasingly to improvements in environmental quality (both physical and social) and the search for new markets. Agarwal (2002) has viewed these responses in terms of restructuring processes, with the focus on product organization and transformation. Such perspectives are useful in two ways. First, they draw attention to the nature of restructuring (involving capital and labour processes) going on in different sectors of the tourism industry (see Chapter 3). This also aids in understanding that resorts are complex tourism spaces, comprising different components and managed in varying ways. Second, these perspectives provide a convenient organizational framework within which to examine the processes of redevelopment. As Table 9.4 shows, it is possible to identify a range of strategies centred on both the reorganization of products and their transformation. Within the context of a case-study of three English resorts, Minehead, Weymouth and Scarborough, Agarwal (2002) has identified nine strategies related to forms of restructuring, namely: product-quality enhancement, diversification, market repositioning, adaption, centralization, preservation (conservation), collaboration, product specialization and technical change. Of course, these strategies are emphasized in different ways by individual resorts, and as the political climate has changed so too have the strategies adopted.

We can illustrate the first point about the nature of restructuring by taking a wider set of examples across the Mediterranean. In those countries touched early by mass tourism, resort restructuring has tended to focus on diversification and preservation concerned with environmental improvement and limits to the physical growth of coastal resorts. More recently, Priestley and Mundet (1998) and Apostolopoulos and Sonmez (2001) have identified the increased importance of collaboration (see Chapter 8). The former noted the growth of collaboration strategies between the private sector and local authorities in Catalan resorts. On a broader scale, Apostolopoulos and Sonmez argue for a restructuring of Mediterranean tourism based on

Table 9.4 **Examples of resort restructuring strategies**

Forms of restructuring	Types of strategies	Examples
Product reorganization	(i) Investment and technical change	Introduction of new facilities, e.g. casino development at Scheveningen, the Netherlands (Van der Weg 1982)
	(ii) Centralization	Formation of 2001 Tourism Summit to support resort regeneration in England
	(iii) Product specialization	Promotion of cultural-event tourism in Sitges, Spain (Priestley and Mundet 1998)
Product transformation	(i) Quality enhancement of product-service	Improved training in mass tourism resorts of the Baltic (Twinning-Ward and Baum 1998)
	(ii) Environmental quality enhancement	The restoration of historic buildings in some English resorts (Turner 1993)
	(iii) Market repositioning	Attempted realignment in the image of resorts toward more special-interest visitors (Agarwal 1999)
	(iv) Diversification	The development of untapped resources to attract new markets, e.g. development of short breaks in Bournemouth and Brighton (Knowles and Curtis 1999)
	(v) Collaboration	Former development of Tourism Development Action Plans (TDAPs) in English resorts
	(vi) Adaption	Market research and the attempts to encourage changing markets (ETC 2001)

Source: Agarwal (2002).

collaborative alliances 'to replace the dead-ended fierce competition of the past' (2001: 283). This would be centred on a cooperative marketing strategy.

Within the context of British and, more especially, English resorts, the response to resort decline has varied and been strongly dictated by state intervention. In turn, the role of the state has changed considerably, and with it the limits of intervention in the resort redevelopment process. It could be argued that the tourism industry, and especially the seaside resort, has been the 'cinderella' of state tourism policy-making. Moreover, the seaside resort has been strongly affected by changing state interest in this economic and social space. The picture is further complicated by the role of sub-state agencies, the local state and the relationship between public and private interests in the tourism industry (Agarwal 1997).

It is possible to identify a series of key shifts in the state's interest in coastal resorts, which, in turn, have conditioned the nature of redevelopment initiatives. At least three phases of engagement with resort regeneration can be recognized within England.

The first, during the 1980s, was associated with a growing awareness of the problems faced by coastal resorts, and solutions were fragmented. State interest and intervention was growing in two distinctive ways. These were:

- The operation of grant aid to elements of the tourism industry in the form of Section Four grants, operated by regional tourist boards (Agarwal 1997; Shaw et al. 1998). Such grants offered financial assistance, which was available to improve accommodation facilities and attractions. In this respect, it did help improve product quality, although this was often taking place in a local policy vacuum. The scheme was abolished in England during 1989, because of cut-backs in government spending. In any case it was not confined to resorts and, increasingly, grants were being given to competing tourism developments in urban and rural areas.

- The rise of area-based strategies during the 1980s, in terms of Tourism Development Action Programmes (TDAPs) and, later, Local Area Initiatives and Strategic Development Initiatives (Bramwell and Broom 1989; Bramwell 1990). These locally based public–private partnerships were designed to promote strategic thinking, although a number failed because of poor collaboration at the local level and through weak leadership. Agarwal (1999) has highlighted their role in resort redevelopment and concluded that their ideas of product improvement and marketing formed a basis for later projects. More tellingly, she concludes from an analysis of three case-studies that the outcome of these TDAPs 'appears to be disappointing, particularly with regard to the halting of local economic decline' (p. 519).

A second phase of intervention occurred from the late 1980s into the 1990s and saw the demise of the TDAP in the face of further changes in government spending. More significantly, the 1990s saw the publication of reports highlighting the need for change in the structure of seaside resorts, with a particular emphasis on the physical environment (Table 9.5). In this context, some resorts were refocusing their development on the notions of heritage, or at least the sense of a Victorian past. This was particularly so in the case of the ideas for redeveloping Weston-Super-Mare in south-west England. Emphasis was also being given to improvements in the economic structure of resorts, with attention directed at the problems of small-scale entrepreneurs.

The third recognizable phase, from the late 1990s to the early part of this century, has witnessed increased attention being directed at the resort product and the problems of local economies dominated by small firms. Moreover, reports by various agencies have increasingly stressed the idea of differentiating the resort product to target different markets. There have also been attempts to learn lessons from other countries, leading to the idea of using new, key attractions to help revitalize resorts. In this context, there has been much debate about the American model of introducing casinos and gambling to increase visitor numbers.

Despite such increased attention, it still remains the case that many British resorts are essentially failing tourist destinations. The plethora of activities by

Table 9.5 **Major reports on the regeneration of English coastal resorts**

Publication	Source	Date
Perspectives on the Future of Resorts	British Resorts Association	1989
The Future of England's Smaller Seaside Resorts	English Tourist Board	1991
Making the most of the Coast	English Tourist Board	1993
Revitalizing the Coast	English Tourist Board	1995
Sea Changes: Creating World-class Resorts in England	English Tourism Council	2001

various agencies and the reports that have appeared have served mainly to highlight problems rather than to put in place workable policies. In part this represents both 'policy failure', and 'policy limitation' in the face of globalization and changes in the mode of regulation (Chapter 2).

SUMMARY: RESORTS IN TRANSITION

This chapter has highlighted the long-term transformation of seaside resorts as pleasure spaces and, in doing so, raised a number of significant themes. The first relates to the spread of the resort into a range of environments related to competition and changes in tourism demand. In addition, the form of the seaside resort is shown to be socially constructed. The resort was historically transformed from an exclusive pleasure space to one catering for mass tourism and, in doing so, produced a range of resort types serving different social groups.

The spread of the resort into the so-called pleasure periphery has ensured that the seaside resort is a global tourism space. The geographies of such resorts are strongly dependent on the roles of capital and the state. Moreover, the global resort has also been recognized as a postmodern space where:

- the built environment of the resort is a key component, with a pastiche style producing an architecture of pleasure

- there is a strongly related commodification of the resort landscape, based on the ideas of consumption spaces (see also Chapter 10)

- the resort is presented as an exclusive space, socially and physically constructed.

The growth of new resorts in the pleasure periphery has produced a switching of demand and investment capital away from older, established resorts. This has led to a decline of such traditional resorts, followed by recent attempts to redevelop such tourism places.

10 Landscapes of Pleasure: the Construction of New Tourism Spaces and Places

THE DEVELOPMENT OF DIVERSE TOURISM SPACES

Tourism spaces are dynamic in that they are constantly being created, abandoned and re-created. Some commentators have viewed such processes via a product life-cycle model (see Chapter 8) and, as we observed in the previous chapter, older coastal resorts have passed through a series of life stages. The early resorts were products of industrialization and related developments in mass consumption – the consequences of modernity. Subsequently, as we have shown, very different resorts have been created in new destinations in the context of postmodernity. Of equal importance, such processes have led to the creation of new and diverse tourism spaces, leading to the commodification of many places (see Chapter 7). The process of tourism commodification, as we have argued, is complex, constituted by, and affecting, spatial and social relationships.

We have already discussed the main factors creating and shaping tourism spaces in Chapter 8. Our aim now is to focus on the more detailed processes involved in the creation of specific types of new tourism spaces. Before examining these, there are three main ideas that require clarification: the conceptualization of tourism spaces, the common features involved in their creation, and their characteristics.

The manner in which space is conceptualized holds the key to understanding the creation and nature of tourism spaces. In this context, we need to highlight the ways in which the sociocultural values of tourists relate to spatial patterns. Zukin (1995) drew attention to a symbolic economy of space and, by doing so, she stressed the important linkages between material and symbolic space. Following the ideas of Lefebvre (1991), Meethan (2001) has argued that the production of tourist space is concerned with material forms along with a symbolic order of meanings. In this context, he suggests that:

- The material order of space may also be viewed as a social one, which embraces both symbolic and material aspects. As we argued in Chapter 7, and will show in this chapter, these symbolic forms are both imposed on material space and derived from it. Such relationships involve 'the creation of coherent spatial representations or narratives' (Meethan 2001: 26). Others have seen such processes as themed environments, i.e. 'products of a cultural production process that seeks to construct spaces as symbols' (Gottdiener 1997: 5).

• Tourism is strongly related to the production and consumption of specific spaces, although, as we shall argue, these are diverse in nature.

• The production of these tourist spaces and their commodification is mediated at different spatial and institutional levels (Chapter 8).

• The creation of such spaces and their relationship to place is part of an open dynamic system.

Our second main theme concerns the processes involved in the creation of these tourist spaces. Zukin, commenting on urban change, drew attention to the 'production of space, with its synergy of capital investment and cultural meanings', alongside the parallel 'production of symbols' (1995: 23). The ways in which such 'production' operates may be viewed at different levels, as Gottdiener (1997) points out. Moreover, Britton (1991) argued that places and sites can be incorporated into the tourism system in two main ways. The first involves the tourism industry giving much more powerful meanings to its products by associating them with particular places and themes. For example, different destinations may be associated with television soap operas or other elements of media culture (see Shaw and Williams 1994; 2002). The second involves the ways in which particular attractions are associated with, or assimilated into, a tourism product: for example, the way that many new shopping malls have incorporated leisure and tourism roles. In this chapter we want to focus on the specific modes of production that have given rise to particular tourist spaces. In doing so, we stress both the more novel aspects of some processes, while at the same time recognizing that these are embedded within more general, common mechanisms. This implies a similarity between the realm of economics – 'capital and commodities – and that of communication – signs and symbols' (Methan 2001: 38) or what we have described as the experience economy. In turn, these two realms can be viewed in terms of production – circulation – and consumption (Featherstone 1991; Meethan 2001).

The final theme concerns the diversity of the new tourism spaces that have been created by what Gottdiener terms 'themes in the circuits of capital' (1997: 48). He claims that there are many such themed environments, which represent 'the melding of material space with the media-scope of television, advertising, movies, cyberspace and commodity marketing' (p. 75). Growing competition has led to increased use of and experimentation in themes and symbols appealing to consumers. These forms of 'symbolic differentiation provide the tourist industry with the raw materials, out of which tourist space can be constructed' (Meethan 2001: 27).

We can recognize, therefore, two aspects of the production of new tourism spaces: one concerns the importance of 'theming', while the other is associated with the way in which such theming is used to construct different tourism spaces. In this context, we can identify a number of themed elements, including theme parks, shopping malls, heritage centres and restaurants (Gottdiener 1997). These must also be viewed in terms of how they have been used in different environments, for example, in different types of urban areas, to help recreate such places. In doing so, it becomes clear that the creation of new tourism spaces has, in many instances, been a goal of the state in conjunction with the private sector. Early work by Harvey (1985) saw this in terms of competition among cities to be major centres of consumption. Hannigan (1996) has drawn attention to the importance of understanding the identity of the gatekeepers who control the evolution of what he terms the 'Fantasy City'. Indeed, we would argue that too much of the discussion on the development of new tourism spaces has focused on the general aspects of production and consumption, while neglecting issues of access and use.

In this chapter, we examine the developments of tourism spaces around the notions of themed environments and symbolic spaces. We do so within the contexts of development processes and the creation of public and private spaces (Cybriwsky 1999). We start with a discussion of themed attractions before going on to consider their use within different environments, as mechanisms for redevelopment, and then move on to consider the wider use of the heritage theme as a means of regenerating redundant industrial spaces. Moreover, in the background to these discussions lies our concern with how tourism contributes to, and is shaped by, places.

FANTASY SPACES: FROM DISNEYLAND TO LAS VEGAS AND BEYOND

The opening of the first Disneyland theme park in California during 1955 offered a completely new leisure space for Americans. It provided 'an encounter for its visitors that was so unique and compelling that it became a new form of commercial enterprise' (Gottdiener 1997: 109). As such, it represents a classic example of disruptive or innovative competition as discussed in Chapter 4. It was fundamentally different from other leisure spaces in three main ways:

- Unlike existing funfairs of the type found in many coastal resorts, visitors had to pay not only for the rides but also to be admitted to the theme park. In this context, visitors were paying to experience the total built environment. To visit Disneyland 'is its own reward' (Gottdiener 1997: 109), as the architecture provides a degree of fantasy and entertainment through its symbolism.

- Visitors at Disneyland enter a carefully constructed, heavily regulated and well-choreographed space. Such regulation in turn engineers the tourist experience (see Chapter 6) and also serves to impose a degree of visitor control within the theme park. As Smith explains, the visitor is immersed in a fantasy environment that 'provides entertainment and excitement, with reassuringly clean and attractive surroundings' (1980: 46). The most dominant feature is, of course, the way Disneyland and its successors are constructed around distinctive themes, or fantasy realms. The original Disneyland contains four different themed areas, each of which represented features developed in films and television programmes made by the Disneyland Corporation. For example, the realm of Frontierland was based on representations of early settlements, as highlighted by old Disney film classics such as 'Davey Crockett' – about the American frontiersman. As the Disneyland theme park became transplanted into new destinations, so the complexity and elaboration of the themed environments changed. Thus, in Disney World, Florida, the Epcot centre, with its laser-light shows, also forms the focal point of a World showcase, represented by mini-themed areas of 11 countries (The Project on Disney 1995).

- Theme parks of the Disneyland type provide a strong contrast to the spaces of daily life. Gottdiener (1997) has highlighted a number of these contrasts in terms of the American way of life, as shown in Table 10.1. In this context, he argues that the success of Disneyland and related theme parks is 'largely because it liberates people from the constraints of everyday life' (p. 114). In saying this, he refers especially to car-orientated, suburban America. This liberation is conditional on the strongly engineered conditions and environment presented by the theme park.

Of course, these spaces are also successful because they are part of a well constructed commercial organization, built on all aspects of the entertainment industry. The extension of Disney movies and their characters into the built environment of the theme park is a clear example of cross-marketing, as is the retailing of Disney merchandise. Williamson (1978) describes these as metastructures, where, in advertising terms, meaning is decoded within one

Table 10.1 **Contrasting environments of the theme park and daily life**

Theme parks offer spaces that can be enjoyed by pedestrians, which stand in sharp contrast to the daily lives of many suburban Americans, where the car is dominant. This same pedestrian experience is also a key element in the themed shopping mall.

Theme parks provide safe, regulated environments free from crime, unlike the wider experience of urban America.

The theme park is a festival-type environment, with both active and passive forms of entertainment. This contrasts with the dominance of passive entertainment in many people's suburban lives.

The theme park provides a seemingly liberating experience, with its illusions of escapism, from the demands of everyday life.

Source: modified from Gottdiener (1997).

structure and then transformed to create another (Goldman and Papson 1996). Such commercial networks are at the core of fantasy spaces, where reality is a construct and both space and time are fabricated as a series of identifiable themes. It is these strong commercialization and fabrication tendencies that have attracted particular criticism (Sorkin 1992; Smoodin 1994). In his wide-ranging discourse, Fjellman (1992) claims that Disney World is the tourist attraction *par excellence* and that the theme park is the 'most ideologically important piece of land in the United States' (p. 10). Such a remarkable claim is based on the influences Disney theme parks have on wider practices and patterns of consumption in America (Hollinshead 1997).

The continual success of Disneyland and its offshoots is based on a number of elements, which MacDonald and Alsford (1995) summarize as:

- high-quality visitor services, which satisfy many of the needs of the postmodern tourist who are 'playful consumers of superficial signs' and spaces (Williams 1998: 189; see also Chapter 5).

- multisensory experiences, which involve:
    (i) simulated environments (natural, cultural, historical and techno-logical)
    (ii) the humanizing of these environments by live interpretations and performances
    (iii) state-of-the-art films
    (iv) themed exhibits and eating places

- a highly structured experience, which attempts to counterbalance the risk of information overload through the structured programming of visitors.

- the constant reinvention and upgrading of experiences, with the latest technologies

Globalizing the theme park

The US theme-park industry has seen a number of changes since the 1990s, the most prominent of which is increased competition, partly within existing, as well as changing, parameters (Chapter 4). Of particular importance has been the acquisitions undertaken by Anheuser Busch to rival the power of Disney World within Florida. In addition, competition has also increased, with the entry of Universal Studios into the theme-park business. As Braun and Soskin (1998) argue, this increased competition has led to major structural changes and shifts in the scope of the product itself. Thus, Universal Studios target the young-adult market by focusing on more high-tech thrill rides set within a greater level of sophistication. It has been able to do this by drawing on scale economies and complementaries in film and television production. In doing so, it has opened up strong competition with Disney and intensified the pace of change.

The popularity and commercial success of the theme park in the United States has contributed to its international diffusion. The transfer of such developments to Europe has proved relatively easy, and Euro-Disney, located outside Paris, was able to utilize the same formula as the American theme parks. Moreover, the French experience also demonstrated the key role of collaboration with the national and local state (d'Hautserre 1999). The global spread of the theme park is a significant trend, with its most recent focus being the Asian-Pacific Rim. Jones (1994), for example, has drawn attention to its development in Japan and argues that the opening of Tokyo-Disney in 1983 led to a rapid expansion of similar, competing theme parks. Tokyo-Disney, like its early predecessors in the United States, was an instant success, attracting 10 million visitors in its first year and rising to over 17 million by 1997. Other similar-style developments include Canal City, Hakata, a large themed space in Fukwoka, which attracted up to 8 million visitors when it opened in 1996. In this instance, a large canal acts as the main thoroughfare through the development, which also includes 'Joypolis', a 55,000 square-foot Sega games arcade and a large 13-screen multiplex cinema (Hannigan 1998). Similarly, in South Korea, two major theme parks were developed in the late 1990s – the Kyongju World Tradition Folk Village, on a 1000-acre site south of Seoul, and Samsung's Everland.

These, and many other developments throughout the region, are being driven by increased consumer purchasing power in parts of South-East Asia. This has created a new middle class, which, in cities such as Bangkok, is set to comprise almost 20% of households by 2010 (Hannigan 1998). In turn, this has led to a growing demand for leisure products and new leisure experiences. These new landscapes of pleasure are being shaped by a mixture of global and local influences, as Hannigan indicates. Moreover, with the exception of Japan, few of these new theme parks are being developed by global companies such as Disney or Universal. Indeed, most are being capitalized by corporate conglomerates and local or regional millionaires, although the developments are often designed by North American companies, such as the Duell Corporation, Landmark, International Theme Park Services and Forrec. This picture emphasizes both the complexity of globalization and the persistence of national regulation (Chapter 2).

Such a complex of capital, development and design factors has produced a mixture of global and local culture. In this context, Lee (1994) talks of the unfolding of a new development paradigm, which stresses ambivalence – where global consumer culture is tailored to the local. At its most extreme, local tradition, lifestyles and culture may be compressed into the themed space and presented in an easily recognized way for visitors (Teo and Yeoh 1997). However, such processes do not happen in a vacuum, since both local people and the state can act to distil global forces. In this context, Chang et al. (1996) demonstrate that local agencies are not merely passive recipients, and that many destinations tend to 'accentuate themes peculiar to their culture and location as a way to differentiate themselves from competitors' (p. 287). For example, many of the recently developed theme parks have reconstructed traditional activities, such as dances and crafts, which, although aimed at the international tourist, may also, according to Hannigan (1998), be attractive to local and regional visitors. He suggests there are two main models of theme-park development within the Asia Pacific rim, namely:

- The so-called 'buffet model', within which the visitor is given a mixture of choices (attractions) covering both global and replicated traditional culture. An example is the Kyongju World Tradition Folk Village, where visitors are faced with a range of themed, cultural zones alongside a recreational centre focusing on Korean history.

- What we can term the 'local model', where a theme park is developed around local culture. This is usually strongly commodified, albeit often around more local themes. Teo and Yeoh (1997) have termed such developments cultural theme parks and have explored their characteristics in terms of Singapore's Haw Par Villa. This original leisure space, developed between 1937 and 1954, was transformed into a cultural theme park with private capital under Singapore's Tourism Product Development Plan, initiated in 1984 (Teo and Huang 1995) (see Box 10.1).

To these two variations of theme-part development we should add a third, which is the extension of the American theme-park model as represented by the Disney Corporation and its emphasis on promoting a global culture. Of course, there are modifications, as commentators such as Hannigan and Tesh and Yesh suggest, but these in no way negate the global reach of the theme-park concept.

Las Vegas: a global pleasure zone

One of the most effective uses and extensions of the theme park is in Las Vegas (Gottdiener et al. 1999). This is both a 'global pleasure zone' and a 'fully themed environment' (Gottdiener 1997: 100). The city's postmodern architecture 'proudly celebrates commercial vulgarity' (Urry 1990: 121). Its transformation during the mid-1970s from a western-style gambling resort to a global pleasure zone is based on the creation of a well-planned themed environment, where fantasy and pleasure are unifying motifs in the landscape. This is a complex environment of themes 'and the rapid-fire transmission of distinct messages' (Gottdiener 1997: 101). This fantasy space is multifunctional, offering visitors gambling, sex, food and nightlife. Moreover, the gambling economy of Las Vegas is embedded within structures that are a complex series of themed resorts, which both compete with and complement one another. As in Disneyland and other theme parks, the 'resorts' of Las Vegas contrive to falsify place and time. These themed resorts are developed around Las Vegas Boulevard, or what is referred to as The Strip (Figure 10.1). The fantasy nature of the place is summed up by the guidebooks, one of which

---

Box 10.1   Singapore's Haw Par Villa: A cultural theme park

Background

Haw Par Villa is located in Singapore's Pasir Panjang Hill on a site fronted by views of the sea. The mansion and its grounds were constructed in 1937 by two millionaire brothers whose business empire included pharmaceutical companies, banks and publishing. The villa's name (in Chinese, 'Haw Par' means 'tiger' and 'leopard') reflects their owners' names and their most famous early product – 'tiger balm'.

Iconography of the Villa's landscape

The grounds of the villa were intended to be a place for relaxation and moral instruction. Between 1937 and 1954 some 1000 statues, along with 150 giant tableaux depicting Chinese folklore, legends and history, were constructed. Most were legends with a moral tone and the gardens became popular with local visitors, who referred to them as the 'Tiger Balm Gardens'.

Creating a Chinese mythological theme park

By the 1970s, the villa and gardens had been taken into public ownership, and in 1984 the Singapore Tourist Promotion Board identified the site as a potential showcase of Chinese heritage, in an attempt to improve the state's tourism industry. Haw Par Villa was to be part of the 'Exotic East' theme, and developed as a theme park. The Disneyfication process was undertaken by public–private partnership, with Battaglia Associates Incorporated (an offshoot of Walt Disney Productions) being commissioned to design the park. It was promoted as a unique Chinese mythological theme park. The emphasis is on two key areas:

- the uniqueness of the villa with its authentic Chinese origins

- the re-writing of the landscape using technological wizardry to create 'Dragon World', so the original Chinese scenes now merely form a backcloth to the new attractions.

Source: based on Teo and Yeoh (1997)

---

proclaims: 'The last time you saw Paris, it was probably right where it always was – in France. Now it's located next to Morocco [sic], just across the street from Italy. If this geography seems a little askew, that's because we are talking about Las Vegas' recent international flavour' (AA Publishing 2002: 6).

The city is one large fantasy space, comprising a series of theme parks, each of which offers varying elements to the visitor, such as hotels, casinos, art galleries, museums, tourist attractions and shopping malls. These, in turn, are a complex series of overlapping symbolic spaces, which have different orders of meaning. According to Gottdiener et al. (1999) these themed areas produce a grand text of a profusion of signs that represent the visitor experience. In manifest material terms, this may be measured by approximately 15,000 miles of neon lights that illuminate the city, and latently by a set of powerful capital accumulation processes. In this highly competitive tourist destination, where image is all important in attracting more consumers, the pleasure environ-

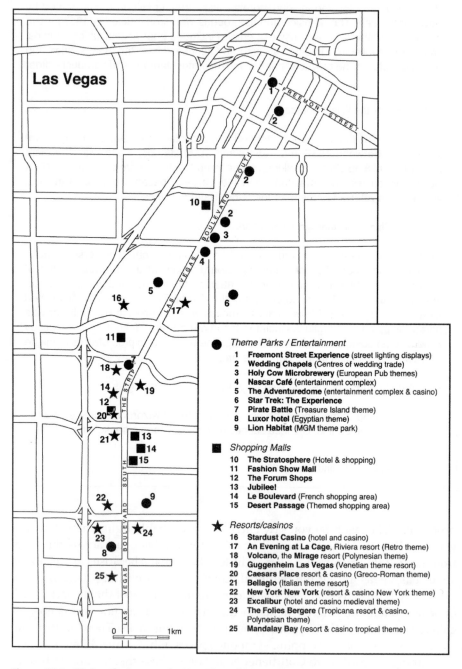

Figure 10.1 **Major themes and attractions in Las Vegas**

ment is reduced to signs and themes, that suggest a differential to the experiences on offer.

Competition, in all its forms, is the key to innovation in this tourism complex. The production of these tourism spaces is the outcome of a complex set of factors, which starts with the competition among casinos. These offer a similar product, but attempt to differentiate themselves by promoting distinctive experiences. As competition increases, so do the scale and production of the themed resorts within the city, and as a consequence, the city is a completely simulated environment, constructed around a set of competing fantasy spaces. For example, New York is represented by a replica of the city's skyline, along with landmarks such as the Statue of Liberty. This development is based around a massive casino, which also includes, within the roof of the gaming areas, the 'Manhattan Express' roller-coaster ride. The visitor is therefore faced with a complex set of overlapping pleasure spaces at a number of different scales.

Over time, the themed environments of Las Vegas have become more elaborate and drawn on a wider set of motifs in response to the global identity being sought by the city. It now claims to be the second most popular tourist destination in the world after Disney World (Florida), attracting some 37 million visitors annually – 11% of whom are from outside the United States. This position of global prominence has been reached through a series of distinct investment stages (linked to innovations), each of which has added to the city's competitiveness and theme-park image.

From the mid-1970s to the early 1990s Las Vegas transformed itself from a western American casino resort to a themed city with global reach. To increase their appeal, casinos invested in themes to create fantasy spaces. The early themes tended to be based on associations with other gambling resorts, such as Monte Carlo, or by falsifying history, especially of ancient Rome and Egypt. Since the 1990s, a second major phase of theming has occurred, based on the mixing of gambling with family-based entertainment and, more especially, shopping malls. This phase has seen the influx of large investment capital from major transnational corporations, such as MGM Grand. Brown and Soskin have outlined the shifts in investment capital and the application of what they term a 'new version of the theme park model' in Las Vegas (1998: 440). This is a 5000-room hotel, casino and entertainment complex, themed around Hollywood's MGM studios. We have, in this particular case, one simulated environment being recreated and represented in another. Despite the global reach of the city, its economic success is also contingent on the particular national and state regulation systems, which facilitate both venture capital along with rapid planning and development decision-making.

CONSTRUCTING NEW LEISURE SPACES: THE RISE OF THE THEMED MEGA-MALL

One of the significant features of the theme park in all its guises is the strong integration of retail outlets. Within Disney's theme parks, for example, cultural values are powerfully reinforced by product-marketing, the various souvenirs reminding the visitors of the mythology of Disney. Similarly, in the themed spaces of Las Vegas, shopping malls are key elements of the city's 'resorts'. For example, visitors to 'Paris Las Vegas' can experience 'Le Boulevard', consisting of 15 French-style boutiques set among winding alleyways, to represent the notion of shopping in Paris (Gottdiener et al. 1999).

Parallel to these developments, most countries have witnessed the growth of the themed shopping-mall, which has increasingly been termed the mega-mall. Once again, this was initially largely an American experience, which was successfully exported. The themed mall has constructed a space for shopping and leisure, with the emphasis on play and fun within a controlled environment. The stress is on engineered, simulated environments, aimed originally at 'recapturing urban ambience' (Gottdiener 1997: 86). Crawford goes further and argues that such malls 'manipulate space and light . . . to create essentially a *fantasy urbanism* devoid of the city's negative aspects' (1992: 22).

Over time, these themed malls have been transformed in two main ways. First, they have become more complex leisure spaces as the importance of theming and entertainment has grown, and this has also increased their size. This development has been in response to increased levels of retail competition and the need to differentiate the shopping experience (Wrigley and Lowe 2002). And such developments have also responded to the changing demands of the consumer and the growing importance of leisure-shopping. The second key transformation relates to the locations of such malls. Early developments in North America and western Europe were located out-of-town and designed to compete with established city-centre retail cores. However, since the late 1980s they have been developed in a range of environments, including city centres. Furthermore, the themed mall has been increasingly used to help regenerate failed industrial areas. It is worthwhile considering both of these important transformations in more detail, as they hold the key to understanding the construction of a wider set of new tourism spaces.

One of the largest and most complex themed shopping malls is the Mall of America, opened in 1992 in Bloomington, Minneapolis. It was developed by the same group that had opened one of the earliest of these mega-malls – the West Edmonton Mall in Canada. Gottdiener claims that its themed environments are merely 'a thinly veiled disguise' (1997: 89) for what is a large retail area. This is only partly the case, as the development is not only a 'mega-retail space' but also an important space for family entertainment. The seven-acre theme park represents an attempt to create a major visitor attraction and what Goss views as the 'apotheosis of the modern mall' (1999: 45); it even operates its own tourism department (see Box 10.2).

In Britain, such themed malls are on a smaller scale, but nevertheless constructed around the same simulated scenes that falsify place and time or which produce facades based on film culture. These malls have been increasingly used to revitalize older industrial areas, as at the Metro-Centre in Gateshead (north-east England), Meadowhall near Sheffield, and the Trafford

Park Centre in Greater Manchester (Chaney 1990; Shaw and Williams 2002). In some cases, as at the Mall of America and the West Edmonton Mall in Canada, the developers planned to create both a shopping experience and a tourist attraction (Goss 1993; Jackson 1995). In doing this, they retain the

---

Box 10.2 The themed mega-mall: the Mall of America

The Mall of America occupies some 78 acres of land in Bloomington, Minneapolis, and was developed as a retail-leisure complex by the Ghermezian brothers. They had rolled out this concept in the West Edmonton Mall in Canada. The Mall of America opened in 1992, adding some 2.5 million square feet of retail space to the Minneapolis area. It contains over 520 stores and the total visitor traffic is estimated at 42.5 million per annum. Of course, this total includes frequent shoppers, although tourists account for an estimated 40% of all visitors. It also has an international clientele, in that some 2.6 million international tourists visit each year, representing around 6% of total visits.

Themed spaces in the Mall

The mall is constructed around a major leisure complex the heart of which is 'Camp Snoopy'. This claims to be the largest indoor themed entertainment park in the United States, covering some seven acres, and it includes 25 rides and attractions. Publicity material describes it as 'an imaginative world of fun, with live, full-sized trees, natural wood buildings, a waterfall, and a stream flowing through the park' (www.campsnoopy.com/general/general.htm2002). Here, Nature is captured and represented as a safe, controlled environment for children. Camp Snoopy is also part of the main theme offered by the Mall, which is one symbolic of the United States itself. This representation begins at the exterior of the Mall, which is emblazoned with the 'Stars and Stripes'. The theme is then partly developed within the four main retail areas, which are:

- North Garden – described as Main Street, USA.

- West Market – which is supposedly symbolic of a European-style market place. In this context, the American theme is largely abandoned in the contrived attempts at representing a different urban scene.

- South Avenue – again this is described in publicity as being representative of European shopping streets.

- East Broadway – this is more of an American street scene, with a focus on up-market stores.

As Gottdiener explains, the 'supercession of reality is typical of themed environments that are also mere simulations of distinct cultural places' (1997: 88). Such themed commercial spaces are mere attempts to create symbols with which to connect these retail areas to 'real' shopping environments.

Sources: www.campsnoopy.com/general/general.htm and Gottdiener (1997)

attention of visitors for longer time periods and thereby attempt to maximize visitor spend. Furthermore, these themed malls allow the reproduction of lifestyles through marketing practices, as they provide ideal environments for replicating signs and motifs important within consumer culture (see Chapter 5). These combine to produce a strongly image-driven environment, where 'fantasy' rules. They encourage the notion of playfulness in terms of the consumption experience, but, of course, always in a strongly engineered setting. For many postmodern consumers, as we argued in Chapter 5, marketing practices have reduced lifestyles to themes, signs or text codes. These malls are postmodernist spaces of consumption, where signs, representations and simulations are all dominant and expected by visitors (Baudrillard 1983). At a more basic level, they are also safer and more attractive environments for consumers, because they are controlled private spaces dominated by the pedestrian. However, while the importance of signs and representation is undeniable, the embedding of these malls in material relationships is also important. A number of them are driven by partnerships between the state and regional, national or international capital. For example, in Britain the Metro-Centre at Gateshead was built by a regional entrepreneur in an enterprise zone (Lowe 1993).

TOURISM SPACES AS THEMED ENVIRONMENTS

As we have argued, theme parks and various shopping malls are important elements in the creation of postmodern tourism spaces, and many authors view them as 'quintessential' postmodern spaces, with their emphasis on mixing different styles of architecture and the 'deliberate confusion of the real with the artificial' (Williams 1998: 189; see also Venturi et al. 1972; Gottdiener 1995). We would go further and argue that the theme park culture, in particular, is an important and widespread phenomenon, which can be recognized in three significant ways:

- It champions the creation of themed environments, through its emphasis on the commodification of leisure and tourism spaces. The business lessons of Disney's success have been widely adopted by competitors.

- As we have seen, it acts as a global carrier of themed attractions, which have now spread into almost all countries, and it creates a space where global culture may be mediated by local traditions.

- It produces a powerful commercial space, which can act to remake the tourism geographies of destination areas in a wide range of environments (Williams 1998).

Theme parks also clearly illustrate the ideas and success of invented spaces and places. As we have argued elsewhere, the importance of themed

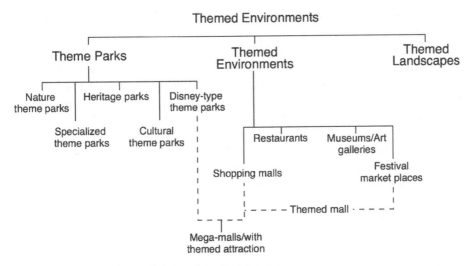

Figure 10.2 **A typology of themed environments**

attractions from the tourist's perspective is that they are 'both locationally and perceptually convenient' (Shaw and Williams 2002: 207). Extending Gottdiener's (1997) ideas, it is possible to recognize a typology of themed spaces, as shown in Figure 10.2. These range from the original theme park through to the theming of different tourism landscapes. For Boorstin (1994) and Urry (1990) these themed spaces are self-perpetuating systems of illusion centred on signs that locate tourist practices.

We can add to these cultural perspectives by viewing the acts of theming as the outcome of capital accumulation processes. These have intensified during recent years, in the face of increased competition and changes in the nature of consumption, as discussed in Chapter 5. These forces have been seen by some as the commodification of landscapes (Boniface and Fowler 1993) and by others as merely some sort of marketing ploy (Williams 1998). In reality, the use of theming relates to the interplay of all these forces and results in a complex system of tourism spaces, characterized by the importance of signs, signifiers and symbols.

The complexity of these themed spaces has partly been revealed by Gottdiener's (1997) work, but we would argue that such ideas can be extended to embrace a wide range of themed spaces. As Figure 10.2 suggests, it is possible to recognize three main subsets of themed spaces: theme parks and related attractions, themed environments, and themed landscapes. In turn, these main types include a range of variations on, and extensions to, the original ideas of theming. For example, there has been increasing experimentation with the original theme park concept, resulting in a variety of offshoots. As we have argued, global–local relationships in parts of South-East Asia have been integral to the spread of the cultural theme park. In contrast, more specialized theme attractions have been developed, such as Dollywood in the United States, which aims to be representative of the lifestyle of the country-and-western music icon Dolly Parton (Gottdiener 1997). A more complex group of visitor attractions has grown around the development of heritage centres, where, in many instances, there has been a degree of borrowing from the theme park 'proper'. As MacDonald and Alsford argue, the worlds of the heritage centre and the theme park 'have made contact and a zone of intersection has become apparent' (1995: 145). It is at this zone of intersection that we can recognize a growth in the use of theme-park technology and theme-park styles of presentation being used, not only in heritage centres, but also in many museums. In this latter context, we can see linkages between the various subsets of themed spaces (Figure 10.2). A greater number of heritage centres and museums are, for example, being promoted around particular themes. Thus, in the UK, a former World War II prisoner-of-war camp – the Eden camp in Yorkshire – is now promoted as the 'Modern History Theme Museum' (MacDonald and Alsford 1995). Furthermore, many heritage centres now use multimedia presentation techniques and actors to represent past lives as at the Ironbridge Museum at Telford, in the West Midlands, or at the Wigan Pier Heritage Centre, in Greater Manchester (Shaw 1992). Orbasli claims that a new generation of high-tech museums are appearing where 'virtual reality is being used to reconstruct the past' (2000: 79). Certainly, there is increasing evidence to suggest a blurring of boundaries between the museum, heritage centre and the theme park (Moore 1997). Both in Norwich and Croydon, in the UK, there are examples of visitor experiences based on the location of local museums in shopping centres. As with theme parks, these represent new forms of competition, and often involve significant public–private collaboration in the context of new forms of governance.

A final set of themed attractions to emerge in recent years have been what we would call Nature or, in some cases, eco-theme parks. In their simplest form, they are constructed around themed gardens, such as the 'Lost Gardens of Heligan' in Cornwall, with the emphasis on a rediscovered garden heritage. At present, possibly in a category of its own, is the eco-theme park known as the Eden Project in Cornwall. This is sold as a unique experience based on the concept of biodiversity and substainable living, in a setting that has as its centrepiece two large biospheres or domes, which house plants from different climate zones (Box 10.3). Here, emphasis is on Nature features, but, of course, these are in artificial settings, so they share some characteristics of the simulated theme-park environment. The outstanding commercial success of this innovative project has, undoubtedly, contributed to changing the parameters of tourism competition in this part of south-west England.

The second main type of themed space consists of those we have termed 'themed environments', in that they encompass a range of settings, from spaces in everyday life to those that act as visitor attractions. Gottdiener (1997) has discussed the emergence of the themed restaurant and café in the American context and we can recognize the spread of such commercial spaces on a global scale. Representative of these developments is the growth of such chains as 'Starbucks' and the 'Hard Rock Café', which offer a particular image based on a cosmopolitan theme. They develop recognizable global brands around particular styles and logos.

Box 10.3   The Eden Project as a themed attraction

Development and location

The Eden Project is located in Cornwall, south-west England, and opened in 2001. It represents a visionary development, aided by Millennium Lottery funding, to create a sustainable tourist attraction within redundant china clay workings. The attraction is sited in an abandoned quarry and based around two large 'biomes', or environmentally controlled domes, that contain vegetation from different climatic regimes. The largest biome measures 240 metres by 110 metres and is 55 metres high. It contains well over 1000 plant species along with constructed waterfalls and variety of created tropical and sub-tropical landscapes. These provide themes within the biome, based on different regimes such as tropical rainforests and agricultural systems. The second, and smaller biome, represents warm temperate climatic regimes, such as the Mediterranean.

Visitors and the themed environment

Unlike other tourist attractions, the Eden Project presents itself as an environmental and sustainable experiment to 'engage and educate the public at large, not just scientists. It will educate, but will do so with a light touch and a style, already emerging, which will delight and amuse as well as inform' (www.edenproject.com/2929.htm). This does not imply that Eden is not a tourist attraction, since it clearly is − attracting almost 2 million visits in its first year. Eden consists of themed environments that closely replicate reality, in that the plant species are real; but they nevertheless fabricate the geography of tropical and warm temperate climates. It has a strong educational programme, similar to many heritage and museum attractions, but for most visitors it represents an exciting day out to a very different tourist attraction.

Source: based on www.edenproject.com

There are other themed environments that have brought together a series of consumption spaces. Of particular note are festival market places. These commercial spaces are themed around speciality shopping, restaurants and entertainment. Early developments were pioneered in the United States by Rouse, who successfully redeveloped the Faneuil Mall and Quincy Market areas in central Boston in 1976 (Whitehall 1977; Law 2002). Such concepts spread rapidly throughout North America, with many developments being funded by combinations of public–private investment capital as part of urban regeneration schemes. Law (2002) estimates that, by the 1990s, there were at least 25 such schemes in North America. Their prime draw is speciality retailing, with an emphasis on products attractive to tourists. The global reach of these festival market places is a testament to their commercial success, and they have been developed in western Europe, Australia and South Africa. Boyer (1992) and Goss (1996), examining specific examples of these developments, conclude that they represent an environment themed around 'illusions of an idealised past' (Law 2002: 169). These pastiches of the past are constructed around old-fashioned speciality stores and are also packaged within architectural settings that emphasize individuality and quality. As in the theme park, private space is regulated, carefully constructed and themed, and ultimately underpinned by a distinctive process of capital accumulation. The festival market place represents a different shopping and leisure complex from that provided by the mega-mall previously discussed.

The final main group of themed spaces, we would argue, are themed landscapes, which help construct the tourist gaze and have increasingly been created by the activities of the tourism industry and the notions of destination- or place-marketing (Gold and Ward 1994; Bramwell and Rawding 1996; Shaw and Williams 2002). These are increasingly themed around popular culture, especially television programmes, films and music, although in many cases there are links between literature and film. In this sense, many places have attempted to develop literary tourism trails, as tourists are drawn to sites associated with novels, especially if these then become popular films. Very often, fictional events and characters generate the strongest images as Pocock (1987) has shown. In a broader context, Herbert (2001) has attempted to summarize the qualities of a literary place, which encompasses the marketing and development of such destinations around specific themes (Figure 10.3). Places and spaces acquire meanings from these imagined worlds (Herbert 2001). Sometimes the links between a novel and its film are themselves continued. The most recent example of this creation of themed landscapes from literary fiction has been the making of J.R.R. Tolkien's *Lord of the Rings* into a series of films, based around locations in New Zealand. Since the release of the first film in 2001, there has been a consistent and successful attempt to theme New Zealand's landscapes around key locations in the film (Box 10.4). The film has provided a powerful set of images for the tourism industry to theme destinations around Tolkien's imagined world. In recalling MacCannell's (1976) account of the semiotics of tourist spaces, Meltzer (2002) emphasizes the formation of these attractions, or what MacCannell called 'site sacralisation'. This covers five main stages, starting with 'naming' or designating the site through to the final stage of 'social reproduction'. It is this latter stage that, we would argue, is becoming especially important, with the naming or labelling of landscapes. This signposting, or theming of the landscape, serves two key purposes. One is to provide convenient guides to tourist about what they are seeing, and the other is to enhance the experience-value of the destination, and is linked to tourism marketing.

PLAYING WITH THE PAST: CREATING TOURISM SPACES IN POSTINDUSTRIAL ECONOMIES

The past, as represented by the heritage industry, has been a powerful attraction to the postmodern tourist. The relationship between heritage and tourism is central to the debates on postmodernity, as outlined in Chapter 5.

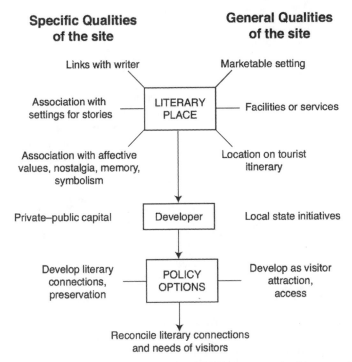

**Specific Qualities
of the site**

**General Qualities
of the site**

Figure 10.3   **The construction of themed literary places (modified from Herbert 2001)**

**Box 10.4   Tolkien's *Lord of the Rings* and the theming of
New Zealand's tourism places**

Background

The adaption of Tolkien's *Lord of the Rings* by New Zealand director Peter Jackson into an ambitious and commercially successful series of films forged the link between literacy and tourism. The decision to use New Zealand as the location for the film established Tolkien's imagined world in identifiable land-scapes. Of course, the original landscapes have, in most cases, been digitally enhanced, in this case by the special effects company Weta, based in Wellington.

Locating 'Middle Earth'

The identification of Tolkien's imagined world of 'Middle Earth' with particular places in New Zealand has been strongly exploited by the tourism industry. This has resulted in the establishment of *Lord of the Rings* tours and the theming or rebranding of locations to fit with places in the film, as shown in Figure 10.4. Large numbers of tour companies are now pursuing marketing based on the film, and especially aimed at adventure tourists and backpackers. Most are strongly promoted on the World Wide Web, which allows companies to firm-up the links between film locations and tourism destinations.

Figure 10.4 **Theming tourism places in New Zealand**

Postmodern consumers show a new awareness of the past, often in the form of nostalgia, which, some argue, has been bred by the 'maelstrom of capitalism' that has contributed to the disruption and displacement of traditional structures (Goldman and Papson 1995). Increasingly, postmodern society is organized around commodity relations, which have facilitated the demand for heritage-based tourism, which, has commodified the past (see Chapter 2). The heritage industry, according to Nuryanti, offers 'opportunities to portray the past in the present' (1996: 250). Additionally, Zukin (1998a) has seen the relationship as part of a broader search for authenticity among the new middle classes, manifest through the processes of gentrification and the reclaiming of urban areas by these social groups.

Our intention is not to rehearse the interest of postmodern tourists in heritage, but rather to discuss the creation of new tourism spaces associated with heritage tourism. Within this context, one significant trend has been the regeneration of 'the redundant spaces of modernity' (Meethan 2001: 22) – old urban and industrial landscapes – through postmodern trends in tourist consumption. The processes producing these changes have been viewed through a variety of lenses. For example, some authors have emphasized the importance of place-identity and place-marketing in the increasing competition among urban areas to be major tourism destinations (Ward 1998). Given that many places attempting to reinvent themselves are past industrial centres, it is hardly surprising that heritage has figured strongly in their strategies (Bramwell and Rawding 1996). Zukin (1995) argues that these are attempts to exploit the uniqueness of accumulated fixed capital – including heritage.

A second perspective views the links between postmodern tourism, heritage and regeneration as part of a series of changes in the patterns of urban management. In turn, this is related to the growth of public–private partnerships, within the context of new forms of governance, which have attempted to reverse economic decline. For example, Ehrlich and Drier (1999) have detailed the redevelopment of Boston as a tourist city through the re-use of historic buildings and the creation of themed retail facilities aimed at the visitor. Similarly, Law (2002) has highlighted the revival of Baltimore as a tourism centre through the efforts of the city authorities and public–private investment strategies. These tourism-based urban regeneration projects lie at various points on the continuum between 'growth machine' and 'growth management strategies' (Gill 2000; see Chapter 8).

These different views are, in fact, emphasizing a set of forces that have combined to aid the transformation of postindustrial landscapes into new playgrounds of consumption. These interlinked forces involve the re-use of industrial spaces, the construction of a heritage industry and the strategies of place promotion. The transformed settings are constructed around new sets of tourism spaces that blur the major distinction between 'signifier and signified, copy and original, and past and present' (Gotham 2002: 1738). However, Gotham, following Gottdiener (2000) and Scott (2000), argues that cultural perspectives have generally failed to expose the links between culture and the political economy. In this respect, they have neglected the profit-making role surrounding the production of the new tourism spaces. To this we should add that this neglect also extends to the role of the state through public-private initiatives. Gotham (2002) has attempted to uncover the different agencies in his study of Mardi Gras in New Orleans. Here he found that 'the production of spectacle and simulation represent very strategic, calculated and methodological campaigns that corporations design to expand markets' (p. 837). Similar findings have been shown by Waitt (2000) in the role played by the Sydney Cove Redevelopment Authority (Australia) in its fashioning of a heritage precinct.

The fact that developments based on heritage tourism are to be found in a range of countries, including both developed and developing economies, is testament to these global influences. According to Green, 'global economic trends are implicated in the pervasiveness of heritage tourism in cities throughout the world' (2001: 191). As with the globalization of the theme park, the debate here is increasingly centred on the global–local nexus. We do not want to focus on this debate in its entirety, but rather to note that increasing evidence shows that 'similarities and differences do exist between urban heritage destinations' (Green 2001: 194). From Green's study and the

Table 10.2 **Examples of disparities in the development of cultural-heritage attractions and the role of local agencies**

Place	Disparities and local agencies
Tilburg	A lack of coordination is evident between the various organizations managing the city's cultural and tourism products. Local agencies operate as autonomous or semi-autonomous organizations, characterized by horizontal disparity.
Bilbao	Complex relationships exist between the Basque government (regional level) and the city council. Local issues, such as heritage products, relate to different administrative levels, leading to vertical disparity in cultural-tourism developments.
Leicester	A degree of ambiguity surrounds the local authorities' cultural-tourism policy. There is potential disparity between the city's promotion agency and the city council in the strategy of promoting Leicester as a quality European city. This is because of a degree of horizontal disparity.

Source: Green (2001).

work of Richards (1996a) and Chang et al. (1996), it appears that the engagement of local agencies is critical in forming the character of heritage spaces (Table 10.2).

The growth of the heritage industry has been reasonably well documented by Law (2002), Shaw and Williams (2002) and Richards (1996b), with the latter claiming that the number of cultural attractions in western Europe increased by 113% between 1970 and 1991. As we have argued, this growth reflects a number of factors, including the use of heritage as a catalyst for economic regeneration. The local state has often been the initial driving force for heritage-based developments, although it is often the private sector that shapes the final product, as the British experience shows (Leslie 2001).

In the UK, there has been extensive use of heritage attractions to reclaim and regenerate redundant industrial spaces. Local authorities have been especially attracted to such strategies, given the relatively high economic leverage from tourism investment (Shaw and Williams 2002). More specifically, in the US context, Strauss and Lord (2001) have monitored the economic performance of a heritage-tourism system of 13 sites in south-western Pennsylvania. These developments were estimated to be worth $33 million in terms of total regional sales impacts, with 74% of that coming from non-resident spend. In addition, Law (2002) has suggested other gains, associated with improvements in the local environment, increased amenities and the creation of positive images through marketing, leading to what Law sees as the gaining of civic pride.

Visitor attractions and the Heritage Lottery

A further major spur to heritage development in the UK since the mid-1990s has been the creation of the Millennium Commission and the Heritage Lottery Fund, both established by the National Lottery Act of 1993. The Commission became a leading fund-giver for tourist attractions as part of its role in assisting communities to help mark the Millennium. It represents the largest single source of non-government public investment for community schemes, many of which are associated with visitor attractions – accounting for £4 billion in 1999 (Waycott 1999). Such investment has aided the creation of a further 25 major visitor facilities, including many museum projects between 1994 and 2002. As Figure 10.5 shows, these are fairly widespread within British cities, including the now defunct Millennium Dome in London.

The 1993 Act also strengthened the role of the National Heritage Memorial Fund, which assumed responsibility for distributing grants from the Heritage Lottery Fund. By 2000, almost £1.5 billion had been distributed to over 5000 projects in the UK (Golding 2000). The net result is that, through various means, there has been a massive increase in the volume of new or upgraded attractions that use heritage as a basis for their development. The initial consequence has been to intensify levels of competition, as more redundant space is brought within the tourism system. This, in turn, has resulted in shifting patterns of visitor demand, with Woodward (2000) showing that while the number of visits had increased, this was mainly accounted for by new sites. Overall, the average number of visitors per site for established heritage attractions was declining, most probably because of the increased competition from newly established sites. This higher level of competition has encouraged heritage sites to attempt to differentiate themselves to attract more visitors, and in doing so they have created new forms of heritage sites that are more innovative in the use of theme park technology.

Of course, the use of heritage attractions to help regenerate past industrial areas is only one of the possible development strategies. Certainly, given the increased levels of competitiveness, visitor attractions have become more complex and more appealing, by offering distinct themes and greater levels of commercialization – usually by the addition of retail facilities. Law sees this as a process of adding more critical mass, in which a 'complex of museums enables the city to project a clear image of itself' (2002: 90). However, this is only part of the change, as it increasingly involves the creation of an assemblage of tourism spaces and the reshaping of places. Examples of this include the Habor place in Baltimore, the Castlefield area in Manchester and the Albert Dock area of Liverpool. At the basic commercial level, such clustering can also help create external economies for specialized retailers selling to tourists (Couch and Farr 2000; see Chapter 4).

The global recognition of cultural spaces

In very different circumstances, Chang (2000) has discussed the attempts by Singapore to present itself as a 'global city for the arts', a phrase coined by the Singapore government in 1992 to help with its attempts to create thriving arts, cultural and entertainment areas. This was part of an economic strategy to attract tourists, especially international visitors, but which also had sociocultural motivations – including 'nation-building'. The strategy saw three key areas of development, creating interlocking consumption spaces, namely: a theatre district to act as a regional centre for South-East Asia, an

Figure 10.5 **Major Millennium-funded visitor attraction projects**

entertainment area specifically for tourists and leisure-seekers, and an art and antique-trading district.

As Chang explains, in 1999 Singapore's prime minister articulated the strategy as creating the 'Renaissance City'; by which 'he meant a fun city with creative people, artistic events and a culturally vibrant environment' (2000: 819).

In most cases, heritage is being played with or manipulated as part of a more general process of re-creating postindustrial and urban places. These processes involve, firstly, the recognition of the need to promote spaces for 'conspicuous consumption, in the shapes of art, food, music, fashion and entertainment' (Jayne 2000: 12). As Hall (1995) and Hubbard (1998) argue, these images are promoted through the enhancement of heritage and the combining of cultural areas that were previously ambiguous. This also involves, in many instances, a second phase of development that is centred on signifying the importance of these new consumption spaces. This is usually achieved by the use of postmodern architecture, which imposes new landscapes, often of a vibrant nature. These constructions serve as identifiable symbols of the postmodern. Some acquire global recognition because of their striking architecture. A good example of this is the Guggenheim Museum building in Bilbao, Spain, opened in 1997. This design by Frank Gehry, of a titanium-clad building at a cost of $100 million, was immediately declared an architectural landmark, and its image became instantly recognizable (Law 2002). This project thus has two major advantages: the spectacle of the building itself, and the high-quality exhibitions it can mount because of its links to the Guggenheim Foundation in New York. Since its opening, the museum has had a marked impact on visitor numbers to this part of Spain, which have increased by almost 35% (Plaza 2000 and Table 10.3). Since the success of Guggenheim-Bilbao, some 60 cities have made contact with the Foundation, offering locations for another museum site.

Guggenheim-Bilbao illustrates the strength of arts and culture in helping create new tourism destinations, but it does so within the important context of postmodern architecture. In this sense, it has parallels with the postmodern

Table 10.3 **Tourist arrivals in Bilbao before and after the opening of the Guggenheim Museum**

	Average per month before Guggenheim: Jan. 1994–Sept. 1997	Average per month after Guggenheim: Oct. 1997–July 1999	Absolute increase	Percentage increase
Total arrivals	83,898	112,887	28,989	34.6
Overseas tourists	22,175	32,058	9,883	44.6
Visitors to Guggenheim	0	97,953	97,953	n/a

Source: Plaza 2000.

forms that are representative of Las Vegas. It is not surprising that many other cities have attempted to develop this critical infrastructure of postmodern buildings, along with arts and culture to help create new, 'vibrant, cosmopolitan, entertaining and happening spaces' (Jayne 2000: 12). Thus, Salford in Greater Manchester has developed the Lowry Centre, which utilizes a dramatic architectural style, as does the nearby Imperial War Museum of the North. Indeed, Manchester has witnessed the opening of three £30 million cultural buildings in all, the other two being the 'museum of the city', or the Urbis Centre, and the extension to the Manchester Art Gallery. Similarly, Glasgow has witnessed the creation of two spectacular buildings along the River Clyde, namely the Scottish Exhibition and Conference Centre, opened in 1987 but extended most dramatically in 1997, and the Scottish National Science Centre, opened in 2001. In Newcastle a new arts and cultural centre has been created in the renovated former Baltic flour mills building, with other postmodern styles of architecture located close by along the River Tyne.

These types of developments bring together, as we have argued, a number of processes that create and shape new tourism spaces. These, in turn, are increasingly forming parts of a cluster of complementary consumption spaces. In detail the focus may vary from festival markets as a hub development, to heritage centres or museums. However, all share some important common features in relation to their stressing of image, signs and themes. If these spaces can be centred on dramatic buildings, then further striking images are created. In total, these elements combine to aid the promotion of place through the production of positive images and significant consumption spaces.

SPACES OF CONSUMPTION OR EXCLUSION?

In the previous chapter we discussed the ways in which new resorts, especially those within the pleasure periphery, had created more exclusive spaces for tourists. Of course recognition of these types of resort enclaves is not new, but it is significant that they are increasing to meet the demands of postmodern trends in tourism consumption. It is also clear that similar trends have emerged in many other tourism destinations and especially within regenerated urban areas. Indeed, the production of the new tourism spaces we have been discussing in this chapter is dependent on a range of interest groups involving public and private actors. These may be seen either as cooperating agencies or as exploitative forces. In terms of the latter perspective, private investment capital combines with state funding to exploit resources in order to create commodities – including those associated with images and themes (Gotham 2002). This view of the production of new tourism spaces sees the processes as leading to the exploitation of labour and the marginalization of particular social groups within the local community.

Following this line of argument, Gotham views these tourism spaces as sites 'of inequality and struggle' (2002: 1739). In his study of New Orleans and the festival of Mardi Gras, he argues that the city leaders and the economic elites used the festival as a strategy to 'refashion the city into a themed landscape of entertainment' (p. 1752). In creating these new landscapes for tourists, there has been a marginalization of parts of the local community and their exploitation in the form of cheap labour. The marginalization process has occurred, in part, because of the increasing privatization of some spaces of consumption, which became more exclusive in character. The same process is to be found in many other cities that have refashioned themselves through the building of festival markets, conference centres, hotels and fashionable apartments.

The creation of more spaces of consumption to fit closely with particular lifestyles inevitably leads to a degree of exclusion, since the postmodern tourist tends to be from the new middle classes (see Chapter 5). These postmodern lifestyles and the postindustrial identities being fashioned by many places are part of consumer society generally, but also need to be grounded in local social relations. In this context, Jayne (2000) argues that if the local state recognizes these issues then these new consumption spaces can create a collective sense of belonging. As yet, success in this area is limited and almost fleeting, in that, in many instances, public–private partnerships may start with an inclusive strategy, but, as the development process takes off, increasing private capital and burgeoning economic elites determine the directions of growth (see Bramwell and Shurma 1999, and Chapter 8 on unequal power relationships within partnerships). We do not argue that civic pride is not part of the process; clearly, there are many examples of this in UK cities (Law 2002). Our argument is more that as postindustrial places are rebranded as new places of spectacle and consumption, the representation of the community becomes narrower. Similarly, Hall (1995), in a related discussion, talks about the festival-led urban regeneration strategy as being shallow.

The use of tourism-led strategies in postindustrial cities may be seen as an exclusive process, producing new spaces of consumption for particular sectors of society. This may bring a form of civic pride and re-image the place, which may help towards investment, but there is limited evidence to suggest these results are widely inclusive. Indeed, in cities such as Bradford, in the north of England, where a tourism-led strategy has been followed since the mid-1980s, tourism has done little to ease racial and social tensions in parts of the city. At an economic level, employment in the tourism, leisure and culture industries grew to almost 13,000 jobs between the mid-1980s and the late 1990s (Hope and Klemm 2001); however, unemployment remained at 8% in the late 1990s. Moreover, the creation of the Bradford festival, a new Imax Cinema, and the Salt Mill and David Hockney Gallery, while adding new cultural spaces to the city, have still been viewed by many marginal economic groups as being irrelevant. And the positive images Bradford had created through its tourism-led strategies were rapidly negated by civil unrest during the summer of 2002.

It seems clear that the creation of new spaces of consumption, while bringing levels of economic and environmental improvement, are also 'pervaded by discourses and issues of enclosure' (Jayne 2000: 20). They are, moreover, both at an individual and a community level, spaces of constraint and, according to Gotham (2002), of conflict. These conflicts demand that our analyses are holistic, embracing economic, political and cultural dimensions.

SUMMARY: CONSTRUCTING TOURISM SPACES

This chapter has focused on the detailed processes involved in the construction of new tourism spaces. In doing so, we highlighted:

- the importance of understanding the ways in which tourism spaces are conceptualised

- the processes involved in creating such spaces, especially the importance of cultural meanings and competition

- the diversity and characteristics of these tourism spaces

We have followed, in part, the ideas of Gottdiener and adopted the notion of themed environments. Within this context, the influence of the theme park – both as a global product and cultural template – has been shown to be critical. Its significance, we argue, operates in three key ways:

- it emphasizes the importance of commodifying tourism spaces

- it acts as a global carrier of the concept

- it operates as a powerful commercial space that can help remake the geographies of tourism destinations.

11 Conclusions

This brief concluding chapter is divided into three parts. First we reflect on the five questions set out in Chapter 1 as informing our approach to this book. Second, we consider some of the issues arising from the need to understand the fundamental changes that characterize tourism. And, third, we identify some of the key challenges that face tourism researchers.

REFLECTING ON TOURISM

In the Introduction, we set out five questions that had informed our approach and which are, to varying degrees, threaded throughout the book. We begin this concluding chapter by reflecting further on these questions.

The first question we posed was: how are tourism structures and flows created? For the early twenty-first century, it is our unavoidable conclusion that most tourism relationships were mediated by market relationships. Increasingly, tourism has been subject to commodification, and this applies as much to pre- and post-tourism experiences as to the actual holiday experiences. Guidebooks, lessons (for example in skiing) and clothing are purchased before the holiday, while holiday experiences may influence the restaurants eaten in and the foods and home furnishings bought after the trip, not to mention longer-term retirement plans, whether to Devon, the Costa del Sol, Florida or elsewhere. Tourism is not, of course, reducible to material relationships, as our discussion of values and behaviour indicated (Chapter 6). But material relationships do fundamentally mediate virtually all tourism experiences. The visiting of friends and relatives involves some form of commodified transport, and probably the purchase of gifts, some eating out, and often entrance fees to attractions. Walking and camping holidays may include the payment of camping fees, and will certainly include the purchase of food and specialist equipment. But even if there are forms of tourism that are only lightly embedded in material relations, most tourism is not only commodified, but also 'fetishised' (Watson and Kopachevsky 1994). Tourism products and experiences have assumed a life of their own, and become transformed into 'the sacred', whereby their exchange value may be detached from the actual costs of production. The economy of signs and symbols is not reducible to the costs of labour, capital and other factors of production.

Tourism flows and structures have to be understood within this framework of material and non-material relations. There is a symbiotic relationship between them, with flows shaping scapes and vice versa, but both being subject to wider shifts in culturally influenced production and consumption. We have argued in this book that regulation theory provides an overarching framework, at least for analysing the material relationships involved in creating or generating structures and flows. The scapes are created by interlocking investments in different types of capital in infrastructures and facilities, and they are also technologically shaped, with the jet engine perhaps being the icon of, as well as a major determinant of, modern tourism. But they are also created by the tourist imagination, fed by place-images and image-makers. These scapes are structural features, although they are not fixed. They determine, but are not deterministic of, tourism flows. Indeed, the tourist resort cycle (Butler 1980) is, in effect, an essay on the evolution of one component of the tourism scape. The history of mass tourism, as outlined in Chapter 9, is the history of the creation of scapes. Mass tourism also presents one of the most intriguing dilemmas for tourism analysts – the question of whether there has been a shift in the regime of accumulation in tourism, from Fordist mass production and consumption to more individualized and flexible post-Fordist forms. The view expressed in this book is that the demise of mass tourism is exaggerated, and that anyway it is more useful to think of the coexistence of different regimes of accumulation, what Ioannides and Debbage term 'the tourism industry polyglot' (1998: 106).

The second question that we posed is: how are tourism structures and flows reproduced? In part this question is a response to the fact that tourism is crisis-prone, as are all forms of capitalist production and consumption. Regulation theory directs our attention to the mode of regulation as a set of time- and place-specific structural forms and institutional arrangements, which regulate economic (and to some extent) social life (Dunford 1990). There are various forms of state intervention, from the local to the international, which regulate travel and hospitality services, but also more fundamentally aim to guarantee that production and consumption are kept in balance. One of the most obvious examples in respect of tourism is the imposition by the state of maximum working hours and minimum holiday entitlements. But equally important are those wider cultural processes that underpin consumption, including cultural capital, the attaching of symbolic values to tourism, and the 'fetishizing' of some forms of consumption. Globalization and neo-liberalism have challenged, but not fundamentally undermined, the role of national states as the key areas of regulation. There remain fundamentally important national differences in how tourism experiences are produced and consumed, as the case-studies in Chapters 9 and 10 illustrate.

The locality is also a key influence on how globalization is mediated in people's (tourism) lives. National and local responses to the globalization of tourism are not preordained, but have to be seen in context of particular historical settings. Local practices and local values shape responses to globalization, but also help to shape globalization. As we argued in Chapter 8, places have to be viewed in context of material relationships but are not reducible to these. They are complex mixes of material objects, companies, workers, local civil societies, the local state with the co-presence of other forms of the state, and a multitude of practices, values, and identities. Tourist practices, and the practices of local residents in response to these, contribute to the shaping of everyday lives, values and identities in places, just as much as place practices and identities contribute to tourism. To some extent the many different interests within a particular place may be incorporated into local governance, although we stress that power relationships within these are highly unequal. Ultimately then, there are complex forms of regulation, operating at different scales, but manifest in both the tourism departure and destination areas, which serve to reproduce tourism in modern societies.

The third question that we posed has long preoccupied tourism policy makers: who and where benefits from, or incurs costs resulting from, these structures and flows of tourism? This is a particularly difficult question to answer. At one level, it should be possible to construct an economic and environmental assessment that produces two outcomes. The first would be an indication of whether particular tourism flows produce net total positive or net total negative impacts. The stress on net total outcomes is important because, too often, analyses of economic and environmental impact focus only on the destination end of the tourism flows. Yet any tourism flow has implications for the place of origin, for example, in terms of displaced leisure trips and expenditure, and their impacts. Second, any such analysis must focus on the distributional question of who benefits where and – given the nature of product cycles – when from tourism flows? Most of the tourism literature focuses on the active participants in tourism – the tourists, the host communities and those with a direct economic interest in the industry. Yet a very large proportion of the world's population has very limited participation in tourism, because of economic, cultural or personal constraints (including many forms of individual disability). And even within the more developed market economies, there are vast inequalities in what may be termed 'tourism welfare' (understood as the individual and collective wellbeing derived from tourism). The sustainable tourism literature tends to focus on intergenerational equity, and pays little more than lip service to the issue of intragenerational equity. This is hardly surprising, for intragenerational inequalities in tourism are born of broader societal inequalities, and it is beyond the scope of local sustainable tourism initiatives to unravel these. The cultural and environmental impacts are also similarly uneven. Hence, the question 'How was tourism for you?' – whether addressed at tourists, business owners or host communities – is ultimately usually answered at the level of the individual. This, however, should not obscure the fact that structural factors do produce regularities in these experiences, and that these are inherently uneven.

This leads us to the fourth question: how do tourists, host communities and other participants experience these flows and structures. As we demonstrated in Chapter 6, notions of the tourist experience are complex and contested. Certainly, the more recent studies have argued that the experiences of tourists are varied and, in different ways, fulfil the desire for authenticity. We have also raised two further issues around this theme. The first concerns the degree to which all tourist experiences are engineered, that is, planned and controlled. There is certainly increasing evidence that the experiences of many 'ecotourists', for example, are no more authentic than those of their counterparts in mass coastal resorts. The second theme extends the debate on the tourist experience into developing countries, in an attempt to move away from an entirely Western perspective of tourist experiences. The ideas of authenticity and the tourist experience are clearly strongly contingent on the processes of commodification. Our discussion in Chapter 7 has shown how these processes impinge on local communities and, equally importantly, how they are negotiated by such communities.

The final question is: to what extent are individuals, communities and states able to contest their relative locations in these structures and flows, and the distribution of costs and benefits that stem from them? As discussed in Chapter 8, and illustrated in Chapters 9 and 10, countries, places and individuals can contest their places in the evolving global tourism scene. Places can mediate their roles – they can seek to reduce or increase tourism flows, they may seek new images to generate different tourism flows, or they may seek to find ways to redistribute the costs and benefits of tourism within communities. Chapter 10, in particular, presented examples of how places have been able to use the (re)theming of the landscape in a variety of ways. The aims are often economic, but they can also be social (for example, bolstering community self-confidence) and cultural (for example, using tourism to support cultural facilities or events, which will also benefit the local community). However, there is no predictable outcome to such contestation: differences in local resources, in local politics (for example, growth-machine versus growth-management policies) and in local cultural milieux all influence these outcomes. Moreover, while the notion of place has arguably become more important in the face of globalization, places are still located within broader structural parameters. National regulatory differences remain important, and there are global shifts in the culture of consumption, and in labour and capital flows, which individual communities are relatively powerless to challenge. Tourism and tourism impacts then need to be seen in the context of the tense knot of relationships among places, between the national and the global, and between human agency and the structural. Moreover, at one level, tourism remains fundamentally a set of market relationships (although this is not to deny that it has meanings at other levels as well), and contestation is mediated by, and through, these.

THE CHALLENGE OF CHANGE

Many tourism books and papers start with a litany of statistics which seems to point to its inexorable growth. Yet the more-than 600 million international tourism trips that are made annually are in fact generated by only a tiny proportion of the world's population. As incomes change, particularly in the emerging market economies, there is the prospect of a sharp shift in the growth of international (and domestic) tourism. This has raised major questions about the resulting impacts, but the focus has been largely on perceived negative environmental impacts, while socio-economic issues have tended to be neglected.

The debate about the growth of tourism often neglects the positive dimension of increased mobility. Yet, as Urry emphasizes, 'the good society would seek to extend the possibilities of co-presence to every social group and regard infringements of this as involving undesirable social exclusion' (2002: 270).

Co-presence (as a result of mobility) is not only desirable *per se*, because of the increased range of social interaction it brings about, but is also related to notions of social capital. Of course, it is also true that because of 'massive resource and environmental constraints, the right to corporeal travel to realize co-presence will never be unlimited' (Urry 2002: 270). Therefore, tourism researchers need to address the needs and rights of non-tourists both now and in the future. Attempts to limit travel, whatever the environmental logic, will have a negative impact on the participation of millions, if not billions, of people in various forms of mobility, including tourism. In this sense, much of the debate about sustainable tourism, with its focus on particular places and on the management of those participating in tourism, misses the much larger question of whether and how tourism can become a right enjoyed by all in the good society.

Other issues pale in comparison to those raised directly or indirectly by the above question. But in practice this question has been, and probably will continue to be, sidelined. Instead, policy-makers and tourism researchers have tended to focus on a set of narrower questions about managing active tourism participation both now and in the near future. One of the most immediate of these questions is whether the growth of tourism really is as inexorable as it has appeared to be in recent years. The world economy was already slowing down before the terrorist attacks of 11 September 2001, while there have also been new and continuing intra- and inter-national tensions and conflicts in many parts of the world. Increased tourism competition, which is one of the most manifest outcomes of globalization, is therefore likely to take place in a context of increasing perceptions of risk and uncertainty. Technological changes, especially in IT, are changing the paradigms of competition (see Chapter 4) more quickly than regulatory systems can keep pace. The result is probably less certainty about the future volume and shape of tourism than at any time since the economic crises of the mid-1970s.

In this context, the challenge faced by places seeking to contest their place in global tourism will become far more complex. Uncertainty and competition have increased, while the nature of tourism is constantly shifting, as illustrated in Chapter 10. For example, a decade ago, many places pinned their faith on cultural tourism as the basis for local tourism or economic development strategies. But this is now in question. Richards writes that:

> Consumption of culture is no longer enough to guarantee success – cities must become centres of creative production as well. Creative production attracts enterprises and individuals involved in the cultural sector, generating important multiplier effects in the local economy, and raising the aesthetic value of creative production locations. (2001a: 64)

He argues that there is a need to look beyond cultural tourism, which is essentially concerned with the past and present, to 'creative' tourism; which is about the past, present and the future, and which, above all, is about experiences. Whether the goals are creative tourism, or the restructuring of mass tourism, places face increasing challenges in achieving these.

With the direct interventionist role of the state in retreat, emphasis has shifted to the notion of the learning economy. The debate about industrial districts (see Chapter 4) has been extended to the concept of learning regions, or learning cities, wherein the capacity of regions to support processes of learning and innovation are identified as the key to competitive advantage (Storper 1997). Much of the research has been in the context of manufacturing and the 'knowledge industries', but it applies equally well to tourism. Research on learning regions and cities focuses on the territorial, social and institutional conditions that shape economic development. Moreover, innovation is conceptualized in this approach as being the outcome of interactive processes within the region rather than being externally driven (Cooke and Morgan 1998). MacKinnon et al. (2000: 301–5) summarize what they consider to be the key propositions concerning learning regions:

- Globalization has not annihilated space but has led to the emergence of new forms of agglomeration economies centred on knowledge creation.

- There is an increasing tendency for non-material advantages based on relations between firms and institutions to be located at the regional rather than the national level.

- Because non-codified knowledge is best transmitted through close interpersonal relations, proximity is important, and this leads to sectoral specialization.

- Collective learning is important, and this is understood as 'cumulative learning processes that take place over time among a community of firms in a locality' (p. 301). Trust is critical to such collective learning.

The shift away from direct state intervention, and greater emphasis on localized responses, drawing on concepts such as learning regions, can be seen to frame the responses of many places to both the restructuring of existing tourism and the creation of new tourism products and experiences, as discussed in Chapters 9 and 10. There are, of course, criticisms of the concept of learning regions. Close interrelations could lead to 'institutional lock in', and failure to be open to outside influences. There is a lack of precision about the scale at which the concept operates. And it downplays the importance of external relationships in stimulating innovation and creating competitive advantage. But the emphasis on knowledge and local institutions does reinforce the importance of the local and the regional as sites of contestation. At the same time, it indicates the emerging lines of differentiation in the abilities of places to contest their engagement with tourism. The one predictable outcome of this is the differentiation of place experiences, and these differences will usually be associated with vast inequalities in the distribution of the costs and rewards of tourism.

THE CHALLENGE FOR TOURISM RESEARCHERS

What is the role of tourism researchers in engaging with the array of questions indicated above? At one level, their role can be nothing less than 'advancing understanding' of tourism, not as a statement of the truth or of a set of invariable laws. Instead, we have to accept that understanding is conditional on place and time. But how are tourism researchers to contribute to understanding? We believe that there are three critical issues here.

- Tourism studies must be critical tourism studies. This is particularly important in a discipline where many researchers have often been closely allied with the needs of a narrow range of research users – the tourism industry, and various levels of government. Only in this way can the broader issues of equity, raised earlier in this chapter, be addressed. Sustainable tourism has at least realigned the tourism research agenda to bring some of these issues within the sights of researchers, but much of the resulting literature has been strong on advocacy, and lacking in the deeper understanding that is offered by critical social-science perspectives.

- Tourism research must be diverse. While this book is strong on political economy, it does not claim that this is the only approach to studying tourism. Such methodological imperialism is to be avoided, and the contributions of different strands of research should be recognized. The problem with tourism research in the past has been that methodology has often been implicit rather than explicit, while some methodological perspectives – notably political economy – have been relatively neglected. If, therefore, we argue for more research on the political economy of tourism, this is to diversify the corpus of understanding, through strengthening an important strand of work, rather than to mark out a claim for hegemony.

- Tourism research needs to be more holistic, in the sense of situating tourism in context of wider social-science debates. There are, of course, distinctive features of tourism (see Chapter 2) that do mark it out from other branches of the service and experience economies. However, we should no more expect an analysis of tourism, than of say manufacturing change or of urban restructuring, to be written in isolation of wider social changes and an understanding of them. Furthermore, while we argue the need to bring mainstream social-science debates about identity, class, etc. to tourism studies, this is not a call for their mechanical application. Moreover, we do not see tourism as just an importer of social-science ideas. Instead, we see tourism as being deeply embedded in all aspects of life. As such, the understanding of tourism contributes to the understanding of society, and in this way tourism researchers should actively seek to contribute to debates in the other social sciences.

There have been significant advances in tourism research in recent decades, and it has become more critical, more diverse and more holistic. But, as Craik argues, 'what is clear is that tourism is a constantly evolving culture and that tourism development changes the dynamics of cultural production in which it is embedded' (1997:135). The challenge for tourism studies is to bring its sharpening perspectives to both a broader and deeper understanding of this rapidly shifting phenomenon.

SUMMARY

- Tourism flows and structures are subject to commodification and have to be seen in a context of capitalist relationships. This does not mean that they are reducible to material relationships, for they also have to be understood in terms of culture and values.

- The reproduction of tourism flows and structures can be understood in terms of the mode of regulation. This is at many levels, including international bodies as well as the national and local states.

- Inevitably, in the context of capitalist societies, there are sharp inequalities in the distribution of the costs and benefits of tourism. We argue the need for a holistic approach to analysing their spatial distribution, as well as the need to ask searching questions about intragenerational social inequalities.

- There is a need for further research to explore how tourists and host communities experience tourism, and this should reinforce awareness of the complexity and diversity surrounding notions such as 'authenticity'.

- Individuals, communities and places can contest the outcome of tourism trajectories, and their places in these. But in order to better understand these opportunities and constraints, there is a need to unravel the tense knot of relationships among places, between the national and the global, and between human agency and structural features.

- Co-presence brings enhanced opportunities for social interaction, whilst tourism is one of the forms of mobility that facilitates this. Debates about the future limitation of tourism and travel must face the equity issues pertaining to the massive current inequalities in access to mobility. Large parts of the world's population lack access to almost any form of tourism.

- Many of the assumptions of a trajectory of 'continuous' tourism growth have been questioned in recent years, and there is now far greater uncertainty about the resulting benefits and costs.

- Knowledge and learning are keys to how individuals, firms and places are able to respond to, and utilize, the opportunities, risks and constraints associated with tourism.

- Tourism research needs to be more critical, more diverse, and more holistic.

References

AA Publishing (2002) *Spiralguide: Las Vegas,* Windsor: AA Publishing.

Abercrombie, N. (1991) 'The privilege of the producer', in R. Keat and N. Abercrombie (eds) *Enterprise Culture,* London: Routledge.

Acott, T.G., La Trobe, H.L. and Howard, S.H. (1998) 'An evaluation of deep ecotourism and shallow ecotourism', *Journal of Sustainable Tourism* 6(3): 238–253.

Adkins, L. (1995) *Gendered Work: Sexuality, Family, and the Labour Market,* Milton Keynes: Open University Press.

Agarwal, S. (1994) 'The resort cycle revisited – implications for resorts', *Progress in Tourism, Recreation and Hospitality Management* 5: 194–207.

Agarwal, S. (1997) 'The public sector: planning for renewal?', in G. Shaw and A. Williams (eds) *The Rise and Fall of British Coastal Resorts,* London: Pinter.

Agarwal, S. (1999) 'Restructuring and local economic development implications for seaside resort regeneration in Southwest Britain', *Tourism Management* 20: 511–22.

Agarwal, S. (2002) 'Restructuring Seaside Tourism: The Resort Lifecycle', *Annals of Tourism Research* 29(1): 25–55.

Agarwal, S., Ball, R., Shaw, G. and Williams, A.M. (2000) 'The geography of tourism production: uneven disciplinary development?', *Tourism Geographies* 2(3): 241–65.

Aitken, C. and Hall, C.M. (2000) 'Migrant and foreign skills and their relevance to the tourism industry', *Tourism Geographies* 2(3): 66–86.

Akis, S., Peristianis, N. and Warner, J. (1996) 'Residents' attitudes to tourism development: the case of Cyprus', *Tourism Management* 17: 481–94.

Allen, J., Cochrane, A. and Massey, D. (1998) *Re-thinking the Region,* London: Routledge.

Allen, L.R., Long, P.T., Perdue, R.R. and Kieselbach, B. (1988) 'The impact of tourism development on residents' perceptions of community Life', *Journal of Travel Research* 27(1): 16–21.

Amin, A. and Thrift, N. (1994) 'Holding down the global', in A. Amin and N. Thrift (eds) *Globalization, Institutions and Regional Development in Europe,* Oxford: Oxford University Press, 257–60.

Amin, A. and Thrift, N. (1997) 'Globalization, socio-economics, territoriality', in R. Lee and J. Wills (eds) *Geographies of Economies,* London: Arnold.

Andrews, H. (2002) 'A theme park for Brits behaving badly', *Times Higher Education Supplement* 19 July: 22.

Ap, J. (1992) 'Residents' perceptions on tourism impacts', *Annals of Tourism Research* 19: 665–90.

Apostolopoulos, Y., Loukissas, P. and Leontidou, L. (2001) 'Tourism, development and change in the Mediterranean', in Y. Apostolopoulos, P. Loukissas and L. Leontidou (eds) *Mediterranean Tourism: Facets of Socioeconomic Development and Cultural Change,* London: Routledge.

Aramberri, J.R. (1991) *'The Nature of Youth Tourism: Concepts, Definitions and Evolution*, unpublished paper presented at WTO conference, New Delhi.

Arbel, A. and Woods, R.H. (1990) 'Debt hitch-hiking: how hotels found low-cost capital', *The Cornell HRA Quarterly* 31(3): 105–10.

Archer, B. (1982) 'The value of multipliers and their policy implications', *Tourism Management* 3: 236–41.

Ascher, B. (1984) Obstacles to international travel and tourism', *Journal of Travel Research* 22: 2–16.

Association of District Councils (1993) *Making the Most of the Coast*, London: Association of District Councils.

Ateljevic, I. (2000) 'Circuits of tourism: stepping beyond the "production/consumption dichotomy" ', *Tourism Geographies* 2: 369–88.

Ateljevic, I. and Doorne, S. (2000) 'Staying within the fence: Lifestyle entrepreneurship', *Journal of Sustainable Tourism* 8: 378–92.

Ateljevic, I. and Doorne S. (2003) 'Unpacking the local: a cultural analysis of tourism entrepreneurship in Murter, Croatia', *Tourism Geographies* vol. 5(2): 123–50.

Atkinson, J. (1984) *Flexibility, Uncertainty and Manpower Management*, Brighton: Institute of Manpower Studies, University of Sussex, Report 89.

Ayala, H. (1991) 'Resort hotel landscape, as an international megatrend', *Annals of Tourism Research* 18(4): 568–87.

Ayala, H. (1993) 'Mexican Resorts: A Blueprint with an Expiration Date', *Cornell Hotel and Restaurant Administration Quarterly* 34(3): 34–42.

Aziz, H. (1999) 'Whose culture is it anyway?' *In Focus*, Spring: 14–15.

Bagguley, P. (1987) *Flexibility, Restructuring and Gender: Employment in Britain's hotels*, Lancaster: Lancaster Regionalism Group, Working Paper 24, University of Lancaster.

Bagloglu, S. and Mangaloglu, M. (2001) 'Tourism destination images of Turkey, Egypt, Greece and Italy as perceived by US-based tour operators and travel agents', *Tourism Management* 22: 1–9.

Bagnell, G. (1996) 'Consuming the past', *Sociological Review* 44(1): 27–47.

Baldacchino, G. (1997) *Global Tourism and Informal Labour Relations: the Small-scale Syndrome at Work*, London: Mansell.

Ballantine, J. and Eagles, P. (1994) 'Defining Canadian ecotourists', *Journal of Sustainable Tourism* 2(4): 210–14.

Banham, R. (1971) *Los Angeles*, New York: Viking Penguin.

Barke, M. and France, L.A. (1996) 'The Costa del Sol', in M. Barke, J. Towner and M.T. Newton, *Tourism in Spain: Critical Issues*, Wallingford: CAB International.

Barkin, D. (2001) 'Strengthening domestic tourism in Mexico: challenges and opportunities', in K.B. Ghimire (ed.), *The Native Tourist: Mass Tourism within Developing Countries*, London: Earthscan.

Barry, J. (1981–82) 'Causal Imagination', *Discourse* 4: 4–31.

Baudrillard, J. (1981) *For a critique of the political economy of sign*, St. Louis: Mo Telos Press.

Baumol, W. (2002) *The Free Market Innovation Machine: Analyzing the Growth Miracle of Capitalism*, Princeton: Princeton University Press.

Beard, J. and Ragheb, M.G. (1983) 'Measuring leisure motivation', *Journal of Leisure Research* 15(3): 219–28.

Beaumont, N. (1998) 'The meaning of ecotourism according to . . . is there now consensus for defining this "natural" phenomenon? An Australian perspective', *Pacific Tourism Review* 2(3/4): 239–50.

Beauregard, R. A. (1998) 'Tourism and economic development policy in US urban areas', in D. Ioannides and K.G. Debbage (eds) *The Economic Geography of the Tourist Industry: A Supply-side Analysis,* London: Routledge.

Belisle, F.J. and Hoy, D.R. (1980) 'The perceived impact of tourism by residents: a case study in Santa Marta, Colombia', *Annals of Tourism Research* 7: 83–101.

Berger, P.L. (1973) 'Sincerity and Authenticity in Modern Society', *Public Interest* 31: 81–90.

Bianchi, R.V. (2000) 'Migrant tourist-workers: exploring the "contact zones" of post-industrial tourism', *Current Issues in Tourism* 3(2): 107–37.

Bianchi, R.V. (2002) 'Towards a new political economy of global tourism', in R. Sharpley and D.J. Telfer (eds) *Tourism and Development: Concepts and Issues,* Clevedon: Channel View Publications, pp. 265–99.

Bird, E., Lynch, P.A. and Ingram, A. (2002) 'Gender and employment flexibility within hotel front offices', *Service Industries Journal* 22(3): 99–116.

Bjorkland, E.M. and Philbrick, A.K. (1975) 'Spatial configurations of mental processes', in M. Belanger and D.G. Janelle (eds) *Building Regions for the Future: Notes of Documents du Recherche No. 6*: Quebec: Department of Geography, University of Laval.

Boden, D. and Molotch, H. (1994) 'The compulsion to proximity', in R. Friedland and D. Boden (eds) *Nowhere: Space, Time and Modernity,* Berkeley: University of California Press.

Boers, H. and Bosch, M. (1994) *The Earth as a Holiday Resort: an Introduction to Tourism and the Environment,* Utrecht: SME.

Boissevain, J. (ed.) (1992) *Revitalising European Rituals,* London: Routledge.

Boissevain, J. (2000) 'Mass tourism' in J. Jafari (ed.) *Encyclopedia of Tourism,* New York; Routledge.

Boniface, P. and Fowler, P. (1993) *Heritage and Tourism in the Global Village,* New York: Routledge.

Boo, E. (1990) *Ecotourism: The Potentials and Pitfalls,* Washington DC: World Wildlife Fund.

Boorstin, D.J. (1964) *The Image: A Guide to Pseudo-Events in America,* New York: Harper.

Bosselman, F.R. (1978) *In the Wake of the Tourist,* Washington DC: Conservation Foundation.

Bosselman, F., Peterson, C. and McCarthy, C. (1999) *Managing Tourism Growth: Issues and Applications,* Washington DC: Island Press.

Bourdieu, P. (1984) *Distinction: A Social Critique of the Judgement of Taste,* London: Routledge and Kegan Paul.

Boyer, C. (1992) 'Cities for sale: merchandising history as South Street', in M. Sortain (ed.) *Variations on a Theme Park: The New American City and the End of Public Space,* New York: Farrar, Strauss and Giroux.

Bramwell, B. (1990), 'Local tourism initiatives', *Tourism Management* 11(2): 176–7.

Bramwell, B. and Broom, G. (1989) 'Tourism Development Action Programmes: an approach to local tourism initiatives', *Insights*: H 9–12.

Bramwell, B. and Lane, B. (2000) 'Collaboration and partnership in tourism planning', in B. Bramwell and B. Lane (eds) *Tourism Collaboration and Partnerships: Politics, Practice and Sustainability,* London: Channel View Publications.

Bramwell, B. and Rawding, L. (1994) Tourism marketing organizations in industrial cities, *Tourism Management* 15(6): 425–35.

Bramwell, B. and Rawding, L. (1996) 'Tourism marketing images of industrial cities', *Annals of Tourism Research* 23(1): 310–20.

Bramwell, B. and Shurma, A. (1999) 'Collaboration in local tourism policies', *Annals of Tourism Research* 26(2): 392–415.

Braun, B.M. and Soskin, M.D. (1998) 'Theme park competitive strategies', *Annals of Tourism Research* 23: 439–43.

Bremner, C. (2002) 'Seine and sand turn Paris into a beach', *The Times* 20 July: 12.

Brennan, F. and Allen, G. (2001) 'Community-based ecotourism, social exclusion and the changing political economy of KwaZulu-Natal, South Africa', in D. Harrison (ed.) *Tourism and the Less Developed World: Issues and Case Studies,* Wallingford: CABI International.

Brewton, C. and Withiam, G. (1998) 'United States tourism policy. Alive but not well', *Cornell Hotel and Restaurant Administration Quarterly* 39(1): 50–9.

British Tourist Authority (1998) *British National Travel Survey,* London: BTA.

British Travel Association (1968) *British National Travel Survey,* London: BTA.

Britton, S. (1989) 'Tourism, dependency and development: a mode of analysis', in T.V. Singh, H.L. Theuns and F.M. Go (eds) *Towards Appropriate Tourism: The Case of Developing Countries,* Frankfurt-am-Main: Peter, pp. 93–116.

Britton, S. (1991) 'Tourism, capital and place: towards a critical geography of tourism', *Environment and Planning D: Society and Space* 9: 452–78.

Brockelman, W. and Dearden, P. (1990) 'The role of nature trekking in conservation: a case study in Thailand', *Environmental Conservation* 17(2) 34–39.

Brougham, J.E. and Butler, R.W. (1981) 'A segmentation analysis of resident attitudes to the social impact of tourism', *Annals of Tourism Research* 8(4): 569–89.

Bruner, E.M. (1989) 'Tourism, creativity and authenticity', *Studies in Symbolic Interaction* 10: 109–14.

Bruner, E.M. (1991) 'Transformation of self in tourism', *Annals of Tourism Research* 18: 238–50.

Bruner, E.M. (1994) 'Abraham Lincoln as authentic reproduction: a critique of postmodernism', *American Anthropologist* 96: 397–415.

Bryden, J.M. (1973) *Tourism and Development: A Case Study of the Commonwealth Caribbean,* London: Cambridge University Press.

Bryman, A. (1995) *Disney and His Worlds,* London: Routledge.

Bryman, A. (1999) 'Theme parks and McDonaldization', in B. Smart (ed.) *Resisting McDonaldization,* London: Sage, pp 101–115.

Budowski, G. (1976) 'Tourism and conservation: conflict, coexistence and symbiosis', *Environmental Conservation* 3(1): 27–31.

Buhalis, D. (1993) 'RICRMS as a strategic tool for small and medium tourism enterprises', *Tourism Management* 14(5): 366–78.

Buhalis, D. (1998) 'Strategic use of information technologies in the tourism industry', *Tourism Management* 19(5): 409–21.

Buhalis, D. (2000) 'Relationship in the distribution channel of tourism: conflicts between hoteliers and tour operators in the Mediterranean region', in J.C. Crott, D. Buhalis and R. March (eds) *Global Alliances and Tourism in Hospitality Management,* New York: Haworth Press, pp. 113–39.

Bull, P.J. and Church, A.P. (1994) 'The geography of employment change in the hotel and catering industry of Great Britain in the 1980s: a subregional perspective', *Regional Studies* 28(1): 13–25.

Burawoy, M. (1979) *Manufacturing Consent: Changes in the Labour Process under Capitalism,* Chicago: University of Chicago Press.

Burch, W.R. (1969) 'The social circles of leisure: competing explanations', *Journal of Leisure Research* 1: 125–47.

Burns, P. (1998) 'Tourism in Russia: background and structure', *Tourism Management* 19 (6): 555–65.

Burns, P. (1999) *An Introduction to Tourism and Anthropology,* London: Routledge.

Burrell, J., Manfredi, S., Rollin, H., Price, L. and Stead, L. (1997) 'Equal opportunities for women employees in the hospitality industry: a comparison between France, Italy, Spain and the UK', *Journal of Hospitality Management* 16: 161–79.

Buswell, R.J. (1996) 'Tourism in the Balearic Islands', in M. Barke, J. Towner and M.T. Newton (eds) *Spain: Critical Issues*, Wallingford: CAB International.

Butler, R.W. (1980) 'The concept of the tourism area cycle of evolution: implications for management of resources', *Canadian Geographer* 24(1): 5–12.

Butler, R.W. (1991) 'West Edmonton Mall as a tourist attraction', *Canadian Geographer* 35: 287–95.

Buultjens, J. and Howard, D. (2001) 'Labour flexibility in the hospitality industry: questioning the relevance of deregulation', *International Journal of Contemporary Hospitality Management* 13(2): 60–9.

Campbell, C. (1987) *The Romantic Ethic and the Spirit of Modern Consumerism*, Oxford: Blackwell.

Canan, P. and Hennessy, M. (1989) 'The Growth Machine: Tourism and the Selling of Culture', *Sociological Perspectives* 32: 227–43.

Carr, N. (1997) 'The holiday behaviour of young tourists: a comparative study' unpublished Ph.D thesis, Department of Geography, University of Exeter.

Carson, T. (1992) 'To Disneyland', *Los Angeles Weekly*, 27 March/2 April, pp. 16–28, quoted in Gottdiener (1997) op. cit.

Cavaco, C. (1995) 'Rural tourism: the creation of new tourist spaces', in A. Montanari and A.M. Williams (eds) *European Tourism: Regions, Spaces and Restructuring,* Chichester: Wiley, pp. 127–50.

Cazes, G. (1992) *Tourisme et tiers-monde: un bilan controversiale: les nouvelles colonies de tourism*, Paris: Editions l'Harmattan.

Chaney, D. (1990) 'Subtopia in Gateshead: the Metro centre as a cultural form', *Theory, Culture and Society* 7: 49–68.

Chang, T.C. (2000) 'Renaissance revisited: Singapore as a global city for the arts', *International Journal of Urban and Regional Research* 24: 818–31.

Chang, T.C., Milne, S.C., Fallon, D. and Pohlmann, C. (1996) 'Urban Heritage Tourism: the Global-Local Nexus', *Annals of Tourism Research* 23(2): 284–305.

Chesshyre, T. (2002) 'A heavy price to pay for paradise', *The Times* (Travel Section) 9 November: 1–2.

Clancy, M. (1998) 'Tourism and development: evidence from Mexico', *Annals of Tourism Research* 26(1): 1–20.

Clancy, M. (2001) *Exporting Paradise: Tourism and Development in Mexico*, Oxford: Pergamon.

Cleverdon, R. (1999) quoted in Holden, A., *Environment and Tourism*, London: Routledge.

Cochrane, A. and Pain, K. (2000) 'A globalizing society', in D. Held (ed.) *A Globalizing World? Culture, Economics and Politics,* London: Routledge.

Cohen, E. (1972) 'Toward a sociology of international tourism', *Social Research* 39: 164–82.

Cohen, E. (1979) 'A phenomenology of tourist experiences', *Journal of British Sociological Association* 13(2): 179–201.

Cohen, E. (1988) 'Authenticity and Commoditization in Tourism', *Annals of Tourism Research* 15: 371–86.

Cohen, E. (1996) 'Touring Tourist in Thailand: Tourist-orientated crime and social structure', in A. Pizam and Y. Mansfeld (eds) *Tourisms, Crime and International Security Issues*, New York: Wiley.

Coleman, S. and Crang, M. (2002) *Tourism: Between Place and Performance*, New York: Berghan.

Conforti, J.M. (1996) 'Ghettos as tourism attractions', *Annals of Tourism Research* 23(4): 830–42.

Cooke, P. and Morgan, K. (1998) *The Associational Economy: Firms, Regions and Innovations,* Oxford: Oxford University Press.

Cooper, C. (1997) 'Parameters and indicators of the decline of the British seaside resort', in G. Shaw and A.M. Williams (eds) *The Rise and Fall of British Coastal Resorts*, London: Pinter.

Cooper, C., Fletcher, J., Gilbert, D., Wanhill, S. and Shepherd, R. (1998) *Tourism: Principles and Practices*, 2nd edition, Harlow: Longman.

Cooper, M.J. (2002) 'Flexible labour markets, ethnicity and tourism-related migration in Australia and New Zealand', in M. Hall and A.M. Williams (eds) *Tourism and Migration: New Relationships between Production and Consumption,* Dordrecht: Kluwer Academic Press, pp. 73–86.

Corbin, A. (1992) *The Lure of the Sea,* Cambridge: Cambridge University Press.

Couch, C. and Farr, S.-J. (2000) 'Museums, galleries, tourism and regeneration: some experiences from Liverpool', *Built Environment* 26: 152–63.

Cox, L.J. , Morton, F. and Bowen, R.L. (1995) 'Does tourism destroy agriculture?', *Annals of Tourism Research* 22: 210–13.

Craik, J. (1997) 'The culture of tourism', in C. Rojek and J. Urry (eds) *Touring Cultures: Transformations of Travel and Theory,* London: Routledge.

Crang, M. (2003) 'Cultural geographies of tourism' in A. Lew, M. Hall and A.M. Williams (eds) *Companion to Tourism Geography*, Oxford: Blackwell.

Crang, P. (1994) 'It's showtime: on the workplace geographies of display in a restaurant in south east England', *Society and Space* 12: 675–704.

Crang, P. (1997) 'Performing the tourist product', in C. Rojek and J. Urry (eds) *Touring Cultures: Transformations of Travel and Theory,* London: Routledge.

Crawford, M. (1992) 'The world in a shopping mall', in M. Sorkin (ed.) *Variations on a Theme Park: The New American City and the End of Public Space*, New York: Noonday Press.

Cressy, R. and Cowling, M. (1996) 'Small business finance', in M. Warner (ed.) *International Encyclopedia of Business and Management,* London: Routledge.

Crick, M. (1989) 'Representations of International Tourism in the Social Sciences: Sun, Sex, Sights, Savings and Servility', *Annual Review of Anthropology* 18: 307–44.

Crompton, J.L. (1979) 'Motivation for pleasure vacation', *Annals of Tourism Research* 6: 408–24.

Crompton, R. and Sanderson, K. (1990) *Gendered Jobs and Social Change,* London: Unwin and Hyman.

Crotts, J.C., Buhalis, D., and March R. (2000) 'Introduction: Global alliances in tourism and hospitality management', in J.C. Crotts, D. Buhalis, and R. March (eds) *Global Alliances in Tourism and Hospitality Management,* New York: Haworth Press.

Cullingworth, B. (1997) Planning in the USA: Policies, Issues and Processes, London: Routledge.

Currie, R. (1997) 'A Pleasure-Tourism Framework', *Annals of Tourism Research* 24(4): 884–97.

Cybriwsky, R. (1999) 'Changing patterns of urban public space: observations and assessments from the Tokyo and New York metropolitan areas', *Cities* 16(4): 223–31.

Dahles, H. (1999a) 'Small businesses in the Indonesian tourist industry: entrepreneurship or employment?', in H. Dahles and K. Bras (eds) *Tourism and Small Entrepreneurs: Development, National Policy and Entrepreneurial Culture: Indonesian Case Studies,* New York: Cognizant Communication Corporation.

Dahles, H. (1999b) 'Tourism and small entrepreneurs in developing countries: a theoretical perspective', in H. Dahles and K. Bras (eds) *Tourism and Small Entrepreneurs: Development, National Policy and Entrepreneurial Culture: Indonesian Case Studies,* New York: Cognizant Communication Corporation.

Dahles, H. and Bras, K. (1999) *Tourism and Small Entrepreneurs: Development, National Policy, and Entrepreneurial Culture: Indonesian Cases,* New York: Cognizant Communication Corporation.

Daniel, Y.P. (1996) 'Tourism Dance Performances', *Annals of Tourism Research* 23(4): 780–97.

Dann, G. (1977) 'Anomie, ego-enhancement and tourism', *Annals of Tourism Research* 41: 184–94.

Dann, G. (1996a) *The Language of Tourism: A Sociolinguistic Perspective,* Oxford: CAB International.

Dann, G. (1996b) 'St Lucia: sociocultural issues', in L. Briguglio, R. Butler, D. Harrison and W.L. Filhó (eds). *Sustainable Tourism in Island and Small States,* London: Pinter.

Dann, G. (1996c) 'The people of tourist brochures', in T. Selwyn (ed.) *The Tourist Image: Myths and Myth Making in Tourism,* London: John Wiley.

David, P.A. and Foray, D. (2002) 'An introduction to the economy of the knowledge society', *International Social Science Journal* 171: 9–24.

Daviddi, R. (ed.)(1995) *Property Rights and Privatization in the Transition to a Market Economy,* Maastricht: European Institute of Public Administration.

Davidson, R. and Maitland, R. (1997) *Tourism Destinations,* London: Hodder and Stoughton.

Davies, D., Allen, J. and Consenza, R.M. (1988) 'Segmenting local residents by their attitudes, interests and opinions toward tourism', *Journal of Travel Research* 27(2): 2–8.

Dawkins, P., Kemo, S. and Cabalu, H. (1995) *Trade and Investment with East Asia in Selected Service Industries: The Role of Immigrants,* Canberra: Bureau of Immigration and Population.

de Albuquerque, K. and McElroy, J. (1999) 'Tourism and Crime in the Caribbean', *Annals of Tourism Research* 26(4): 968–84.

Decrop, A. (2000) 'Tourists' decision-making and behaviour processes', in A. Pizam and Y. Mansfeld (eds) *Consumer behaviour in travel and tourism,* New York: Haworth Hospitality Press.

d'Mautserre, A. (1999) 'The French mode of social regulation and sustainable tourism development: the case of Disneyland Paris', *Tourism Geographics* 1(1): 86–107.

de Kadt, E. (ed.) (1979) *Tourism: Passport to Development,* Oxford: Oxford University Press.

Demetriadi, J. (1997) 'The golden years: English seaside resorts 1950–1974', in G. Shaw and A.M. Williams (eds), *The Rise and Fall of British Coastal Resorts,* London: Pinter.

Desforges, L. (2000) 'State tourism institutions and neo-liberal development: a case study of Peru', *Tourism Geographies* 2(2): 177–92.

Dewhurst, P. and Horobin, H. (1998) 'Small business owners', in R. Thomas (ed.) *The Management of Small Tourism and Hospitality Firms,* London: Cassell.

D'Hautserre, A.-M. (1999) 'The French model of social regulation and sustainable tourism development: the case of Disneyland Paris', *Tourism Geographies* 1(1): 86–107.

Dicken, P. (1998) *Global Shift,* 3rd edition, London: Paul Chapman.

Dicken, P. and Thrift, N. (1992) 'The organisation of production and the production of organisation: why business enterprises matter in the study of geographical industrialisation', *Transactions of the Institute of British Geographers* 17: 270–91.

Dicken, P., Peck, J. and Tickell, A. (1997) 'Unpacking the global', in R. Lee and J. Wills (eds) *Geographies of Economies*, London: Arnold.

Dieke, P.U.C. (2000) 'The nature and the scope of the political economy of tourist development in Africa', in P.U.C. Dieke (ed.) *The Political Economy of Tourism Development n Africa*, New York: Cognizant Communication Corporation.

Dieques, A.C. (2001) 'Regional and domestic mass tourism in Brazil: an overview', in K.B. Ghimire (ed.) *The Native Tourist: Mass Tourism within Developing Countries*, London: Earthscan.

Dinan, C. (1999) 'A marketing geography of sustainable tourism, with special reference to Devon, England', unpublished Ph.D thesis, Department of Geography, University of Exeter.

Doherty, L. and Manfredi, S. (2001) 'Women's employment in Italian and UK hotels', *Hospitality Management* 20: 61–76.

Doxey, G.V. (1976) 'When Enough's Enough: The Natives are Restless in Old Niagara', *Heritage Canada* 2(2): 26–27.

Drucker, P.F. (1992) 'The new productivity challenge', *Harvard Business Review* 696: 69–79.

Duffus, D. and Dearden, P. (1990) 'Non-consumptive wildlife-orientated: a conceptual framework', *Biological Conservation* 53(3): 213–31.

Dundjerovic, A. (1999) 'Consortia: how far can they go towards helping independent hotels compete with chains?', *Tourism and Hospitality Research* 1(4): 370–4.

Dunford, M. (1990) 'Theories of regulation', *Society and Space* 8: 297–321.

Dunning, J.H. and McQueen, M. (1982) 'The eclectic theory of the multinational enterprise and the international hotel industry', in A.M. Rugman (ed.) *New Theories of the Multinational Enterprise*, London: Croom Helm.

Dwyer, L. and Forsyth, P. (1994) 'Foreign tourism investment: motivation and impact', *Annals of Tourism Research* 21(3): 512–37.

Dyess, R. (1997) 'Adventure travel or ecotourism?' *Adventure Travel Business*, 2 April.

Eagles, P. (1992) 'The travel motivations of Canadian ecotourists', *Journal of Travel Research* 31(2): 3–7.

Earthscan (2002) Press Release on *The Good Alternative Travel Guide*, June.

Easterlin, R. (2001) 'Income and happiness: towards a unified theory', *Economic Journal* 111(473): 465–84.

Eaton, M. (1995) 'British expatriate service provision in Spain's Costa del Sol', *Service Industries Journal* 15(2): 251–66.

Economist Intelligence Unit (1990) *Spain*, International Tourism Quarterly Report 4, London: Economist Intelligence Unit.

Ecotourism Society (1998) *Ecotourism Statistical Fact Sheet*, North Bennington, Vermont: Ecotourism Society.

Edwards, T. (2000) *Contradictions of Consumption*, Buckingham: Open University Press.

Ehrlich, H. and Dreier, P. (1999) 'The new Boston discovers the old: tourism and the struggle for a livable city', in D.R. Judd and S.S. Fainstein (eds) *The Tourist City*, New Haven: Yale University Press.

English Tourism Council (2001) *Sea Changes: Creating World-Class Resorts in England*, London: English Tourism Council.

Esping-Andersen, G. (1990) *The Three Worlds of Welfare Capitalism*, Cambridge: Polity Press.

Evans, G. (2000) 'Contemporary crafts as souvenirs, artefacts and functional goods, and their role in local economic diversification and cultural development', in M. Hitchcock and K. Teague (eds) *Souvenirs: The Material Culture of Tourism*, Aldershot: Ashgate.

Evans, N. (2001) 'Collaborative strategy: an analysis of the changing world of international airline alliances', *Tourism Management* 22: 229–43.

Fainstein, S.S. (1983) *Restructuring the City: the Political Economy of Urban Redevelopment*, London: Longman.

Falk, P. and Campbell, C. (eds) (1997) *The Shopping Experience*, London: Sage.

Featherstone, M. (1991) *Consumer Culture and Postmodernism*, London: Sage.

Feifer, M. (1985) *Going Places*, London: Macmillan.

Fennell, D. (1999) *Ecotourism: An Introduction*, London: Routledge.

Finn, A. and Erdem, T. (2001) 'The economic impact of a mega-multi-mall: estimation issues in the case of the West Edmonton Mall', *Tourism Management* 16: 367–73.

Fjellman, S. (1992) *Vinyl Leaves: Walt Disney World and America*, Boulder, Colorado: Westview.

Fodness, D. (1994) 'Measuring Tourist Motivation', *Annals of Tourism Research* 21: 555–81.

Freitag, T.G. (1994) 'Enclave tourism development: for whom the benefits roll?', *Annals of Tourism Research* 21: 538–54.

Frideres, J. and Goldenberg, S. (1982) 'Ethnic identity: myth and reality in Western Canada', *International Journal of Intercultural Relations* 6, 137–51.

Friedman, J. (1999) 'The hybridisation of roots and the abhorrence of the bush', in M. Featherstone and S. Lash (eds) *Spaces of Culture: City, Nation, World,* London: Sage.

Friedmann, H. (1980) 'Household production and the national economy: concepts for the analysis of agrarian formations', *Journal of Peasant Studies* 7: 158–84.

Garland, A. (1997) *The Beach*, London: Penguin.

Gartner, W.C. (1999) 'Small scale enterprises in the tourism industry in Ghana's central region', in D.G. Pearce and R. Butler (eds) *Contemporary Issues in Tourism Development,* London: Routledge, pp. 158–75.

Gee, C., Makens, J. and Choy, D. (1989) *The Travel Industry*, 2nd edition, New York: Van Nostrand Reinhold.

Gertler, M.S. (1997) 'The invention of regional culture', in R. Lee and J. Wills (eds) *Geographies of Economies,* London: Arnold.

Getz, D. (1987) *Tourism Planning and Research: Traditions, Models and Futures,* conference paper at the Australian Travel Research Workshop, Banbury, Australia, 5–6 November; quoted in Hall (2000), *Tourism Planning: Policies, Processes and Relationships*, Harlow: Pearson.

Getz, D. (1994) 'Residents' attitude towards tourism', *Tourism Management* 15(4): 247–58.

Getz, D., and Carlsen, J. (2000) 'Characteristics and goals of family and owner-operated businesses in the rural tourism and hospitality sectors', *Tourism Management* 21: 547–60.

Ghimine, K.B. (2001) 'The growth of national and regional tourism in developing countries: an overview', in K.B. Ghimine (ed.) *The Native Tourist: Mass Tourism within Developing Countries*, London: Earthscan.

Giddens, A. (1996) 'Affluence, poverty and the idea of a post-scarcity society', *Development and Change* 27: 365–77.

Gilbert, E.M. (1954) *Brighton: Old Ocean's Bauble*, London: Methuen.

Gill, A. (2000) 'From growth machine to growth management: the dynamics of resort development in Whistler, British Columbia', *Environment and Planning A* 32: 1083–103.

Gnoth, J. (1997) 'Tourism Motivation and Expectation Formation', *Annals of Tourism Research* 24: 283–304.

Go, F.M. and Pine, R. (1995) *Globalization Strategy and the Hotel Industry,* London: Routledge.

Goffee, R. and Scase, R. (1983) 'Class entrepreneurship and the service sector: towards a conceptual clarification', *Services Industries Journal* 3: 146–60.

Gold, J.R. and Ward, S.V. (1994) *Place Promotion: The Use of Publicity and Marketing to Sell Towns and Regions,* Chichester: John Wiley.

Goldman, R. and Papson, S. (1996) *Sign Wars: The Cluttered Landscape of Advertising,* London: Guilford Press.

Goldthorpe, J.H., Lackwood, D., Bechhofer, F. and Platt, J. (1968) *The Affluent Worker: Industrial Attitudes and Behaviour,* Cambridge: Cambridge University Press.

Gonçalves, V.F. da C., and Aguas, P.M.R. (1997) 'The concept of life cycle: an application to the tourist product', *Journal of Travel Research* 36(2): 12–22.

Goodwin, M. and Painter, J. (1996) 'Local governance, the crises of Fordism and the changing geographies of regulation', *Transactions of the Institute of British Geographers* 21: 635–48.

Goossens, C. (1998) 'Tourism Information and Pleasure Motivation', *Annals of Tourism Research* 25: 301–20.

Gordon, C. (1991) 'Sustainable leisure', *Ecos* 12(1): 7–13.

Gordon, I. and Goodall, B. (2000) 'Localities and tourism', *Tourism Geographies* 2(3): 290–311.

Goss, J. (1993) ' "The Magic of the Mall": an analysis of form, function and meaning in the contemporary retail built environment', *Annals of the Association of American Geographers* 83: 18–47.

Goss, J. (1996) 'Disquiet on the waterfront: reflections on nostalgia and utopia in the urban archetypes of festival marketplaces', *Urban Geography* 17: 221–47.

Goss, J. (1999) 'Once-upon-a-time in the commodity world: an official guide to the Mall of America', *Annals of the Association of American Geographers* 89: 45–75.

Gotham, K. (2002) 'Marketing Mardi Gras: commodification, spectacle and the political economy of tourism in New Orleans', *Urban Studies* 39(10): 1735–56.

Gottdiener, M. (1997) *The Theming of America: Dreams, Visions and Commercial Spaces,* Boulder, Colorado: Westview Press.

Gottdiener, M. (2000) 'The consumption of space and the spaces of consumption', in M. Gottdiener (ed.) *New Forms of Consumption: Consumers, Culture and Commodification,* Lanham, Maryland: Rowman and Littlefield.

Gottdiener, M. (2000) Approaches to Consumption: classical and contemporary perspectives, in M. Gottdiener (ed.) *New Forms of Consumption: Consumers, Culture and Commodification,* Lanham, Maryland: Rowman and Littlefield: 3–32.

Gottlieb, A. (1982) 'Americans' Vacations', *Annals of Tourism Research* 9: 165–87.

Graburn, N. (1976) 'The Eskimos and airport art', *Trans-Action* 4: 28–33.

Graburn, N. (1983) 'The Anthropology of Tourism', *Annals of Tourism Research* 10(1): 9–33.

Granovetter, M. (1985) 'Economic action and social structure: the problem of embeddedness', *American Journal of Sociology* 91(3): 481–510.

Gratton, C. (1990) 'Consumer behaviour in tourism: a psycho-economic approach', paper presented at 'Tourism Research into the 1990s' conference, University of Durham.

Green, M. (2001) 'Urban Heritage Tourism: Globalisation and Localisation', in G. Richards (ed.) *Cultural Attractions and European Tourism,* Wallingford: CAB International.

Greenwood, D.J. (1977) 'Culture by the pound: an anthropological perspective on tourism as cultural commoditization', in V. Smith (eds) *Host and Guests: The Anthropology of Tourism,* Oxford: Blackwell, pp. 129–39.

Greenwood, J. (1992) 'Producer interest groups in tourism policy: case studies from Britain and the European Community', *American Behavioural Scientist* 36(2): 236–56.

Gregson, N. (1995) 'And now it's all consumption', *Progress in Human Geography* 19: 135–41.

Grindle, M.S. (1986) *State and Countryside: Development Policy and Agrarian Politics in Latin America*, Baltimore: Johns Hopkins University Press.

Gursoy, D., Jurowski, C. and Uysal, M. (2002) 'Resident attitudes: a structural modelling approach', *Annals of Tourism Research* 29(1): 79–105.

Hall, C.M. (1994) 'Gender and economic interests in tourism prostitution: the nature, development and implications of sex tourism in South-east Asia' in V. Kinnaird and D. Hall (eds) *Tourism: A Gender Analysis*, New York: John Wiley, pp.142–63.

Hall, C.M. (1996) *Tourism and Politics: Policy, Power and Place*, Chichester: John Wiley.

Hall, C.M. (1998) 'The institutional setting – tourism and the state', in D. Ioannides and K.G. Debbage (eds) *The Economic Geography of the Tourist Industry: A Supply-Side Analysis*, London: Routledge, pp. 199–219.

Hall, C.M. (2000) *Tourism Planning: Policies, Processes and Relationships*, Harlow: Pearson.

Hall, C.M. and Jenkins, J.M. (1995) *Tourism and Public Policy*, London: Routledge.

Hall, C.M. and Page, S. (1999) *The Geography of Tourism and Recreation: Environment, Place and Space*, London: Routledge.

Hall, T. (1995) 'The Second Industrial Revolution: cultural reconstructions of industrial regions', *Landscape Research* 20: 112–23.

Handler, R. and Saxton, W. (1988) 'Dissimulation: reflexivity, narrative, and the quest for authenticity in "Living History" ', *Cultural Anthropology* 3: 242–60.

Hann, C. (2000) *The Tragedy of the Privates? Postsocialist Property Relations in Anthropological Perspective*, Working Paper 2. Halle: Max Planck Institute for Social Anthropology.

Hannerz, U. (1996) *Transnational Connections: Culture, People, Places*, London: Routledge.

Hannigan, J. (1998) *Fantasy City: Pleasure and Profit in the Postmodern Metropolis*, London: Routledge.

Harrison, D. (1992a) 'The background', in D. Harrison (ed.) *Tourism and Less Developed Countries*, London: Belhaven Press.

Harrison, D. (ed.) (1992b) *Tourism and the Less Developed Countries*, London: Belhaven Press.

Harvey, D. (1985) *The urbanisation of capital*, Oxford: Blackwell.

Harvey, D. (1988) 'Vodoo cities', *New Statesman and Society*, 30 September: 33–5.

Harvey, D. (1989a) *The Condition of Postmodernity*, Oxford: Blackwell.

Harvey, D. (1989b) 'From managerialism to entrepreneurialism: the transformation in urban governance in late capitalism', *Geografiska Annaler* 71B: 3–17.

Harvey, D. (1996) *Justice, Nature and the Geography of Difference*, Oxford: Blackwell.

Haywood, M. (1998) 'Economic business cycles and the tourism life-cycle concept', in D. Ioannides and K.G. Debbage (eds) *The Economic Geography of the Tourist Industry: A Supply-Side Analysis*, London: Routledge, pp. 273–84.

Held, D. (2000) 'Introduction', in D. Held (ed.) *A Globalizing World? Culture, Economics, Politics*, Routledge, London, pp. 1–12.

Hennessy, S. (1994) 'Female employment in tourism development in South-West England', in V. Kinnaird and D. Hall (eds) *Tourism: A Gender Analysis*, Chichester: John Wiley.

Herbert, D. (2001) 'Literary Places, Tourism and the Heritage Experience', *Annals of Tourism Research* 28: 312–33.

Hermans, D. (1981) 'The encounter of agriculture and tourism: a Catalan case', *Annals of Tourism Research* 8: 462–79.

Hiernaux, N.D. (1999) 'Cancun Bliss' in D.R. Judd and S.S. Fainstein (eds) *The Tourist City*, New Haven: Yale University Press.

Hirschman, E.C. and Holbrook, M.B. (1982) 'Hedonic Consumption: Emerging Concepts, Methods and Propositions', *Journal of Marketing* 46: 92–101.

Hirst, P. and Thompson, G. (1996) *Globalization in Question,* Cambridge: Polity Press.

Hjalager, A.-M. (2000) 'Tourism destinations and the concept of industrial districts', *Tourism and Hospitality Research* 2(3): 199–213.

Hodson, M. (2002) 'Thailand to die for', *The Sunday Times*, Travel section, 7 July: 57.

Holcomb, B. (1993) 'Revisioning place: de- and re-constructing the image of the industrial city', in G. Kearns and C. Philo (eds) *Selling Places: the City as Cultural Capital, Past and Present*, Oxford: Pergamon Press.

Holden, A. (2000) *Environment and Tourism*, London: Routledge.

Hollinshead, K. (1997) 'Heritage Tourism under Post-modernity: Truth and the Past', in C. Ryan (ed.) (1997) *The Tourist Experience: A New Introduction*, London: Cassell.

Holloway, J.C. (1998) *The Business of Tourism,* Harlow: Longman.

Holt, D.B. (1995) 'How Consumers Consume: A Typology of Consumption Practices', *Journal of Consumer Research* 22: 1–16.

Hope, C.A. and Klemm, M.S. (2001) 'Tourism in difficult areas revisited: the case of Bradford', *Tourism Management* 22: 629–35.

Horner, A.E. (1993) 'Tourist arts in Africa before tourism', *Annals of Tourism Research* 20: 52–63.

Horner, S. and Swarbrooke, J. (1996) *Marketing Tourism, Hospitality and Leisure in Europe*, London: Thomson Business Press.

Hotels (2000) 'Corporate 300', *Hotels* July: 49–58.

Hubbard, P. (1998) 'Introduction, representation, culture and identities', in T. Hall and P. Hubbard (eds), *The Entrepreneurial City: Geographies of Politics, Regime and Representation*, Chichester: Wiley.

Hudson, R. (1997) 'The end of mass production and of the mass collective worker? Experimenting with production, employment and their geographies', in R. Lee and J. Wills (eds) *Geographies of Economies,* London: Edward Arnold.

Hudson, R. (2001) *Producing Places,* New York: Guilford Press.

Hudson, R. and Townsend, A. (1992) 'Trends in tourism employment and resulting policy choices for local government', in P. Johnson and B. Thomas (eds) *Perspectives on Tourism Policy*, London: Mansell.

Hudson, R. and Williams, A.M. (1995) *Divided Britain,* Chichester: John Wiley.

Huong, N.T. and King, B. (2002) 'Migrant communities and tourism consumption: the case of the Vietnamese in Australia', in M. Hall and A.M. Williams (eds) *Tourism and Migration: New Relationships between Production and Consumption,* Dordrecht: Kluwer Academic Press.

Hvenegaard, G. (1994) 'Ecotourism: a status report and conceptual framework', *Journal of Tourism Studies* 5(2): 24–35.

ILO (1983) *Social Problems and Employment in the Hotel and Catering Trade, Restaurants and Similar Establishments in Developing Countries,* Geneva: ILO.

ILO (2001) *ILO Yearbook 2000*, Geneva: ILO.

Ioannides, D. (1998) 'Tour operators: the gatekeepers of tourism', in D. Ioannides and K.G. Debbage (eds) *The Economic Geography of the Tourist Industry: A Supply-Side Analysis,* London: Routledge.

Ioannides, D. and Debbage, K.G. (1997) 'Post-Fordism and flexibility: the travel industry polyglot', *Tourism Management* 18(4): 229–41.

Ioannides, D. and Debbage, K.G. (1998) 'Neo-Fordism and flexible specialization in the travel industry: dissecting the polyglot', in D. Ioannides and K.G. Debbage (eds) *The Economic Geography of the Tourist Industry: a Supply-Side Analysis*, London: Routledge.

Ireland, M. (1993) 'Gender and class relations in tourism employment', *Annals of Tourism Research* 20: 666–84.

Iso-Ahola, S. (1982) 'Towards a social psychology of tourism motivation: a rejoinder', *Annals of Tourism Research* 9: 256–61.

Iuoto (1974) 'The role of the state in tourism', *Annals of Tourism Research* 1(3): 66–72.

Jackson, I. (1986) 'Carrying capacity in small tropical Caribbean islands', *Industry and Environment* 9(1): 7–10.

Jackson, P. (1995) 'Changing geographies of consumption', *Environment and Planning A* 27: 1875–6.

Jafari, J. (1989) 'Socio-cultural dimensions of tourism: and English language literature review', in J. Bustrzanowski (ed.) *Tourism as a Factor of Change: a Sociocultural Study*, Vienna: Economic Coordination Centre for Research and Documentation in Social Sciences.

Jamal, T.B. and Getz, D. (1995) 'Collaboration theory and community tourism planning', *Annals of Tourism Research* 22: 186–204.

Jayne, M. (2000) 'Imag[in]ing a post-industrial Potteries', in D. Bell and A. Haddour (ed.) *City Visions*, Harlow: Pearson Education.

Jeans, D.N. (1990) 'Beach resort morphology in England and Australia: a review and extension', in P. Fabbri (ed.) *Recreational Uses of Coastal Areas*, Dordrecht: Kluwer.

Jeffries, D. (2001) *Government and Tourism*, Oxford: Butterworth-Heinemann.

Jenkins, C.L. (1982) 'The effect of scale in tourism projects in developing countries', *Annals of Tourism Research* 9: 229–49.

Jesitus, J. (1993) 'Megaresort Expected to Boost Mexico's Tourism', *Hotel and Motel Management* 208(2): 3–5.

Jessop, B. (1994) 'Post-Fordism and the state: a reader', in A. Amin (ed.) Oxford: Blackwell, pp. 251–79.

Jessop, R. (2001) 'Institutional re(turns) and the strategic-relational approach', *Environment and Planning A* 33: 1213–35.

Johanson, J., Lars, H. and Nazeem, S.M. (1991) 'Interfirm adaptation in business relationships', *Journal of Marketing* 55(2): 29–37.

Johnson, J.D., Snepenger, D.J. and Akis, S. (1994) 'Residents' perceptions of tourism development', *Annals of Tourism Research* 21: 629–42.

Jones, T.S.M. (1994) 'Theme parks in Japan', *Progress in Tourism, Recreation and Hospitality Management* 6: 111–25.

Jordan, F. (1997) 'An occupational hazard? Sex segregation in tourism employment', *Tourism Management* 18(8): 525–34.

Judd, D.R. (1995) 'Promoting tourism in US cities', *Tourism Management* 16(3): 175–89.

Judd, D.R. and Fainstein, S.S. (1995) *The Tourist City*, New Haven: Yale University Press.

Kaosa-ard, M. Bezic, D. and White, S. (2001) 'Domestic tourism in Thailand: supply and demand', in K.B. Ghimine (ed.) *The Native Tourist*, London: Earthscan.

Kayat, K. (2002) 'Power, social exchanges and tourism in Langkawi: rethinking resident perceptions', *International Journal of Tourism Research* 4: 171–91.

Kerin, P. and Peterson, R.A. (1980) *Perspectives on Strategic Marketing Management,* Boston: Allyn and Bacon.

Kim, J. (1991) 'A study on Korean perceptions toward overseas travel', *Journal of Tourism Sciences* 15, 29–42.

King, B. (1997) *Creating Island Resorts*, London: Routledge.

King, B. (2001) 'Resort-based Tourism on the Pleasure Periphery', in D. Harrison (ed) *Tourism and the Less Developed World: Issues and Case Studies*, Wallingford: CABI International.

King, B., Pizam, A. and Milman, A. (1993) 'Impacts of Tourism: Host Perceptions', *Annals of Tourism Research* 20: 650–5.

King, R. (1995) 'Tourism, labour and international migration', in A. Montanari and R. King (eds) *European Tourism: Regions, Spaces, and Restructuring*, Chichester: John Wiley.

Kirchner, E.J. (1992) *Decision Making in the European Community: The Council Presidency and European Integration,* Manchester: Manchester University Press.

Knowles, T. and Curtis, S. (1999) 'The market viability of European mass tourist destinations: a post-stagnation lifecycle analysis', *International Journal of Tourism Research* 1(1): 87–96.

Knowles, T., Diamantis, D. and El-Mourhabi, J.B. (2001) *The Globalization of Tourism and Hospitality: A Strategic Perspective,* London: Continuum.

Kotkin, J. (1993) *Tribes: How Race, Religion, and Identity Determine Success in the New Global Economy,* New York: Random House.

Kotler, P, Haider, D.H. and Rein, I. (1983) *Marketing Places: Attracting Investment, Industry, and Tourism to Cities, States, and Nations,* New York: Free Press.

Kousis, M. (1989) 'Tourism and family life in a rural Cretan community', *Annals of Tourism Research* 16: 318–32.

Krippendorf, J. (1987) *The Holiday Makers*, London: Heinemann.

Kumar, K. (1995) *From Post-Industrial to Post-Modern Society: New Theories of the Contemporary World,* Oxford: Blackwell.

Kusler, J. (1991) 'Ecotourism and resource conservation: introduction to issues', in J. Kusler (ed.) *Ecotourism and Resource Conservation: A Collection of Papers*, vol. 1, Madison: Omnipress.

Langer, F.J. and Piper A.T. (1987) 'The prevention of mindlessness', *Journal of Personality and Social Psychology* 52: 269–78.

Langford, S.V. and Howard, D.R. (1994) 'Developing a tourism impact attitude scale', *Annals of Tourism Research* 21: 121–39.

Lankford, S.V. (1994) 'Attitudes and Perceptions Toward Tourism and Rural Regional Development', *Journal of Travel Research* 32(3): 35–43.

Lash, S. (1991) *Sociology of Postmodernism*, London: Routledge.

Lash, S. and Urry, J. (1989) *The End of Organised Capital*, Cambridge: Polity Press.

Lavery, P. and van Doren, C. (1990) *Travel and Tourism: A North American European Perspective*, Huntingdon: ELM Publications.

Law, C.M. (1993) *Urban Tourism: Attracting Visitors to Large Cities*, London: Mansell.

Law, C.M. (2002) *Urban Tourism: The Visitor Economy and the Growth of Large Cities*, 2nd edition, London: Continuum.

Laxson, J. (1991) 'How "we" see "them": Tourism and native American Indians', *Annals of Tourism Research* 18(3): 365–89.

Lea, S., Kemp, S. and Willetts, K. (1994) 'Residents' Concepts of Tourism', *Annals of Tourism Research* 21: 406–09.

Lee, M. (1993) *Consumer Culture Reborn: The Cultural Politics of Consumption*, London: Routledge.

Lee, R. (1994) 'Modernisation, postmodernism and the third world', *Current Sociology* 42(2): 1–66.

Lee, R. and Wills, J. (eds) (1997) *Geographies of Economies*, London: Edward Arnold 1997.

Lee-Ross, D. (1999) 'Seasonal hotel jobs: an occupation and a way of life', *International Journal of Tourism Research* 1: 239–53.

Lefebvre, H. (1991) *The Production of Space*, Oxford: Blackwell.

Lehtonen, T.-K. and Mäenpää, P. (1997) 'Shopping in the East Centre Mall', in P. Falk and C. Campbell (eds) *The Shopping Experience*, London: Sage.

Leiper, N. (1990) 'Partial industrialization of tourism systems', *Annals of Tourism Research* 17: 600–05.

Lencek, L. and Basker, G. (1998) *The Beach: The History of Paradise on Earth*, London: Secker and Warburg.

Leslie, D. (2001) 'Urban Regeneration and Glasgow's Galleries with particular reference to the Burrell Collection', in G. Richards (ed.) *Cultural Attractions and European Tourism*, Wallingford: CAB International.

Lett, J.W. (1983) 'Ludic and liminoid aspects of charter yacht tourism in the British Virgin Islands', *Annals of Tourism Research* 10: 35–56.

Lew, A. and Wong, A. (2002) 'Tourism and the Chinese diaspora', in M. Hall and A.M. Williams (eds) *Tourism and Migration: New Relationships between Production and Consumption,* Dordrecht: Kluwer Academic Press.

Lewis, R. (1980) 'Seaside holiday resorts in the United States and Britain', *Urban History Yearbook*: 44–52.

Lewis, S. (1996) 'Young People's Holidays', *Travel Weekly* 11 December: 33–36.

Li, Y. (2000) 'Geographical Consciousness and Tourism Experience', *Annals of Tourism Research* 27(4): 863–83.

Liebfried, S. (1993) 'Conceptualising European social policy: the EC as a social actor', in L. Hantrais and S. Mangen (eds) *The Policy Making Process and the Social Actors,* Loughborough: Loughborough University European Research Centre.

Light, D. and Prentice, R.C. (1994) 'Who Consumes the Heritage Product: implications for European heritage', in P. Larkham and G. Ashworth (eds) *Building a New Heritage: Tourism, Culture, and Identity in the New Europe*, London: Routledge.

Lindberg, K. (1991) *Politics for Maximising Nature Tourism's Ecological and Economic Benefits*, North Bennington, Washington, D.C.: World Resources Institute.

Lindberg, K. and Hawkins, D. (eds) *Ecotourism: A Guide for Planners and Managers*, vol. 1, Vermont: Ecotourism Society.

Lindberg, K. and Johnson, R.L. (1993) 'Modelling resident attitudes toward tourism', *Annals of Tourism Research* 24(2): 402–24.

Lindblom, C.E. (1980) *The Policy Making Process*, 2nd edition, Englewood Cliffs, New Jersey: Prentice-Hall.

Liu, J.C. and Var, T. (1986) 'Resident attitudes toward tourism impacts in Hawaii', *Annals of Tourism Research* 13: 193–214.

Lockwood, A. and Guerrier, Y. (1989) 'Flexible working in the hospitality industry: current strategies and future potential', *Journal of Contemporary Hospitality Management* 1: 11–6.

Logan, J.R. and Molotoch, H. (1987) *Urban Fortunes: The Political Economy of Place,* Berkeley: University of California Press.

Loker-Murphy, L. and Pearce, P.L. (1995) 'Young Budget Travellers: Backpackers in Australia', *Annals of Tourism Research* 22(4): 819–43.

Loverseed, H. (1994) 'Theme parks in North America', *Travel and Tourism Analyst* 4: 51–63.

Lowe, M. (1993) 'Local hero! An examination of the role of the regional entrepreneur in the regeneration of Britain's regions', in G. Kearns and C. Philo (eds) *Selling Places: The City as Cultural Capital, Past and Present*, Oxford: Pergamon Press.

Lumsdon, L. and Swift, J. (2001) *Tourism in Latin America*, London: Continuum.

Lundgren, J. (1973) 'Tourist impact/island entrepreneurship in the Caribbean', conference paper quoted in Mathieson and Wall, *Tourism: Economic, Physical and Social Impacts*, London: Longman.

Lury, C. (1996) *Consumer Culture*, Cambridge: Polity Press.

Luzer, E., Diagne, A., Gan, C. and Henning, B. (1995) 'Evaluating nature-based tourism using the new environmental paradigm', *Journal of Agricultural and Applied Economics* 27(2): 544–55.

MacCannell, D. (1973) 'Staged authenticity: arrangements of social space in tourist settings', *American Journal of Sociology* 79(3): 589–603.

MacCannell, D. (1976) *The Tourist: A New Theory of the Leisure Class*, New York: Sulouker Books; revised edition, 1989.

McCracken, G. (1990) *Culture and Consumption*, Bloomington: Indiana University Press.

MacDonald, G.F. and Alsford, S. (1995) 'Museums and theme parks: worlds in collision?' *Museum Management and Curatorship* 14: 129–47.

Macdonald-Wallace, D. (1999) 'UK tourism and the Internet: the slow stumble towards success', *Insights* : 139–42.

McDowell, L. (1997) 'A tale of two cities? Embedded organisations and embodied workers in the City of London', in R. Lee and J. Wills (eds) *Geographies of Economies,* London: Edward Arnold.

McElroy, J.L. and Albuquerque, K. (1998) 'Tourism Penetration Index in Small Caribbean Islands', *Annals of Tourism Research* 25(1): 145–68.

McIntosh, A.J. and Prentice, R.C. (1999). 'Affirming authenticity: consuming cultural heritage', *Annals of Tourism Research* 26(3): 589–12.

MacKay, J. (1994) 'Eco tourists take over', *The Times* 17 February.

McKercher, B. (2002) 'Towards a classification of cultural tourists', *International Journal of Tourism Research* 4: 29–38.

MacKinnon, D., Cumbers, A. and Chapman, K. (2002) 'Learning, innovation and regional development: a critical appraisal of recent debates', *Progress in Human Geography* 26(3): 293–11.

Mackun, P. (1998) 'Tourism in the Third World: labor and social-business networks', in D. Ioannides and K.G. Debbage (eds) *The Economic Geography of the Tourist Industry: A Supply-Side Analysis,* London: Routledge.

McLaren, D. (1998) *Rethinking Tourism and Ecotravel: The Paving of Paradise and What You Can Do to Stop It*, West Hartford, Connecticut: Kumarian Press.

MacLeod, G. (2001) 'Beyond soft institutionalism: accumulation, regulation and their geographical fixes', *Environment and Planning A* 33:1145–67.

McMullan, W. and Long, W.A. (1990) *Developing New Ventures: The Entrepreneurial Option',* San Diego: Harcourt Brace Jovanovich.

Macnaghten, P. and Urry, J. (1998) *Contested Natures*, Sage: London.

McTaggart, W. D. (1988) 'Hydrologic management in Bali', *Singapore Journal of Tropical Geography* 9: 96–111.

Madrigal, R. (1993) 'A tale of tourism in two cities', *Annals of Tourism Research* 20: 336–53.

Madrigal, R. (1995) 'Resident perceptions and the role of government', *Annals of Tourism Research* 22: 86–102.

Mansperger, M.C. (1995) 'Tourism and cultural change in small-scale societies', *Human Organisation* 54: 87–94.

Markwick, M.C. (2001) 'Tourism and the development of handicraft production in the Maltese islands', *Tourism Geographies* 3(1): 29–51.

Marshall, G. (1986) 'The workplace culture of a licensed restaurant', *Theory, Culture and Society* 3: 33–48.

Martin, B. (1982) *A Sociology of Contemporary Popular Culture*, Oxford: Blackwell.

Martin, P. (2002) 'Easy Jet takes a difficult route', *Financial Times* 7 May.

Martin, R., Sunley, P. and Wills, J. (1994) 'Unions and the politics of deindustrialisation: comments on how geography complicates class analysis', *Antipode* 26(1): 59–76.

Martinez, D.P. (1998) *The Worlds of Japanese Popular Culture: Gender, Shifting Boundaries and Global Cultures*, Cambridge: Cambridge University Press.

Maslow, A. (1970) *Motivation and Personality*, 2nd edition. New York: Harper.

Mason, P. (2002) 'The Big OE: New Zealanders' overseas experience in Britain', in M. Hall and A.M. Williams (eds) *Tourism and Migration: New Relationships between Production and Consumption*, Dordrecht: Kluwer Academic Press.

Massey, D. (1995) *Spatial Divisions of Labour: Social Structures and the Geography of Production*, second edition, London: Macmillan.

Mathieson, A. and Wall, G. (1982) *Tourism: Economic, Physical and Social Impacts*, London: Longman.

Mazanec, J.A. and Zins, A.H. (1994) 'Tourist behaviour and the new European lifestyle typology', in W. Theobold (ed.) *Global Tourism: The Next Decade*, Oxford: Butterworth-Heinemann.

Meethan, K. (2001) *Tourism in Global Society: Place, Culture, Consumption*, Basingstoke: Palgrave.

Meltzer, E. (2002) 'Performing Place: A Hyperbolic Drugstore in Wall, South Dakota, in S. Coleman and M. Crang (eds) *Tourism: Between Place and Performance*, New York: Berghahn Books.

Meyer-Arendt, K.J. (1990) 'Gulf of Mexico seaside resorts', *Journal of Cultural Geography* 11(1): 39–55.

Middleton, V. and Clark J. (2001) *Marketing in Travel and Tourism*, 3rd edition, Oxford: Butterworth-Heinemann.

Miles, S. (1998) *Consumerism as a Way of Life*, London: Sage.

Milman, A. and Pizam, A. (1988) 'Social impacts of tourism on central Florida', *Annals of Tourism Research* 15: 191–204.

Milne, S. and Gill, K. (1998) 'Distribution technologies and destination development: myths and realities', in D. Ioannides and K.G. Debbage (eds) *The Economic Geography of the Tourist Industry: A Supply-Side Analysis*, London: Routledge.

Milne, S. and Pohlmann, C. (1998) 'Continuity and change in the hotel sector: some evidence from Montreal', in D. Ioannides and K.G. Debbage (eds) *The Economic Geography of the Tourist Industry: A Supply-Side Analysis*, London: Routledge.

Mo, C.M., Howard, D.R. and Havitz, M.E. (1993) 'Testing an international tourist role typology', *Annals of Tourism Research* 20(2): 319–25.

Mok, C. and DeFranco, A.L. (1999) 'Chinese cultural values: their implications for travel and tourism marketing', *Journal of Travel and Tourism Marketing* 8(2): 99–114.

Moore, K. (2000) *Museums and Popular Culture*, London: Cassell.

Morgan, M. (1991) 'Dressing up to survive: marketing Majorca anew', *Tourism Management* 11: 15–20.

Morgan, N. (1997) 'Seaside resort strategies: the case of inter-war Torquay', in S. Fisher (ed.) *Recreation and the Sea*, Exeter: Exeter University Press.

Morgan, N. and Pritchard, A. (1999) *Tourism, Promotion and Power: Creating Images, Creating Identities*, Chichester: John Wiley.

Morgan, N., Pritchard, A. and Pride, R. (2002) *Destination Branding: Creating the unique destination proposition*, Oxford: Butterworth-Heinemann.

Morrell, P.S. (1998) 'Airline sales and distribution channels: the impact of new technology', *Tourism Economics* 4(1): 5–19.

Morrison, A., Rimmington, M. and Williams, C. (1999) *Entrepreneurship in the Hospitality, Tourism and Leisure Industries*, Oxford: Butterworth-Heinemann.

Mowforth, M. (1993) *Ecotourism: Terminology and Definitions*, occasional paper, Department of Geographical Sciences, University of Plymouth.

Mowforth, M. and Munt, I. (1998) *Tourism and Sustainability: New Tourism in the Third World*, London: Routledge.

Mowforth, M. and Munt, I. (2003) *Tourism and Sustainability: Development and Tourism in the Third World*, 2nd edition, London: Routledge.

Müller, D.K. (2001) 'German second home development in Sweden', in M. Hall and A.M. Williams (eds) *Tourism and Migration: New Relationships between Production and Consumption,* Dordrecht: Kluwer Academic Press.

Munt, I. (1994) 'The "other" postmodern tourism: culture travel and the new middle classes', *Theory, Culture and Society* 11(3): 101–25.

Murphy, P.E. (1983) 'Perceptions and attitudes of decision-making groups in tourism centers', *Journal of Travel Research* 21(3): 8–12.

Murphy, P.E. (1985) *Tourism: A Community Approach*, London: Methuen.

Murphy, P.E. (1988) 'Community driven tourism planning', *Tourism Management* 9: 96–104.

Nielson, C. (2001) *Tourism and the Media*, Melbourne: Hospitality Press.

Nimmonratana, T. (2000) 'Impacts of tourism on a local community: a case study of Chiang Mai', in K.S. Chon (ed.) *Tourism in Southeast Asia*, New York: Haworth Hospitality Press.

Noronha, F. (1999) 'Culture Shocks', *In Focus* Spring: 4–5.

Nūnez, T. (1989) 'Touristic studies in anthropological perspective', in V. Smith (ed.) *Hosts and Guests: The Anthropology of Tourism*, 2nd edition, Philadelphia: University of Pennsylvania Press.

Nuryanti, W. (1996) 'Heritage and Postmodern Tourism', *Annals of Tourism Research* 23(2): 249–60.

OECD (1974) *Government Policy in the Development of Tourism*, Paris: Organisation for Economic Co-operation and Development.

Ohmae, K. (1990) *The Borderless World,* London: Collins.

Olding, S. (2000) 'Funding Heritage', *Insights* July: A9–A12.

Orams, M. (2000) 'Towards a more desirable form of ecotourism', in C. Ryan and S.J. Page (eds) *Tourism Management: Towards the New Millennium*, Oxford: Pergamon Press.

Orbasli, A. (2000) *Tourist in Historic Towns: Urban Conservation and Heritage Management*, London: Spon.

Outhart, T., Taylor, L., Barker, R. and Marvell, A. (2000) *Travel and Tourism*, London: HarperCollins.

Page, S. and Dowling, K. (2002) *Ecotourism*, London: Prentice-Hall.

Page. S., Forer, P. and Lawton, G.R. (1999) 'Small business development and tourism: terra incognita?', *Tourism Management* 20(4): 435–60.

Painter, J. and Goodwin, M. (1995) 'Local governance and concrete research: investigating the uneven development of regulation', *Economy and Society* 24: 334–56.

Painter, M. (1992) 'Participation in power', in M. Munro-Clarke (ed.) *Participation on Government*, Sydney: Male and Iremonger.

Palacio, V. and McCool, S. (1997) 'Identifying ecotourists in Belize through benefit segmentation: a preliminary analysis', *Journal of Sustainable Tourism* 5(3): 234–43.

Parry, K. (1983) *Resorts on the Lancashire Coasts*, Newton Abbot: David and Charles.

Passariello, P. (1983) 'Never on Sunday? Mexican tourists at the beach', *Annals of Tourism Research* 10: 109–22.

Pattinson, M. (1993) 'Seaside towns cry for help', *Planning* 59(7): 24–25.

Pearce, D. (1995) *Tourism Today: A Geographical Analysis*, 2nd edition, London: Longman.

Pearce, J.A. (1980) 'Host community acceptance of foreign tourists: strategic considerations', *Annals of Tourism Research* 7: 224–33.

Pearce, P.L. (1988) *The Ulysses Factor: Evaluating Visitors in Tourist Settings*, New York: Soringer.

Pearce, P.L. (1993) 'Fundamentals of tourist motivation' in D.G. Pearce and R.W. Butler (eds) *Tourism Research: Critiques and Challenges*, London: Routledge.

Pearce, P.L. and Moscardo, G.M. (1986) 'The concept of authenticity in tourist experiences', *Australian and New Zealand Journal of Sociology* 22: 121–32.

Pearce, P.L., Moscardo, G. and Ross, G.F. (1996) *Tourism Community Relationships*, Oxford: Pergamon Press.

Perdue, R.R., Long, P.T. and Allen, L. (1987) 'Rural resident tourism perceptions and attitudes', *Annals of Tourism Research* 14: 420–29.

Philo, C. and Kearns, G. (1993) 'Culture, history, capital: a critical introduction to the selling of places', in G. Kearns and C. Philo (eds) *Selling Places: The City as Cultural Capital, Past and Present,* Oxford: Pergamon Press.

Picard, M. (1996) *Bali: Cultural Tourism and Touristic Culture*, 2nd edition, Singapore: Archipelago Press.

Pine, B.J. and Gilmore, J.H. (1999) *The Experience Economy: Work is Theatre and Every Business a Stage,* Cambridge, Massachusetts: Harvard University Press.

Pizam, A. (1978) 'Tourist impacts: the social costs to the destination community as perceived by its residents', *Journal of Travel Research* 16(4): 8–12.

Pizam, A., Reichel, A., and Shieh, C.F. (1982) 'Tourism and crime: is there a relationship?', *Journal of Travel Research* 20: 7–11.

Pizam, A. and Milman, A. (1986) 'The social impacts of tourism', *Tourism Recreation Research* 11: 29–32.

Plaza, B. (2000) 'Guggenheim Museum's effectiveness to attract tourism', *Annals of Tourism Research* 27: 1055–58.

Pleumaron, A., (1992) 'Course and effect: golf tourism in Thailand', *Ecologist* 22(3): 104–10.

Plog, S. (1977) 'Why destination areas rise and fall in popularity', in E.M. Kelly (ed.) *Domestic and International Tourism*, Wellesley: Institute of Certified Travel Agents.

Pocock, D.C.D. (1987) 'Haworth: the experience of a literary place', in W.E. Mallony and P. Simpson-Harsley (eds) *Geography and Literature*, Syracuse, NY: Syracuse University Press.

Podilchak, W. (1991) 'Distinctions of fun, enjoyment and leisure', *Leisure Studies* 10(2): 133–48.

Pollard, J. and Rodriguez, R.D. (1993) 'Tourism and Torremolinos: recession or reaction to environment?' *Tourism Management* 14: 247–58.

Poon, A. (1993) *Tourism, Technology and Competitive Strategies,* Wallingford: CAB International.

Porter, M.E. (1980) *Competitive Strategy,* New York: Free Press.

Porter, M.E. (1985) *Competitive Advantage: Creating and Sustaining Superior Performance,* New York: Free Press.

Prentice, R. (1993a) 'Community-driven tourism planning and residents' perceptions', *Tourism Management* 14(3): 218–227.

Prentice, R. (1993b) *Tourism and Heritage Attractions,* London: Routledge.

Priestley, G. and Maundet, L. (1998), 'The post-stagnation phase of the resort lifecycle', *Annals of Tourism Research* 25(1): 85–111.

Prideaux, B. (2000a) 'Analysing bilateral tourism flows in the case of Thailand and Australia', paper presented to the 4th International Conference on Tourism in Southeast Asia and Indochina, Chiang Mai, Thailand, 24–26 June, quoted in Cooper 2002, op. cit.

Prideaux, B. (2000b) 'The resort development spectrum: a new approach to modelling resort development', *Tourism Management* 21: 225–40.

Project on Disney (1995) *Inside the Mouse: Work and Play at Disney World,* London: Rivers Oram Press.

Prunster, J. and Socher, K. (1983) 'The world recession and the future of tourism', *Association Internationale d'Experts Scientifique du Tourisme* 24: 145–56.

Purcell, K. (1997) 'Women's employment in UK tourism', in M.T. Sinclair (ed.) *Gender, Work and Tourism,* London: Routledge.

Rao, N. and Suresh, K.T. (2001) 'Domestic tourism in India' in K.B. Ghimine (ed.) *The Native Tourist: Mass Tourism within Developing Countries,* London: Earthscan.

Richards, G. (1996) 'Production and Consumption of European Cultural Tourism', *Annals of Tourism Research* 23(2): 9–13.

Richards, G. (1996a) 'The policy context of cultural tourism', in G. Richards (ed.) *Cultural Tourism in Europe,* Wallingford: CAB International.

Richards, G. (1996b) 'Production and consumption of European cultural tourism', *Annals of Tourism Research* 23: 2–11.

Richards, G. (2001a) 'The market for cultural attractions' in G. Richards (ed.) *Cultural Attractions and European Tourism,* Wallingford: CAB International.

Richards, G. (2001b) 'The experience industry and the creation of attractions', in G. Richards (ed.) *Cultural Attractions and European Tourism,* Wallingford: CAB International.

Riley, M. (1984) 'Hotels and group identity', *International Journal of Tourism Management* 5(2): 102–9.

Riley, M., Ladkin, A. and Szivas, E. (2002) *Tourism Employment: Analysis and Planning,* Clevedon: Channel View Publications.

Rimmer, M. and Zappala, J. (1988) 'Labour market flexibility and the second tier', *Australian Bulletin of Labour* 14(4): 564–91.

Ritchie, J.R.B. (1988) 'Consensus policy formulation in tourism: measuring resident views via survey research', *Tourism Management* 9: 199–212.

Ritzer, G. (1995) *Expressing America: A Critique of the Global Credit Card Society,* Newbury Park, California: Pine Forge.

Ritzer, G. (1998) *The McDonaldization Thesis,* London: Sage.

Ritzer, G. (1999) *Enchanting a Disenchanted World: Revolutionizing the Means of Consumption,* Thousand Oaks, California: Pine Forge.

Ritzer, G. and Liska, A. (1997) ' "McDisneyization" and "Post Tourism": complementary perspectives on contemporary tourism', in C. Rojek and J. Urry (eds) *Touring Cultures,* London: Routledge.

Ritzer, G. and Ovadia, S. (2000) 'The process of McDonaldization is not uniform, nor are its settings, consumers or the consumption of its goods and services', in

M. Gottdiener (ed.) *New Forms of Consumption: Consumers, Culture and Commodification*, Lanham, Maryland: Rowman and Littlefield.

Rivers, P. (1973) 'Tourist troubles', *New Society* 23: 250.

Robinson, W. (1996) 'Globalisation: nine theses on our epoch', *Race and Class* 38(2): 13–31.

Rojek, C. (1990) 'Baudrillard and leisure', *Leisure Studies* 9: 7–20.

Rojek, C. (1993) *Ways of Escape: Modern Transformations in Leisure and Travel*, Basingstoke: Macmillan.

Rojek, C. (1997) 'Indexing, dragging and the social construction of tourist sights', in C. Rojek and J. Urry (eds) *Touring Cultures: Transformations of Travel and Theory*, London: Routledge.

Rojek, C. (2000) 'Mass tourism or the re-enchantment of the world? Issues and contradictions in the study of travel', in M. Gottdiener (ed.) *New Forms of Consumption: Consumers, Culture and Commodification*, Lanham, Maryland: Rowman and Littlefield.

Ross, G.F. (1992) 'Resident perceptions of the impact of tourism on an Australian city', *Journal of Travel Research* 30(3): 13–17.

Rowe, H. (2002) 'The heritage of heritage tourism: a case study of Devon, 1940–2000', unpublished thesis, Department of Geography, University of Exeter.

Royle, T. (2000) *Working for McDonald's in Europe: The Unequal Struggle?*, London: Routledge.

Russo, A.P. (2002) 'The "vicious circle" of tourism development in heritage cities', *Annals of Tourism Research* 29(1): 165–82.

Ryan, C. (1993) 'Crime, violence, terrorism and tourism: an accidental or intrinsic relationship', *Tourism Management* 14: 173–83.

Ryan, C. (ed.) (1997) *The Tourist Experience: A New Introduction*, London: Cassell.

Ryan, C. (ed.) (2002) *The Tourist Experience: A New Introduction*, 2nd edition, London: Continuum.

Ryan, C. and Glendon, I. (1998) 'Application of leisure motivation scale to tourism', *Annals of Tourism Research* 25: 169–84.

Ryan, C. and Montgomery, D. (1994) 'The attitudes of Bakewell residents to tourism and numbers in community responsive tourism', *Tourism Management* 15: 358–69.

Sack, R.D. (1992) *Place, Modernity and the Consumer's World*, Baltimore: Johns Hopkins University Press.

Sadi, M.A. and Henderson, J.C. (2001) 'Tourism and foreign direct investment in Vietnam', *International Journal of Hospitality and Tourism Administration* 2(1): 67–90.

Salamone, F.A. (1997) 'Authenticity in tourism: The San Angel Inns', *Annals of Tourism Research* 24: 305–21.

Sàlva-Tomàs, P.A. (2002) 'Foreign immigration and tourism development in Spain's Balearic Islands', in M. Hall and A.M. Williams (eds) *Tourism and Migration: New Relationships between Production and Consumption*, Dordrecht: Kluwer Academic Press.

Sanghera, B. (2002) 'Microbusiness, household and class dynamics: the embedding of minority ethnic petty commerce', *Sociological Review* 50(2): 241–57.

Santagata, W. (2002) 'Cultural districts, property rights and sustainable economic growth', *International Journal of Urban and Regional Research* 26(1): 9–23.

Saunders, P. (1981) *Social Theory and the Urban Question*, London: Hutchinson.

Scarborough Borough Council (1997–98) *Scarborough Tourism Economic Activity Monitor*, Scarborough: Scarborough Borough Council.

Scheyvens, R. (2002) 'Backpacker tourism and Third World development', *Annals of Tourism Research* 29: 144–64.

Schiller, N.G., Basch, L. and Blanc-Szanton, C. (1992) 'Transnationalism: a new analytic framework for understanding migration', *Annals of the New York Academy of Sciences* 645: 1–24.

Schofield, P. (1996) 'Cinematographic images of a city', *Tourism Management* 17: 333–40.

Schott, C. (2002) 'Motivation and lifestyles amongst young holidaymakers: a case study of Exeter', unpublished Ph.D thesis, University of Exeter.

Schumpeter, J. (1919) *The Theory of Economic Development,* Cambridge, Massachusetts: Harvard University Press.

Schumpeter, J.A. (1939) *Business Cycles: A Theoretical, Historical and Statistical Analysis of the Capitalist Process,* New York: McGraw Hill.

Scott, A.J. (2000) *The Cultural Economy of Cities: Essays on the Geography of Image-Producing Industries,* London: Sage.

Seaton, A.V. (1992) 'Social stratification in tourism choice and experience since the War, part 1', *Tourism Management* 13: 106–11.

Selwyn, T. (1996) *The Tourist Image,* Chichester: John Wiley.

Serageldin, I. (1999) *Very Special Places: The Architecture and Economics of Intervening in Historic Cities,* New York: World Bank.

Sharma, P., Chrisman, J. and Chua, J. (1996) *A Review and Annotated Bibliography of Family Business Studies,* Boston: Kluwer.

Sharpley, R. (1994) *Tourism, Tourists and Society,* Huntingdon: ELM Publications.

Shaw, G. and Curtin, S. (2001) 'Contesting tourism destinations: resident reactions to tourists in south Devon resorts', unpublished paper presented to Royal Geographical Society Annual Conference, Plymouth.

Shaw, G. and Williams, A.M. (1990) 'Tourism, economic development and the role of entrepreneurial activity', *Progress in Tourism, Recreation and Hospitality Management* 2: 67–81.

Shaw, G. and Williams, A.M. (1994) *Critical Issues in Tourism: A Geographical Perspective,* Oxford: Blackwell.

Shaw, G. and Williams, A.M. (1997) 'The private sector: tourism entrepreneurship – a constraint on resource', in G. Shaw and A.M. Williams (eds) *The Rise and Fall of British Coastal Resorts,* London: Pinter.

Shaw, G. and Williams, A.M. (1998) 'Entrepreneurship and small business culture and tourism development', in D. Ioannides and K.G. Debbage (eds) *The Economic Geography of the Tourist Industry: A Supply-Side Analysis,* London: Routledge.

Shaw, G. and Williams, A.M. (2002) *Critical Issues in Tourism: A Geographical Perspective,* 2nd edition, Oxford: Blackwell.

Shaw, G., Agarwal, S., and Bull, P. (2000) 'Tourism consumption and tourist behaviour: a British perspective', *Tourism Geographies* 2(3): 264–89.

Shaw, G., Thornton, P., and Williams, A.M. (1998) 'The UK: market trends and policy responses', in A.M. Williams and G. Shaw (eds) *Tourism and Economic Development: European Experiences,* 3rd edition, Chichester: John Wiley.

Sheldon, P. (1986) 'The tour operator industry: an analysis', *Annals of Tourism Research* 13: 349–65.

Sheldon, P. (1994) 'Information technology and computer systems', in S. Witt and L. Mountinho (eds) *Tourism Marketing and Management Handbook,* London: Prentice-Hall.

Sheldon, P.J. and Var, T. (1984) 'Resident attitudes to tourism in north Wales', *Tourism Management* 5: 40–47.

Shields, R. (1991) *Places on the Margin: Alternative Geographies of Modernity,* London: Routledge.

Short, J.R. and Kim, K.-H. (1999) *Globalization and the City,* London: Longman.

Silberberg, T. (1995) 'Cultural tourism and business opportunities for museums and heritage sites', *Tourism Management* 16(3): 361–5.

Simmons, D.G. (1994) 'Community participation in tourism planning', *Tourism Management* 15: 98–108.

Simms, J., Hales, C., and Riley, M. (1988) 'Examination of the concept of internal labour markets in UK hotels', *Tourism Management* 9: 3–12.

Simpson, P. and Wall, G. (1999) 'Consequences of resort development: a comparative study', *Tourism Management* 20: 283–96.

Sinclair, M.T. (1997) 'Issues and theories of gender and work in tourism', in M.T. Sinclair (ed.) *Gender, Work and Tourism,* London: Routledge.

Sinclair, M.T. and Stabler, M. (1997) *The Economics of Tourism,* London: Routledge.

Sindiga, I. (1999) *Tourism and African Development: Change and Challenge of Tourism in Kenya,* Aldershot: Ashgate.

Singer, J. and Donahu, C. (1992) 'Strategic management planning for the successful family business', *Journal of Business and Entrepreneurship* 4(3): 39–51.

Sklair, L. (1995) *Sociology of the Global System,* 2nd edition, Baltimore: Johns Hopkins University Press.

Slater, D. (1997) *Consumer Culture and Modernity,* Cambridge: Polity Press.

Smith, D. (1980) *New to Britain: A Study of Some New Developments in Tourist Attractions,* London: English Tourist Board.

Smith, S. (1998) 'Tourism as an industry: debates and concepts', in D. Ioannides and K. Debbage (eds) *The Economic Geography of the Tourist Industry: A Supply-Side Analysis,* London: Routledge.

Smith, V. and Hughes, H. (1999) 'Disadvantaged families and the meaning of holiday', *International Journal of Tourism Research* 1(2): 123–33.

Smoodin, G.E. (1994) *Disney Discourse: Producing the Magic Kingdom,* New York: Routledge.

Snepenger, D., Johnson, J.D. and Rasker, R. (1995) 'Travel-stimulated entrepreneurial migration', *Journal of Travel Research* 34(1): 40–4.

Sorkin, M. (1992) *Variations on a Theme Park,* New York: Hill and Wang.

Spreitzhofer, G. (1998) 'Backpacking tourism in South-East Asia', *Annals of Tourism Research* 25: 979–83.

Stallinbrass, C. (1980) 'Seaside resorts and the hotel accommodation industry', *Progress in Planning* 13: 103–74.

Stansfield, C.A. (1978) 'Atlantic City and the resort cycle: background to the legalisation of gambling', *Annals of Tourism Research* 5(2): 238–51.

Stebbins, R.A. (1996) 'Cultural tourism as serious leisure', *Annals of Tourism Research* 23: 948–50.

Stone, C.N. (1987) 'Summing up urban regimes, development policy and political arguments', in C.N. Stone and H.T. Saunders (eds) *The Politics of Urban Development,* Lawrence: University of Kansas Press.

Storey, D.J. (1994) *Understanding the Small Business Sector,* London: Routledge.

Storper, M. (1995) 'The resurgence of regional economies, ten years later: the region as a nexus of untraded interdependencies', *European Urban and Regional Studies* 2(3): 191–222.

Storper, M. (1997) *The Regional World: Territorial Development in a Global Economy,* New York: Guilford Press.

Storper, M. and Walker, R. (1983) 'The theory of labour and the theory of location', *International Journal of Urban and Regional Research* 7: 1–41.

Strauss, C.H. and Lord, B.E. (2001) 'Economic impacts of a heritage tourism system', *Journal of Retailing and Consumer Services* 8: 199–204.

Sutcliffe, W. (1999) *Are You Experienced?,* London: Penguin.

Szivas, E. and Riley, M. (2002) 'Labour mobility and tourism in the post 1989 transition in Hungary', in M. Hall and A.M. Williams (eds) *Tourism and Migration: New Relationships between Production and Consumption,* Dordrecht: Kluwer Academic Press.

Takaki, R. (1994) *Ethnic Islands: The Emergence of Urban Chinese America,* New York: Chelsea House.

Tamamura, K. (2002) *Package Tourism: A Comparative Study between the UK and Japan,* unpublished Ph.D thesis, University of Exeter.

Taylor, J.S. (2000) 'Tourism and "embodied" commodities: sex tourism in the Caribbean', in S. Clift and S. Carter (eds) *Tourism and Sex: Culture, Commerce and Coercion,* London: Pinter, pp. 41–53.

Taylor, M. and Thrift, N. (eds) (1986) *Multinationals and the Restructuring of the World Economy,* London: Croom Helm.

Telfer, D.J. and Wall, G. (1996) 'Linkages between tourism and food production', *Annals of Tourism Research* 23: 635–53.

Telfer, D.J. and Wall, G. (2000) 'Strengthening backward economic linkages: local food purchasing by three Indonesian hotels', *Tourism Geographies* 2(4): 421–47.

Teo, P. and Huang, S. (1995) 'Tourism and Heritage Conservation in Singapore', *Annals of Tourism Research* 22: 589–615.

Teo, P. and Yeoh, B.S.A. (1997) 'Remaking local heritage for tourism', *Annals of Tourism Research* 24: 192–213.

Thlognfeldt, T. (2001) 'Second home ownership: a sustainable semi-migration', in M. Hall and A.M. Williams (eds) *Tourism and Migration: New Relationships between Production and Consumption,* Dordrecht: Kluwer Academic Press.

Thomas, G. and Fernandez, T.V. (1994) 'Mangrove and tourism: management strategies', *Indian Forester* 120(5): 406–12.

Thomas, R. (ed.)(1998) *The Management of Small Tourism and Hospitality Firms,* London: Cassell.

Thomas, R. (2000) 'Small firms in the tourism industry: some conceptual issues', *International Journal of Tourism Research* 2: 345–53.

Thomason, P., Crompton, J.L. and Kamp, B.D. (1979) 'A study of the attitudes of impacted groups within a host community towards prolonged stay tourist visitors', *Journal of Travel Research* 17(3): 2–6.

Thomson Holidays (2002) *Club Freestyle,* 3rd edition, March–October, London: Thomson.

Thrift, N. and Olds, K. (1996) 'Refiguring the economic in economic geography', *Progress in Human Geography* 20(3): 311–37.

Thurot, J. and Thurot, G. (1983) 'The ideology of class and tourism', *Annals of Tourism Research* 10: 173–89.

Tickell, A. and Peck, J. (1992) 'Accumulation, regulation and the geographies of post-Fordism: missing links in regulationist research', *Progress in Human Geography* 16(2): 190–218.

Timmerman, G.F. (1992) 'Media analysis: the image of tourism as portrayed by the print news media', unpublished Hons thesis, Department of Tourism, James Cook University.

Timothy, D. (2002) 'Tourism and the growth of urban ethnic islands', in M. Hall and A.M. Williams (eds) *Tourism and Migration: New Relationships between Production and Consumption,* Dordrecht: Kluwer Academic Press.

Todd, S. (1999) 'Examining tourist motivation methodologies', *Annals of Tourism Research* 26: 1022–24.

Travel and Tourism Analyst (1998) *Travel and Tourism Analyst* 4.

Travel and Tourism Analyst (2000) *Travel and Tourism Analyst* 2.

Travis, J. (1997) 'Continuity and change in English sea-bathing 1730–1900: a case of swimming with the tide', in S. Fisher (ed.) *Recreation and the Sea*, Exeter: Exeter University Press.

Trilling, L. (1972) *Sincerity and Authenticity*, London: Oxford University Press.

Turner A. (1996) 'Water World', *Leisure Management* 16(8): 66–9.

Turner, C. and Manning, P. (1988) 'Placing authenticity – on being a tourist: a reply to Pearce and Moscardo', *Australia and New Zealand Journal of Sociology* 24: 136–9.

Turner, G. (1993) 'Tourism and the environment: the role of the seaside', *Insights*: A 125–31.

Turner, L. and Ash, J. (1975) *The Golden Hordes: International Tourism and the Pleasure Periphery*, London: Constable.

Turner, V. (1974) 'Liminal to liminoid play flow and ritual: an essay in comparative symbology', *Rice University Studies* 60: 53–92.

Turok, I. (1993) 'Inward investment and local linkages: how deeply embedded is "Silicon Glen"', *Regional Studies* 27: 401–18.

2wentys Holidays (2002) *2wentys Holidays*, Alton, Hampshire: First Choice Group.

Twinning-Ward, L. and Baum, T. (1998) 'Dilemmas facing mature island destinations: cases from the Baltic', *Progress in Tourism, Recreation and Hospitality Management* 4: 131–40.

Tyler, D. and Dinan, C. (2001) 'The role of interest groups in England's emerging tourism policy network', *Current Issues in Tourism* 4(2–4): 210–52.

Tyrell, T. and Spaulding, I.A., (1984) 'A survey of attitudes toward tourism growth in Rhode Island', *Hospitality Education and Research Journal* 8(2): 22–33.

Um, S. and Crompton, J.L. (1987) 'Measuring residents' attachment levels in a host community', *Journal of Travel Research* 26(1): 27–29.

UNESCO (1976) 'The effects of tourism on sociocultural values', *Annals of Tourism Research* 4: 74–105.

United Nations Environment Programme (1983) *Workshop on Environmental Aspects of Tourism*, New York: United Nations.

Uriely, N. (2001) ' "Travelling workers" and "working tourists": variations across the interaction between work and tourism', *International Journal of Tourism Research* 3: 1–8.

Uriely, N., Yonay, Y. and Simchai, D. (2002) 'Backpacking experiences: a type and form analysis', *Annals of Tourism Research* 29(2): 520–38.

Urry, J. (1990) *The Tourist Gaze: Leisure and Travel in Contemporary Societies*, London: Sage.

Urry, J. (1995) *Consuming Places*, London: Routledge.

Urry, J. (1997) 'Cultural change and the seaside resort', in G. Shaw and A.M. Williams (eds) *The Rise and Fall of British Coastal Resorts*, London: Pinter.

Urry, J. (2000) *Sociology beyond Societies: Mobilities for the Twenty-First Century*, London: Routledge.

Urry, J. (2001) 'Globalizing the tourist gaze', Department of Sociology, Lancaster University: www.comp.lancs.ac.uk/sociology/soc079ju.html

Urry, J. (2002) 'Mobility and proximity', *Sociology* 36(2): 255–74.

Valenzuela, M. (1998) 'Spain: from the phenomenon of mass tourism to the search for a more diversified model', in A.M. Williams and G. Shaw (eds) *Tourism and Economic Development: European Experiences,* 3rd edition, Chichester: John Wiley.

Van der Weg, H. (1982) 'Trends in design and development facilities', *Tourism Management* 3(2): 303–7.

Var, T. (2002) 'The State, the private sector, and tourism policies in Turkey', in Y. Apostolopoulos, P. Loukissas and L. Leontidou (eds) *Mediterranean Tourism: Facets of Socioeconomic Development and Cultural Change*, London: Routledge.

Var, T., Kendall, K.W. and Tarakcioglu, E. (1985) 'Resident attitudes towards tourists in a Turkish resort town', *Annals of Tourism Research* 12: 652–7.

Varley, C.G. (1978) *Tourism in Fiji: Some Economic and Social Problems,* Cardiff: University of Wales Press.

Vellas, F. and Bécherel, L. (1995) *International Tourism: An Economic Perspective*, Basingstoke: Macmillan.

Ventures Consultancy (1989) *Seaside Resorts in England Market Profile*, London: BTA/ETB Research Services.

Venturi, R., Brown, D.S. and Izenour, S. (1972) *Learning from Las Vegas*, Cambridge, Massachusetts: MIT Press.

Wagner, U. (1977) 'Out of time and place: mass tourism and charter trips', *Ethnos* 42: 38–52.

Waitt, G. (2000) 'Consuming heritage: perceived historical authenticity', *Annals of Tourism Research* 27(4): 835–62.

Waitt, G. (2001) 'The Olympic spirit and civic boosterism: the Sydney 2000 Olympics', *Tourism Geographies* 3(3): 249–78.

Wales Tourist Board (1992) *Perspectives for Coastal Resorts: a Paper for Discussion*, Cardiff: Wales Tourist Board.

Waller, J. and Lea, S.E.G. (1998) 'Seeking the real Spain? Authenticity in motivation', *Annals of Tourism Research* 26(1): 110–29.

Walton, J.K. (1981) 'The demand for working-class seaside holidays in Victorian England', *Economic History Review* 34(2): 249–65.

Walton, J.K. (1983) *The English Seaside Resort: A Social History, 1750–1914*, Leicester: Leicester University Press.

Walton, J.K. (1997a) 'The seaside resorts of western Europe, 1750–1939', in S. Fisher (ed.) *Recreation and the Sea*, Exeter: Exeter University Press.

Walton, J.K. (1997b) 'The seaside resorts of England and Wales, 1900–1950: growth, diffusion and the emergence of new forms of coastal tourism', in G. Shaw and A.M. Williams (eds) *The Rise and Fall of British Coastal Resorts*, London: Pinter.

Walton, J.K. (2000) *The British Seaside: Holiday and Resorts in the Twentieth Century*, Manchester: Manchester University Press.

Wang, N. (1999) 'Rethinking authenticity in tourism experience', *Annals of Tourism Research* 26(2): 349–70.

Wang, Z.H., Kandampully, J. and Ryan, C. (1998) 'Taiwanese visitors to New Zealand: an analysis of attitudes,' *Pacific Tourism Review* 2(1): 29–41.

Ward, S.V. (1998) *Selling Places: The Marketing and Promotion of Towns and Cities, 1850–2000*, London: Spon.

Warde, A. (1990) 'Production, consumption and social change: reservations regarding Peter Saunders' sociology of consumption', *International Journal of Urban and Regional Research* 14(2): 228–48.

Waters, M. (1995) *Globalization*, London: Routledge.

Watson, G.L. and Kopachevsky, J.P. (1994) 'Interpretations of tourism as commodity', *Annals of Tourism Research* 21(3): 643–60.

Waycott, R. (1999) 'Marvels of the millennium or millennium madness', *Insights* November: D15–D20.

WCED (1987) *Our Common Future* [the Brundtland Report] London: Oxford University Press for the World Commission on Environment and Development.

Weaver, D.B. (1998) *Ecotourism in the Less Developed World*, Wallingford: CAB International.

Weissinger, E. and Bandalos, D.L. (1995) 'Development and validity of a scale to measure intrinsic motivation in leisure', *Journal of Leisure Research* 27: 379–400.

Welsch, W. (1999) 'Transculturality: the puzzling form of cultures today', in M. Featherstone and S. Lash (eds) *Spaces of Culture: City, Nation, World*, London: Sage.

Wheatcroft, S. (1998) 'The airline industry and tourism', in D. Ioannides and K. Debbage (eds) *The Economic Geography of the Tourist Industry: A Supply-Side Analysis,* London: Routledge.

Wheatcroft, S. and Seekings, J. (1995) *Europe's Youth Travel Market*, Brussels: Commission of the European Union.

Whitehall, W.M. (1977) 'Recycling Quincy Market', *Ekistics* 256: 155–77.

Wilkinson, P.F. (1997) *Tourism Policy and Planning: Case Studies from the Commonwealth Caribbean,* New York: Cognizant Communication Corporation.

Williams, A.M. (1994) *The European Community: The Contradictions of Integration,* 2nd edition, Oxford: Blackwell.

Williams, A.M. (1995) 'Capital and the internationalisation of tourism', in A. Montanari and A.M. Williams (eds) *European Tourism: Regions, Spaces and Restructuring,* Chichester: John Wiley.

Williams, A.M. (1996) 'Mass tourism and international tour companies' in M. Barke, J. Towner and M.T. Newton (eds) *Tourism in Spain: Critical Issues*, Wallingford: CAB International.

Williams, A.M. (2001) 'Tourism on the fabled shore', in R. King, P. de Mas and J.M. Beck (eds) *Geography, Environment and Development in the Mediterranean,* Brighton: Sussex Academic Press.

Williams, A.M. (2002) 'Mobility and culture: issues for cultural tourism', paper presented to The Tourist Historic City conference, Brugge, March.

Williams, A.M. and Balaz, V. (2000a) 'Privatisation and the development of tourism in the Czech Republic and Slovakia: property rights, firm performance and recombinant property', *Environment and Planning A* 2000, 32: 715–34.

Williams, A.M. and Balaz, V. (2000b) *Tourism in Transition: Economic Change in Central Europe*, London: I.B. Tauris.

Williams, A.M. and Balaz, V. (2001) 'From collective provision to commodification of tourism?', *Annals of Tourism Research* 28(1): 27–49.

Williams, A.M. and Hall, C.M. (2000) 'Tourism and migration: new relationships between production and consumption', *Tourism Geographies* 2(3): 5–27.

Williams, A.M. and Hall, M. (2002) 'Tourism, migration, circulation and mobility: the contingencies of time and place', in M. Hall and A.M. Williams (eds) *Tourism and Migration: New Relationships between Production and Consumption,* Dordrecht: Kluwer Academic Press.

Williams, A.M. and Shaw, G. (1988) 'Tourism: candyfloss industry or job generator', *Town Planning Review* 59: 81–104.

Williams, A.M. and Shaw, G. (1994) 'Tourism and the EC challenge', in M. Blacksell and A.M. Williams (eds) *The European Challenge: Geography and Development in the European Community*, Oxford: Oxford University Press.

Williams, A.M. and Shaw, G. (1998a) 'Tourism policies in a changing economic environment', in A.M. Williams and G. Shaw (eds) *Tourism and Economic Development: European Experiences,* 3rd edition, Chichester: John Wiley.

Williams, A.M. and Shaw, G. (1998b) 'Tourism and uneven economic development', in A.M. Williams and G. Shaw (eds) *Tourism and Economic Development: European Experiences*, 3rd edition, Chichester: John Wiley.

Williams, A.M. and Shaw, G. (1999) 'Tourism and the environment: sustainability and economic restructuring', in C.M. Hall and A. Lew (eds) *Sustainable Tourism: A Geographical Perspective,* London: Longman.

Williams, A.M., Shaw, G. and Greenwood, J. (1989) 'From tourist to tourism entrepreneur, from consumption to production: evidence from Cornwall, England', *Environment and Planning A* 21: 1639–53.

Williams, A.M., King, R., Warnes, A.M. and Patterson, G. (2000) 'Tourism and retirement migration: new forms of an old relationship in Southern Europe', *Tourism Geographies* 2(3) 28–49.

Williams, J. and Lawson, R. (2001) 'Community issues and resident opinions of tourism', *Annals of Tourism Research* 28(2): 269–90.

Williams, S. (1998) *Tourism Geography*, London: Routledge.

Williamson, J. (1978) *Decoding Advertisments*, London: Marion Boyars.

Wilson, K. (1998) 'Market/industry confusion in tourism economic analyses', *Annals of Tourism Research* 25: 803–17.

Witt, S.F. (1991) 'Tourism in Cyprus: Balancing the benefits and costs', *Tourism Management* 12: 37–46.

Wood, R. (1993) 'Tourism, culture and the sociology of development', in M. Hitchcock, V. King and M. Parnwell (eds) *Tourism in South East Asia*, London: Routledge.

Wood, R.C. (1992) *Working in Hotels and Catering,* London: Routledge.

Wood, R.C. (1997) *Working in Hotels and Catering*, 2nd edition, London: International Thomson Business Press.

Wood, R.E. (2000) 'Caribbean cruise tourism: globalization at sea', *Annals of Tourism Research* 27(2): 345–70.

Woodward, S. (2000) 'The market for industrial heritage sites', *Insights* January: D21–D30.

World Tourism Organization (1994) *Compendium of Tourism Statistics, 1988–1992,* Madrid: World Tourism Organization.

Wright, E.O. (1989) 'Rethinking once again the concept of class structure', in E.O. Wright, U. Becker, J. Brenner, M. Burawoy, V. Burris, G. Carchedi, G. Marshall, P.F. Meiksins, D. Rose, A. Stinchcombe and P. van Parijs, *The Debate on Classes,* London: Verso.

Wrigley, N. and Lowe, M. (2002) *Reading Retail: A Geographical Perspective on Retailing and Consumption Spaces*, London: Edward Arnold.

Yiannais, A. and Gibson, H. (1992) 'Roles tourists play', *Annals of Tourism Research* 19: 287–303.

Zukin, S. (1995) *The Culture of Cities*, Oxford: Blackwell.

Zukin, S. (1998a) 'From Coney Island to Las Vegas', *Urban Affairs Review* 33: 627–54.

Zukin, S. (1998b) 'Urban lifestyles: diversity and standardisation in spaces of consumption', *Urban Studies* 35: 825–40.

Index